METHODS IN MOLECULAR BIOLOGY™

AF773643

Series Editor
John M. Walker
School of Life Sciences
University of Hertfordshire
Hatfield, Hertfordshire, AL10 9AB, UK

For further volumes:
http://www.springer.com/series/7651

Molecular Chaperones

Methods and Protocols

Edited by

Stuart K. Calderwood

Molecular and Cellular Radiation Oncology, Beth Israel Deaconess Medical Center, Harvard Medical School, Boston, MA, USA

Thomas L. Prince

Urologic Oncology Branch, National Cancer Institute, National Institutes of Health, Bethesda, MD, USA

Humana Press

Editors
Stuart K. Calderwood, Ph.D.
Molecular and Cellular Radiation Oncology
Beth Israel Deaconess Medical Center
Harvard Medical School
Boston, MA
USA
scalderw@bidmc.harvard.edu

Thomas L. Prince
Urologic Oncology Branch
National Cancer Institute
National Institutes of Health
Bethesda, MD
USA
thomas.prince@nih.gov

ISSN 1064-3745 e-ISSN 1940-6029
ISBN 978-1-4939-6170-2 ISBN 978-1-61779-295-3 (eBook)
DOI 10.1007/978-1-61779-295-3
Springer New York Dordrecht Heidelberg London

Printed on acid-free paper

Humana Press is part of Springer Science+Business Media (www.springer.com)

Dedication

I would like to dedicate the book to my beautiful wife Laura and to the memory of my mother and father Albert and Gwyneth Calderwood.

Preface

The proteome consists of a complex mixture of proteins each of which need to be folded correctly for their function and for the health of the organism. Many of these proteins require molecular chaperones to reach the correct conformation and in some cases to remain in a folded form. Although many proteins fold spontaneously under in vitro conditions, under the crowded confines of the cell these molecular chaperones are required to prevent loss of function and aggregation [1–3]. Proteome quality is mediated by a number of mechanisms, including facilitated folding by the chaperones, degradation of the protein, or disposal of aggregates by asynchronous cell division. Cells have a repertoire of folding molecules consisting of multiple cohorts of molecular chaperones with subtly different properties in the protein quality control pathways. Under some conditions, such as heat shock, protein unfolding reaches crisis conditions and overwhelms these quality control mechanisms [4]. Cells respond by deploying the heat shock response and synthesizing new molecular chaperones, the heat shock proteins (HSPs) [4]. This response is regulated by heat shock transcription factors (HSFs), proteins that activate the promoters of the HSP genes, and rapidly produce new HSPs to address the folding crisis [5].

We have commenced this volume by addressing approaches to the study of HSF in vitro and in vivo. In Chapter 1, we have examined mouse models to study HSF. There are at least five HSF genes expressed in mammalian cells: Hsfl, Hsf2, Hsf3, Hsf4, and Hsf5. To understand the physiological roles of Hsfl, Hsf2, and Hsf4 in vivo, knockout mouse lines for these factors were generated. The chapter describes the design of the targeting vectors, the plasmids used, and the successful generation of mice lacking the individual genes.

In Chapter 2, we overview techniques for the study of HSF in vitro. HSF proteins bind the heat shock elements of *HSP* genes and mediate transcription in trans. We discuss methods for the study of HSP binding to *HSP* promoters and the consequent increases in HSP gene expression.

We next examine the roles and biological properties of the molecular chaperones themselves, commencing with Hsp90 (Chapter 3). The historical path of Hsp90 research within each class of client is investigated (kinase, hormone receptor, and nonsignal transduction clients) and current areas of active investigation highlighted. Chapter 4 focuses on the role of Hsp90 and its cofactors (cochaperones) in promoting the folding and activation of a diverse group of proteins with critical roles in cellular signaling and function while in Chapter 5 Hsp90 regulation by posttranslational modification is discussed. An important cochaperone for Hsp90 is the protein Cdc37/p50, and this protein has the capacity of targeting Hsp90 to protein kinases (Chapter 6).

Chapter 7 discusses the study of another family of molecular chaperones: the Hsp70 family. Hsp70 binds a number of cochaperones that mediate its ability to bind and hydrolyze ATP and interact with clients. Nucleotide exchange factors (NEFs) accelerate ADP release from Hsp70, which results in the rebinding of ATP and the release of the substrate. This chapter describes several methods suitable to study NEFs of Hsp70 chaperones. On the one hand, steady-state ATPase assays provide information on how the NEF influences progression of the Hsp70 through the entire ATPase cycle. On the other hand, nucleotide release can be measured directly using labeled nucleotides, which enable identification and further characterization of NEFs.

A very specialized cochaperone for Hsp70 is CHIP. CHIP, the *carboxyl terminus of Hsp70-interacting protein*, is both an E3 ubiquitin ligase and an Hsp70 cochaperone which is implicated in the degradation of cytosolic quality control and numerous disease substrates. CHIP contains a TPR domain, which allows for interaction with the C-terminus of Hsp70, as well as a U box domain, which is responsible for the E3 ubiquitination activity. CHIP has both Hsp70 dependent and independent activities, both of which are important for maintaining cellular homeostasis. For example, CHIP regulates HSF activation in a U box-independent manner as well as the turnover of Hsp70 after a heat shock response [6], thereby playing a crucial role in fine-tuning the stress response. The assays described in this chapter allow for the biochemical measurement of both Hsp70 dependent and independent ubiquitination activity of CHIP.

Chapter 9 discusses a third family of chaperones: the small Hsps typified by Hsp27 or HspB1. This chapter describes procedures for testing the oligomeric and phosphorylation pattern of HspB1 as well as its interaction with specific partner/client polypeptides using tissue culture cells genetically modified to express different levels of this protein. In addition, Hsp27 plays a causal role on many types of cancer, and, in Chapter 10, Kaur et al. discuss approaches to targeting Hsp27 in these diseases. Methods for quantification of the molecular chaperone activities, extracellular Hsp70 levels, and HSP localization are described in Chapters 11–14.

Understanding protein quality control requires a synthetic approach to determine the multiple interactions between the chaperone families and the proteome. Chapter 15 describes a computational protocol to identify functional modules and pathway interrelationships of chaperones based on physical interaction data derived from high-throughput proteomic experiments. The protocol first identifies interacting proteins shared by the different chaperone systems to organize the chaperones into functional modules. The chaperone functional modules represent groups of chaperones that are involved in mediating the folding of the shared interacting proteins. The chaperones in a module can either function along a single folding pathway of a given substrate protein or the substrate protein might have two or more different folding pathways that the chaperones act on independently. As described in the computational protocol, probabilities of these pathway relationships between two chaperones in a two-component chaperone module can be determined using whole-genome expression and cellular pathways as reference. This protocol is potentially useful for identifying functional modules and pathway relationships in other biological systems that involve multiple proteins with many identified interactions.

In addition to their functions in protein quality control, molecular chaperones function to directly inhibit cell inactivation by inhibiting programmed cell death pathways. Thus, cells may be maintained in a viable form by molecular chaperones until the proteome is healed by protein folding or degradation [7]. Chapter 16 discusses the role of molecular chaperones in these pathways, and Chapter 17 describes the methods to quantitate cell survival and death.

Hsp60 is a chaperonin with essential functions for cell physiology and survival. Additionally, its involvement in the pathogenesis of a number of diseases (e.g., some autoimmune disorders and cancer) is becoming evident with new research. For example, the distribution and levels of Hsp60 in cells and tissues have been found altered in many pathologic conditions, and the significance of these alterations is being investigated in a number of laboratories. Chapter 18 discusses methods to study Hsp60 in tissues. In this chapter, they describe in detail how to perform immunohistochemical reactions for detecting the chaperonin, determining its location, and measuring its levels of expression.

Finally, recent studies have uncovered a profound role for molecular chaperones in the extracellular spaces [8, 9]. Extracellular HSPs play important roles in cell signaling and immunity. Many of these effects are mediated by cell surface receptors expressed on a wide range of cell types. Chapters 19, 20, and 21 describe approaches to using the chaperone function of HSPs in the design of vaccines. Chapter 22 follows from this by describing approaches to study HSP receptors by cloning candidate receptors into cells (CHO-K1) with the rare property of being null for HSP binding. Using this approach, it has been shown that Hsp70 binds to at least two classes of receptor: c-type lectin receptors (CLR) and scavenger receptors (SR). However, the nature of the receptor–ligand interactions is not yet clear. In this chapter, methods for determining HSP receptors, approaches to study individual receptors in cells that contain multiple such receptors, and methods for investigating HSP receptor function in vivo are described. Finally, HSPs and HSF1 have been shown to play roles in cancer and metastasis, and Kouspou et al. (Chapter 23) describe methods to study cell migration (a key step in transformation) in cancer mediated by components of the HSP response.

Boston, MA ***Stuart K. Calderwood***
Bethesda, MD ***Thomas L. Prince***

References

1. Mayer MP, Bukau B. Hsp70 chaperones: cellular functions and molecular mechanism. Cell Mol Life Sci 2005;62(6):670–84.
2. Ellis RJ. Protein misassembly: macromolecular crowding and molecular chaperones. Adv Exp Med Biol 2007;594:1–13.
3. Bukau B, Weissman J, Horwich A. Molecular chaperones and protein quality control. Cell 2006;125(3):443–51.
4. Lindquist S, Craig EA. The heat shock proteins. Ann Rev Genet 1988;22:631–7.
5. Wu C. Heat shock transcription factors: structure and regulation. Ann Rev Cell Dev Biol 1995;11:441–69.
6. Qian SB, McDonough H, Boellmann F, Cyr DM, Patterson C. CHIP-mediated stress recovery by sequential ubiquitination of substrates and Hsp70. Nature 2006;440(7083):551–5.
7. Beere HM. 'The stress of dying': the role of heat shock proteins in the regulation of apoptosis. J Cell Sci 2004;117(Pt 13):2641–51.
8. Murshid A, Gong J, Calderwood SK. Heat-shock proteins in cancer vaccines: agents of antigen cross-presentation. Expert Rev Vaccines 2008;7(7):1019–30.
9. Mambula SS, Calderwood SK. Heat shock protein 70 is secreted from tumor cells by a nonclassical pathway involving lysosomal endosomes. J Immunol 2006;177(11):7849–57.

Acknowledgments

We thank Ayesha Murshid, Yue Zhang, and Ketty Chou for help and advice and Roxanne Calderwood for carrying out the preliminary editing of the book.

Contents

Contributors

ANDRÉ-PATRICK ARRIGO • *Stress, Chaperons, and Cell Death Laboratory, CNRS UMR 5534, Claude Bernard University Lyon 1, Villeurbanne, France*
ALEXZANDER ASEA • *Division of Investigative Pathology, College of Medicine, Scott & White Memorial Hospital and Clinic, Temple, TX, USA; Texas A&M Health Science Center, Temple, TX, USA*
DANIEL N. BOLON • *Department of Biochemistry and Molecular Biology, University of Massachusetts Medical School, Worcester, MA, USA*
BERND BUKAU • *Zentrum für Molekulare Biologie, der Universität Heidelberg, DKFZ-ZMBH Alliance, Heidelberg, Germany*
STUART K. CALDERWOOD • *Molecular and Cellular Radiation Oncology, Department of Radiation Oncology, Beth Israel Deaconess Medical Center, Harvard Medical School, Boston, MA, USA*
AVROM J. CAPLAN • *Department of Biology, The City College of New York, New York, NY, USA*
FRANCESCO CAPPELLO • *Department of Experimental Biomedicine and Clinical Neuroscience, University of Palermo, Palermo, Italy*
GISELA CASTRO • *Laboratory of Oncology, Institute of Experimental Medicine and Biology of Cuyo, Scientific and Technological Center of Mendoza, National Research Council (CONICET), Mendoza, Argentina; Argentina Foundation for Cancer Research, Mendoza, Argentina*
NIUBYS CAYADO-GUTIÉRREZ • *Laboratory of Oncology, Institute of Experimental Medicine and Biology of Cuyo, Scientific and Technological Center of Mendoza, National Research Council (CONICET), Mendoza, Argentina; Argentina Foundation for Cancer Research, Mendoza, Argentina*
SHIUH-DIH CHOU • *Molecular and Cellular Radiation Oncology, Beth Israel Deaconess Medical Center, Harvard Medical School, Boston, MA, USA*
DANIEL R. CIOCCA • *Laboratory of Oncology, Institute of Experimental Medicine and Biology of Cuyo, Scientific and Technological Center of Mendoza, National Research Council (CONICET), Mendoza, Argentina; Argentina Foundation for Cancer Research, Mendoza, Argentina*
EVERLY CONWAY DE MACARIO • *Department of Microbiology and Immunology, School of Medicine, University of Maryland at Baltimore, Baltimore, MD, USA*
MARC B. COX • *Department of Biological Sciences, University of Texas at El Paso, El Paso, TX, USA; Border Biomedical Research Center, El Paso, TX, USA*
FERNANDO D. CUELLO-CARRIÓN • *Laboratory of Oncology, Institute of Experimental Medicine and Biology of Cuyo, Scientific and Technological Center of Mendoza, National Research Council (CONICET), Mendoza, Argentina; Argentina Foundation for Cancer Research, Mendoza, Argentina*
DOUGLAS M. CYR • *Department of Cell and Developmental Biology, School of Medicine, University of North Carolina at Chapel Hill, Chapel Hill, NC, USA*

NINA C. DEMPSEY • *Chester Centre for Stress Research, University of Chester, Chester, UK*
BINNUR EROGLU • *Center for Molecular Chaperone/Radiobiology and Cancer Virology, Medical College of Georgia, Augusta, GA, USA*
MARIEL A. FANELLI • *Laboratory of Oncology, Institute of Experimental Medicine and Biology of Cuyo, Scientific and Technological Center of Mendoza, National Research Council (CONICET), Mendoza, Argentina; Argentina Foundation for Cancer Research, Mendoza, Argentina*
VLADIMIR L. GABAI • *Department of Biochemistry, Boston University Medical School, Boston, MA, USA*
CARMEN GARRIDO • *Faculty of Medicine and Pharmacy, INSERM, U-866, Dijon, France*
CAROLINE GAUCHER-DI-STASIO • *Faculty of Medicine and Pharmacy, INSERM, U-866, Dijon, France*
JESSICA GOBBO • *Faculty of Medicine and Pharmacy, INSERM, U-866, Dijon, France*
JIANLIN GONG • *Department of Medicine, Boston University School of Medicine, Boston, MA, USA*
YUNCHEN GONG • *Centre for the Analysis of Genome Evolution and Function, University of Toronto, Toronto, ON, Canada*
JEAN GUZZO • *Institut Jules Guyot, Université de Bourgogne, Equipe de Recherche en vigne et vin (REVV), BP, Dijon*
WALID A. HOURY • *Department of Biochemistry, King's College Circle, Medical Sciences Building, University of Toronto, Toronto, ON, Canada*
CHRISTINE HOYLE • *Betsi Cadwaladr University Health Board, Glan Clwyd Hospital, Bodelwyddan, North Wales, UK*
H. ELYSE IRELAND • *Chester Centre for Stress Research, University of Chester, Chester, UK*
GAËTAN JEGO • *Faculty of Medicine and Pharmacy, INSERM, U-866, Dijon, France*
XIONGJIE JIN • *Center for Molecular Chaperone/Radiobiology and Cancer Virology, Medical College of Georgia, Augusta, GA, USA*
JILL L. JOHNSON • *Department of Microbiology, Molecular Biology, and Biochemistry, University of Idaho, Moscow, ID, USA*
ALEXANDER E. KABAKOV • *Department of Radiation Biochemistry, Medical Radiology Research Center, Obninsk, Russia*
PUNIT KAUR • *Division of Investigative Pathology, College of Medicine, Scott & White Memorial Hospital and Clinic, Temple, TX, USA; Texas A&M Health Science Center, Temple, TX, USA*
MICHELLE M. KOUSPOU • *Department of Biochemistry and Molecular Biology, School of Biomedical Sciences, Monash University, Clayton, VIC, Australia*
VLADIMIR A. KUDRYAVTSEV • *Department of Radiation Biochemistry, Medical Radiology Research Center, Obninsk, Russia*
FRANCESCA LEONI • *Chester Centre for Stress Research, University of Chester, Chester, UK*
ALBERTO J.L. MACARIO • *Department of Microbiology and Immunology, School of Medicine, University of Maryland at Baltimore, Baltimore, MD, USA*

Atin K. Mandal • *Department of Biology, City College New York, New York, NY, USA*
Matthias P. Mayer • *Zentrum für Molekulare Biologie der Universität Heidelberg, DKFZ-ZMBH Alliance, Heidelberg, Germany*
Parul Mishra • *Department of Biochemistry and Molecular Biology, University of Massachusetts Medical School, Worcester, MA, USA*
Nahid F. Mivechi • *Charlie Norwood VA, Medical Center, One Freedom Way, Augusta, GA, USA; Center for Molecular Chaperone/Radiobiology and Cancer Virology, Georgia Health Sciences University, Augusta, GA, USA*
Mehdi Mollapour • *Urologic Oncology Branch, Center for Cancer Research, National Cancer Institute, National Institutes of Health, Bethesda, MD, USA*
Demetrius Moskophidis • *Center for Molecular Chaperone/Radiobiology and Cancer Virology, Medical College of Georgia, Augusta, GA, USA*
Ayesha Murshid • *Molecular and Cellular Radiation Oncology, Beth Israel Deaconess Medical Center, Harvard Medical School, Boston, MA, USA*
Ganachari M. Nagaraja • *Division of Investigative Pathology, College of Medicine, Scott & White Memorial Hospital and Clinic, Temple, TX, USA; Texas A&M Health Science Center, Temple, TX, USA*
Len Neckers • *Urologic Oncology Branch, Center for Cancer Research, National Cancer Institute, National Institutes of Health, Bethesda, MD, USA*
Nadinath B. Nillegoda • *Department of Biology, City College New York, New York, NY, USA*
Cam Patterson • *Department of Cell and Developmental Biology, School of Medicine, University of North Carolina at Chapel Hill, Chapel Hill, NC, USA*
John T. Price • *Department of Biochemistry and Molecular Biology, School of Biomedical Sciences, Monash University, Clayton, VIC, Australia*
Thomas L. Prince • *Urologic Oncology Branch, National Cancer Institute, National Institutes of Health, Bethesda, MD, USA*
Heike Rampelt • *Zentrum für Molekulare Biologie, der Universität Heidelberg, DKFZ-ZMBH Alliance, Heidelberg, Germany*
Hong Yu Ren • *Department of Cell and Developmental Biology, School of Medicine, University of North Carolina at Chapel Hill, Chapel Hill, NC, USA*
Anne-Laure Rérole • *Faculty of Medicine and Pharmacy, INSERM, U-866, Dijon, France*
Meredith F.N. Rosser • *Department of Cell and Developmental Biology, School of Medicine, University of North Carolina at Chapel Hill, Chapel Hill, NC, USA*
Sheila Schreiner • *Molecular and Cellular Radiation Oncology, Beth Israel Deaconess Medical Center, Harvard Medical School, Boston, MA, USA*
Mary Ann Stevenson • *Department of Radiation Oncology, Beth Israel Deaconess Medical Center, Harvard Medical School, Boston, MA, USA*
John R. Subjeck • *Department of Cellular Stress Biology, Roswell Park Cancer Institute, Buffalo, NY, USA*
Maria A. Theodoraki • *Department of Biology, City College New York, New York, NY, USA*

JIMMY THERIAULT • *Molecular and Cellular Radiation Oncology, Beth Israel Deaconess Medical Center, Harvard Medical School, Boston, MA, USA*
XIANG-YANG WANG • *Department of Human and Molecular Genetics, Massey Cancer Center, Institute of Molecular Medicine, Virginia Commonwealth University, Richmond, VA, USA*
NATALIE WAYNE • *Department of Biochemistry and Molecular Biology, University of Massachusetts Medical School, Worcester, MA, USA*
STÉPHANIE WEIDMANN • *Institut Jules Guyot, Université de Bourgogne, Equipe de Recherche en vigne et vin (REVV), BP, Dijon*
DESHENG WENG • *Department of Medicine, Boston University School of Medicine, Boston, MA, USA*
JOHN H.H. WILLIAMS • *Chester Centre for Stress Research, University of Chester, Chester, UK*
HUANFA YI • *Department of Human and Molecular Genetics, Massey Cancer Center, Institute of Molecular Medicine, Virginia Commonwealth University, Richmond, VA, USA*
XIAOFEI YU • *Department of Human and Molecular Genetics, Massey Cancer Center, Institute of Molecular Medicine, Virginia Commonwealth University, Richmond, VA, USA*
YUE ZHANG • *Molecular and Cellular Radiation Oncology, Beth Israel Deaconess Medical Center, Harvard Medical School, Boston, MA, USA*
ZHAOLEI ZHANG • *Banting and Best Department of Medical Research, Terrence Donnelly Centre for Cellular and Biomolecular Research, University of Toronto, Toronto, ON, Canada; Department of Molecular Genetics, Terrence Donnelly Centre for Cellular and Biomolecular Research, University of Toronto, Toronto, ON, Canada*
GIOVANNI ZUMMO • *University of Palermo, Department of Experimental Biomedicine and Clinical Neurosciences, Palermo, Italy*
DAMIN ZUO • *Department of Human and Molecular Genetics, Massey Cancer Center, Institute of Molecular Medicine, Virginia Commonwealth University, Richmond, VA, USA*

Chapter 1

Targeted Deletion of Hsf1, 2, and 4 Genes in Mice

Xiongjie Jin, Binnur Eroglu, Demetrius Moskophidis, and Nahid F. Mivechi

Abstract

Heat-shock transcription factors (Hsfs) regulate transcription of heat-shock proteins as well as other genes whose promoters contain heat-shock elements. There are at least five Hsfs in mammalian cells, Hsf1, Hsf2, Hsf3, Hsf4, and Hsfy. To understand the physiological roles of Hsf1, Hsf2, and Hsf4 in vivo, we generated knockout mouse lines for these factors. In this chapter, we describe the design of the targeting vectors, the plasmids used, and the successful generation of mice lacking the individual genes. We also briefly describe what we have learned about the physiological functions of these genes in vivo.

Key words: Hsf1, Hsf2, Hsf4, Knockout mice, Targeting vector, Hsf4-EGFP

1. Introduction

1.1. Mammalian Cells Possess Multiple Hsfs with Diverse Functions

Transcription of the genome is controlled by a class of proteins known as transcription factors. Transcription factors bind to specific DNA sequences and enhance (or repress) expression of specific genes. These factors usually share a high level of overlap in their DNA recognition sequence. The heat-shock transcription factor (Hsf) family members (Hsf1, 2, and 4) bind to heat-shock elements (HSEs) (5′-nGAAn-3′ units) and regulate transcription of Hsps and other molecular chaperones (1, 2). Comparisons between Hsf protein sequences between different organisms indicate the presence of a conserved DNA-binding domain and three hydrophobic heptad repeats known as the trimerization domain. These domains are located within the amino-terminal portion of the protein.

Stuart K. Calderwood and Thomas L. Prince (eds.), *Molecular Chaperones: Methods and Protocols*, Methods in Molecular Biology, vol. 787, DOI 10.1007/978-1-61779-295-3_1, © Springer Science+Business Media, LLC 2011

The transcriptional activation domain is located toward the carboxyl-terminus end of the molecule. The intramolecular interactions between the amino- and carboxyl-terminal domains keep Hsf1 in an inactive state under nonstress conditions (1). The expression of Hsf2 and Hsf4 in the cell correlates with their increased DNA-binding activity (2–4).

Gene targeting in mice by homologous recombination in embryonic stem (ES) cells has become a routine procedure (5–8). Gene targeting alters the mouse genome at specific selective locus. A targeting vector that carries a specific portion of the gene to be targeted is normally flanked by a neomycin gene, most often containing its own promoter sequence. Other strategies, where neomycin expression is under the control of an endogenous gene, have also been used. The neomycin gene is used as a positive selectable marker for isolation of embryonic stem cells that carry the targeted allele. Neomycin gene may be flanked by loxP sites so that it can be removed following generation of the knockout mouse line (7, 8). The targeting vector may also contain one or two thymidine kinase (TK) genes that can be used as a negative selectable marker. Following electroporation of the targeting vector into the ES cells, if the targeting vector is randomly inserted into the genome by nonhomologous recombination, the TK genes will also be inserted and the gene is expressed. Treatment of the ES cells with ganciclovir ensures removal of the cells containing the random integration of the targeting constructs. In contrast, if the targeting vector is inserted into the genome by homologous recombination, the TK genes will not be inserted into the genome and the ES clones will survive the treatment with ganciclovir. The strategy we used was to generate mutant *hsf1*, *hsf2*, and *hsf4* targeting vectors containing a neomycin gene flanked by two loxP sites and two TK genes.

In this chapter, we describe the targeting vectors and generation of *hsf1*-, *hsf2*-, and *hsf4*-deficient mouse lines. We also briefly describe the phenotype of the *hsf* knockout mice generated in our laboratory.

2. Materials

2.1. Genomic DNA Identification, Isolation, and Analyses

At the time when we began constructing targeting vectors for the *hsf* genes, the mouse genome had not been entirely sequenced. Therefore, we cloned these genes as described in the following sections. However, currently, a BAC clone has been identified that encodes the murine *hsf* genes. To identify the BAC clone number that contains specific gene, one can go to the NCBI Web site "Clone Registry" and search for the gene of interest. After clicking the BAC clone number, the Web sites for the *hsf1*, *2*, and *4* clones are indicated.

The BAC clone containing the *hsf1* gene can be found at: http://www.ncbi.nlm.nih.gov/genome/clone/clname.cgi?

stype=Name&list=RP23-266H9. The BAC clone containing the *hsf2* gene can be found at: http://www.ncbi.nlm.nih.gov/genome/clone/clname.cgi?stype=Name&list=RP23-212L2, and the BAC clone containing the *hsf4* gene can be found at: http://www.ncbi.nlm.nih.gov/genome/clone/clname.cgi?stype=Name&list=RP23-118P12.

Once the clone is identified, the clone can be obtained from http://bacpac.chori.org.

One method of using a BAC clone to construct the targeting vector is from the available sequencing information, first design the targeting vector, and then fragments that are needed can be amplified by PCR. Any fragment that is amplified via PCR must be sequenced entirely to detect the presence of any errors. For PCR of genomic DNA, use of a high-fidelity TAQ polymerase is highly recommended. For the conventional knockout strategy, presence of more than one or two base pair (bp) differences per kb pairs of DNA may reduce homologous recombination. For the conditional cre-loxP techniques, even one change in the DNA bases may be detrimental in proper expression of the gene under study and should be avoided.

2.2. Plasmids and Phages

Conventional plasmids can be used for manipulation of the DNA fragments. For the final targeting vector, we used Lamda DASH II-254-2TK Phage (Stratagene, and a modified version was provided by Drs. NR Manley and B. Condi, Department of Genetics, University of Georgia, Athens, GA 30602, USA). A map of Lamda DASH II-254-2TK has been provided in Fig. 1. It contains stuffer sequences that can be removed by restriction enzyme (XhoI) digestion, and the gene fragments (in two or three pieces) of interest can then be inserted. If three fragments are inserted into the final phage, it is advisable to use different restriction enzymes for each fragment to avoid excessive self-ligation.

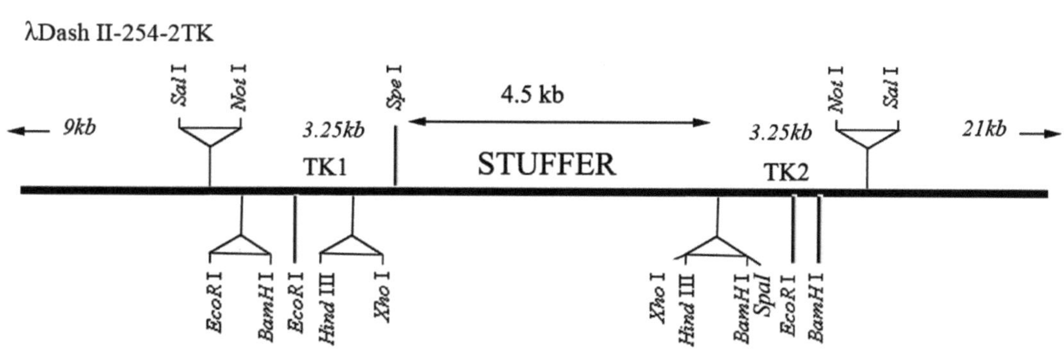

Fig. 1. Schematic presentation of Lambda Dash II-254-2TK Vector. The portion of the map of lambda-Dash II-254-2TK is presented. The stuffer sequence can be removed and the targeting vector can be ligated into the phage DNA in two or three fragments.

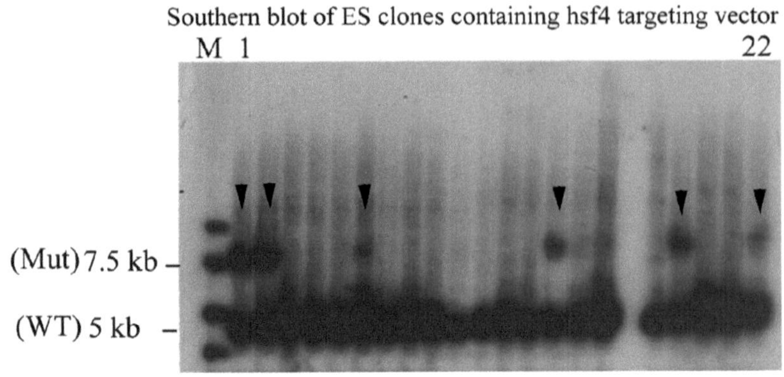

Fig. 2. Southern blot of ES cells electroporated with *hsf4* targeting constructs. From 22 clones presented, 6 clones carried the correct targeting for the *hsf4* mutant (*arrow* heads). WT is wild-type band (5 kb), and Mut is the *hsf4* mutant band at 7.5 kb.

2.3. ES Electroporation, Southern Blotting of ES Clones, and Identification of Positively Targeted ES Clones

For identification of ES clones with correct targeting, Southern blotting from a small amount of DNA (may be as low as 2 μg) is essential. For Southern blotting, we follow the procedure in the Molecular Cloning, Laboratory Manual (9). Figure 2 represents Southern blotting of 22 clones of ES cells obtained for targeting of the *hsf4* gene. As the data indicate from 22 clones, 6 clones were found to be positive. Other methods of this section have been briefly described for each knockout mouse line; however, the methodology can be found in detail in the following references (5–8, 10).

2.4. ES Cell Microinjection into Blastocysts, Generation of Chimeras, and Germline Transmission

This section has been briefly described for each knockout mouse line; however, the methodology can be found in more detail in the following references (5–8, 10).

3. Methods

3.1. Knockout of hsf1 Gene

Hsf1 gene structure: The mouse Hsf1 gene (*hsf1*) is located on chromosome 15 and encodes an Hsf1 protein that contains 503 amino acids (11, 12). The mouse *hsf1* gene has 12 exons. Exon 1 contains 274 bp, and the start codon (ATG) is located at 158 bp, i.e., exon 1 encodes 39 amino acids. The distance between exon 1 and exon 2 (intron 1) is 18 kb (18, 254 bp). The remaining exons (exon 2 through exon 12) are located compactly in a 4.6-kb region. The entire *hsf1* gene is 23.5 kb.

Design of the targeting vector *to delete hsf1*: To target the *hsf1* gene, the cloned fragment containing *hsf1* was sequenced and analyzed for the presence of unique restriction enzyme sites. In our targeting strategy, we selected to delete a portion of exon 2 of the *hsf1* gene for the reasons provided here: (1) In the *hsf1* gene, exon 1 is located 18 kb apart from other exons. (2) Intron 1 encodes the promoter of another gene (known as Bop1), as we have previously reported (12). Disruption of exon 1 (plus insertion of neomycin gene in this exon) could potentially disrupt the expression of the Bop1 gene. (3) Since the distance between exon 1 and exon 2 is 18 kb, deletion of both exons would have been impossible. (4) If only exon 1 was deleted, there would be a possibility that a truncated Hsf1 mRNA and protein encoded by other exons (from exon 2 to exon 12) would be generated. This truncated Hsf1 would potentially contain amino acid residues 40–503 due to the presence of an ATG at amino acid 40. (5) For the *hsf1* targeting construct, we planned that a LacZ gene could be inserted under control of the *hsf1* promoter. The best strategy would have been to insert the LacZ gene at the first ATG. However, for the reasons noted above, we inserted the LacZ gene into the *hsf1* exon 2.

The amino acid sequence encoded by exon 2 is critical for the DNA-binding domain of Hsf1 protein. If exon 2 was deleted, the Hsf1 DNA-binding domain would be disrupted and Hsf1 could not bind to the DNA. Furthermore, if exon 2 was deleted, the *hsf1* open reading frame (i.e., a cDNA-encoding exon 1 and exons 3–12) will shift and a stop codon will appear immediately at amino acid residue 49 in exon 2. Three more stop codons are also present within the next 100 bp. Therefore, deletion of exon 2 completely disrupts the Hsf1 protein structure and function. According to these criteria, we designed an *hsf1* targeting vector with deletion of 55 bp of exon 2. The final targeting vector contains a 3.2 kb proximal fragment with homology to *hsf1*, a 3.2 kb internal ribosomal entry site (IRES)-LacZ-neomycin fragment, and a 3.7 kb distal fragment with homology to *hsf1* (Fig. 3) (11).

Targeting vector: Cloning of the *hsf1* gene: An 18 kb clone containing a portion of the *hsf1* gene fragment (exon 2–exon 12) was isolated following screening a 129/SvJ mouse genomic library in Lambda Fix II vector (Stratagene) using Hsf1 cDNA as a probe (12). This 18 kb DNA fragment was used as a template for constructing the *hsf1* targeting vector.

External probe for detection of correct targeting: To identify homologous recombination, an external probe was designed to screen the *hsf1* mutant ES clones and eventually the *hsf1*-deficient mice. The external probe should hybridize to the DNA region that is close to, but not included in the targeting vector. According to our targeting strategy, we designed a 1 kb probe that hybridized to the 5′ region of the targeting vector as presented in Fig. 3a. The probe was tested on genomic DNA prepared from wild-type mouse tail that was digested by the restriction enzyme PstI. PstI was selected because of its suitability of detecting the correct targeting of ES cell clones.

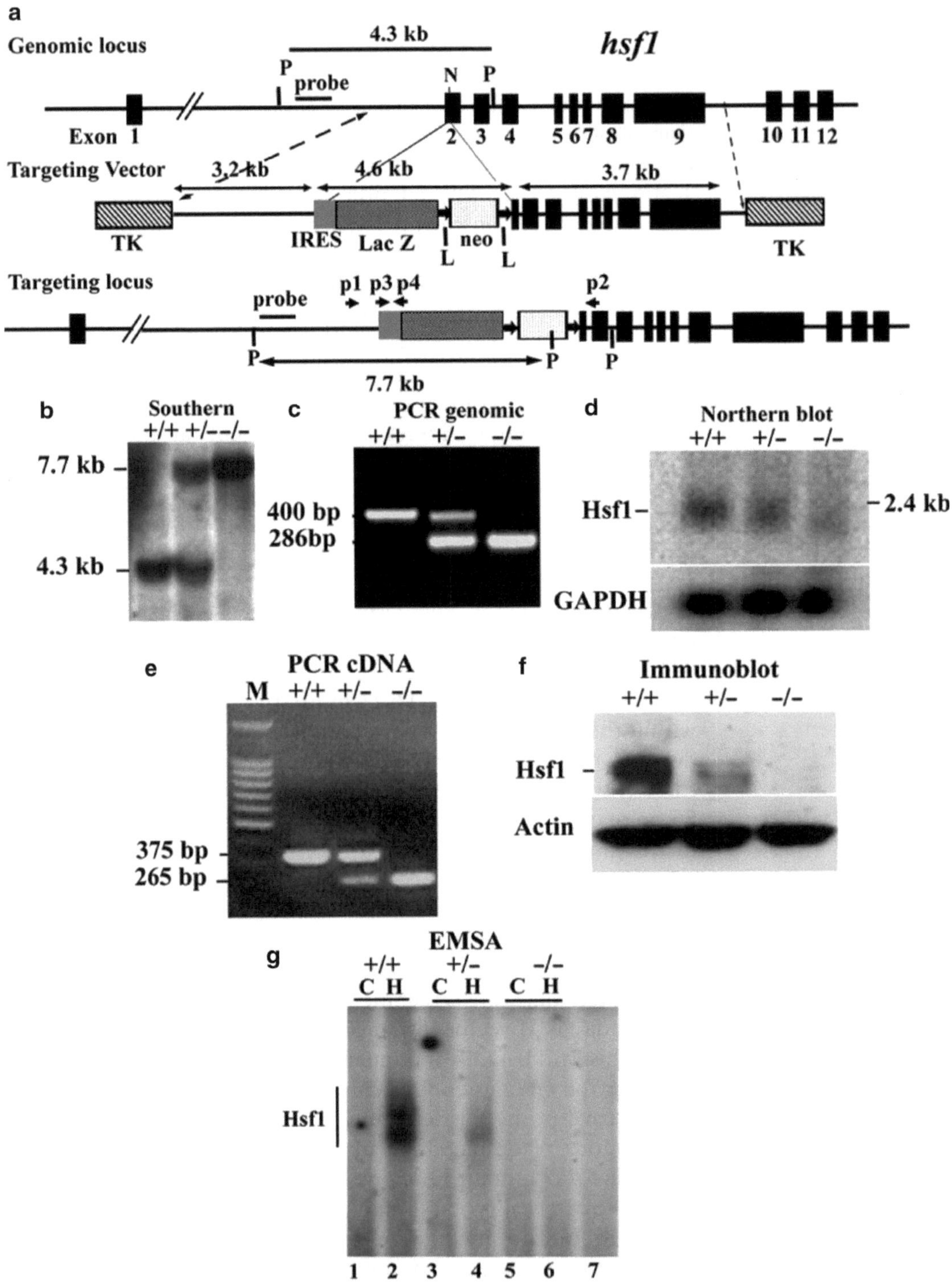

Fig. 3. Targeted disruption of *hsf1* by homologous recombination. (**a**) Schematic of segment of the *hsf1* locus, targeting construct, and targeted locus. Coding exons are *boxes in black*, beginning with exon 2 (12). Targeting construct replaces the coding region for 55 bp of exon 2. *LoxP* flanked, PGK-neomycin cassette with upstream *IRES-LacZ* is indicated. Two TK genes were used for negative selection. "Outside probe" was used for screening ES cell clones to distinguish between endogenous and targeted alleles. The 3.2 and 3.7 kb fragments are the proximal and distal *hsf1* homologous segments in

Vector construct: A 3.2 kb proximal fragment with homology to *hsf1* was amplified by PCR using the isolated 18 kb *hsf1* clone as a template using forward primer: 5′-CTG CAG AAC CAA TGC ATT GGC GGC CGC TCG AGA ACA CAG CAT TC TTG AAA GAA A-3′ that included BstXI, NotI, EagI, and XhoI restriction enzyme sites and a reverse primer: 5′-GAA TCG GCC GTG GTC AAA CAC GTG GAA GCT GTT-3′ that included an EagI restriction enzyme site. The PCR product was sequenced to confirm DNA sequence fidelity. The PCR product was digested by EagI for subcloning.

A plasmid containing an IRES-lacZ-neomycin cassette was used to insert an IRES-lacZ-neomycin fragment in the targeting vector (Fig. 3a) (11). The neomycin gene (used as a positive selectable marker) was driven by the phosphoglycerate kinase (PGK) promoter and contained an SV40 poly-(A) signal and a stop codon. The neomycin gene was flanked by Cre recombinase recognition sequences (loxP) to allow removal of the neomycin gene in the mutant mice. The lacZ gene contained sequences of the picornaviral IRES at its 5′-end and a poly-(A) signal and a stop codon at its 3′-end (clones encoding sequences for lacZ and IRES were the gift of Dr. A. Smith (Univ. of Edinburgh, Scotland)). Since the decision was made to insert the lacZ gene (containing its own ATG) into exon 2, it was possible that Hsf1 transcripts that start from exon 1 would interfere with the lacZ expression. Therefore, an IRES was inserted before the lacZ gene to direct translation of the reporter gene. However, as we have noted in our previous publication (11), in *hsf1*-deficient mice, the lacZ gene (plus the entire exon 2) is spliced out, fusing exon 1 directly to exon 3 and making a shorter transcript that excluded exon 2. Since we had predicted this may occur in vivo, we still went with such a design since fusion of exon 1 to exon 3 would generate an Hsf1 transcript that would

Fig. 3. (continued) the targeting vector (11). The final insert between the two TK genes is 13 kb. N and P represent NheI and PstI sites, respectively. Primers P1, P2, P3, and p4 are used for genotyping. (**b**) Southern blot analysis of genomic DNA prepared from tails of wild-type (+) and targeted mutant mice (–). The 7.7 and 4.3 kb bands are PstI-digested fragments corresponding to the targeted (–) and wild-type (+) alleles, respectively. (**c**) PCR analysis of tail DNA derived from wild-type (+) and targeted mutant mice (–) showing 420 and 890 bp fragments derived from wild-type and targeted alleles, respectively. (**d**) Northern blot analysis of total RNA derived from mouse embryo fibroblasts (MEFs) of wild-type (+) or mutant (–) mice. Full-length murine *hsf1* cDNA was used as a probe. Hsf1 generates an approximately 2.4 kb fragment. GAPDH is shown for equal loading. (**e**) cDNA from wild-type (+) and *hsf1* mutant (–) mice were amplified using forward primer located in exon 1 and reverse primers located in exon 3. Sequencing the 375 bp (+) and 265 bp (–) fragment indicated normal splicing of exons 1, 2, and 3 and splicing of exon 1 to exon 3, respectively. (**f**) Immunoblot analysis of extracts of MEFs derived from wild-type (+) or mutant (–) *hsf1* analyzed by SDS-PAGE using antibody to Hsf1. Actin is shown as an indicator of loading. (**g**) Electrophoretic mobility shift assays (EMSAs) (11). Nuclear extracts of wild-type (+) or *hsf1* mutant (–) MEFs were prepared from untreated control (C) or heated (43°C for 20 min plus 30 min recovery at 37°C to ensure Hsf1 activation). *Lanes 1* and *2*, *3* and *4*, *5* and *6* represent untreated control (C) and heated (H) samples, respectively. *Lane 7* is the same extract as in lane *2*, but with 200× excess cold HSE to show specificity.

be out-of-frame and no protein could be generated from this transcript (as noted above). As such, cells deficient in the *hsf1* gene do not express the lacZ gene (11).

The proximal 3.2 kb fragment was subcloned into the IRES-lacZ-neomycin plasmid. The resulting plasmid was then digested with XhoI/NruI to release the 7.8 kb proximal fragment. For the distal 3.7 kb fragment with homology to *hsf1*, the 18 kb genomic clone was digested with NheI to release a 7.5 kb fragment, which was subcloned into plasmid pBlueScript at an EcoRV site. This plasmid was then digested with HindIII to remove a 3.8 kb fragment. The remaining 3.7 kb fragment (portion of exons 2–9) was subsequently released by SmaI/XhoI digestion.

The proximal fragment containing the IRES-lacZ-neomycin cassette and the 3.7 kb distal fragment were then ligated into the phage DNA vector λDASHII-254-2TK at an XhoI site (11). The targeting construct is flanked by two TK genes, which are used as a negative selectable marker. The vector was packaged into phage, and the positive phage clones were selected by PCR and restriction enzyme digestion. Several positive phage DNA clones were digested by NotI (see Fig. 1) to release the final vector that contained the targeting vector and the two TK genes. This final vector can be recircularized into a plasmid that could be amplified in bacteria. After amplification, the final vector was linearized with NotI digestion for ES cell transfection.

ES cell electroporation: ES (D3; Incyte Genomics, St. Louis, MO) cells were cultured as described previously (10, 11). ES cells were electroporated (BioRad Gene Plus, 250 V, 950 μF) with the linearized targeting vector and cultured in the presence of neomycin (200 μg/ml) and ganciclovir (2 μM) for 10 days. Following Southern blotting using the external probe, two doubly resistant ES cell clones (Fig. 3b) (from 167 clones tested) were selected and expanded. Generation of *hsf1* mutant mice: The two ES clones that were found to be positive by Southern blotting were microinjected into C57BL/6 blastocysts, and germline-transmitting chimeric mice were obtained. The chimeric mice were then crossed with C57BL/6 mice to obtain *hsf1*$^{+/-}$ mice which were interbred to generate *hsf1*$^{-/-}$ mice.

Genotyping of mutant mice: Southern blotting: ES clones, germline-transmitting chimeric mice, and the first several litters of mice were genotyped by Southern blotting. Mouse genomic DNA was isolated and digested with the restriction enzyme PstI. Southern blotting was performed using an external probe (indicated in Fig. 3b) (11). This generates a 7.7 kb fragment for the targeted locus and a 4.3 kb fragment for the wild-type locus.

PCR: When the *hsf1* mutant mouse line was established after screening by Southern blotting (Fig. 3b), the mice were subsequently routinely genotyped by PCR (Fig. 3c). Genomic DNA isolated from mouse tail was used as template. Two sets of primers were used to identify wild-type and mutant alleles: for wild type, forward primer (P1): 5′-GAG ATG ACC AGA ATG CTG TGG GTG-3′ and reverse primer (P2): 5′-GCA AGC ATA GCA TCC TGA AAG AG-3′; for mutant alleles (the primers to amplify IRES region): forward primer (P3): 5′-ACT GGC CGA AGC CGC TTG GAA TAA-3′ and reverse primer (P4): 5′-ATA CAC GTG GCT TTT GGC CGC AGA-3′. These PCR reactions generated a 400 bp fragment for wild type and a 285 bp fragment for mutants (Fig. 3) (11).

Figure 3d–g shows analyses of *hsf1*$^{-/-}$ tissues confirming that no Hsfl protein was produced.

3.2. Knockout of hsf2 Gene

Hsf2 gene structure: The mouse *hsf2* gene (*mhsf2*) is located on chromosome 10 and encodes Hsf2 protein that contains 517 amino acids. The mouse *hsf2* gene contains 12 exons. Exon 1 contains 117 bp and the start codon (ATG) begins at 25 bp. Exon1 and exon 2 are separated by a 9.4 kb intron (13).

Design of the targeting vector to delete *hsf2*: To disrupt the mouse *hsf2* gene, we designed a targeting vector in which 67 bp from the start codon were deleted. An EGFP reporter gene with a start and stop codon were inserted into this region of the *hsf2* gene. Removal of the first 67 bp from exon I of the *hsf2* gene results in an out-of-frame shift in the cDNA. Since EGFP inserted in exon 1 of the *hsf2* gene contained a stop codon, it is therefore unlikely that the truncated Hsf2 cDNA could be translated. Therefore, deletion of 67 bp from exon I of the *hsf2* gene completely disrupts the *hsf2* gene. The final targeting vector contained a 2.8 kb proximal fragment with homology to the *hsf2* gene, a 2.2 kb EGFP-neomycin fragment, and a 6.1 kb distal fragment with homology to *hsf2* (Fig. 4) (13).

Targeting vector: Isolation of the *hsf2* gene: A 22 kb DNA fragment containing 3.8 kb of the promoter region and the first 6 of the 12 exons of the murine *hsf2* gene were isolated from a 129/SvJ mouse genomic library in Lambda FixII vector (Stratagene, La Jolla, CA) by hybridization with a mouse *hsf2* exon 1 cDNA as a probe. This genomic clone was used to construct a targeting vector (13).

External probe for detection of correct targeting: To identify the homologous recombination, a 500 bp external probe was generated by PCR using primers: 5′-GTT TCT GCA CTG AGC CCT TG-3′ and 5′-CAA GGA TTC AAT AAT CGT GAC AC-3′. This probe hybridizes to a fragment of the *hsf2* gene that is located in the 5′ region of the targeting vector (Fig. 4a). The probe was tested on wild-type genomic DNA digested with PvuII restriction enzyme.

A 2.8 kb proximal fragment, including part of the *hsf2* promoter and the *hsf2* start codon, was amplified by PCR using the following primers: 5-AGT CCG CTC GAG GAG AGG TGG TAT ACA TAA ACA AGG (included a *Xho*I site, underlined) and 5-GAA CTC GGA TCC ATT GTT AGC CCG GTG CAG GGA TTC CAA ATT CTA CTA CCG AAC GCG GAG GTC GCA

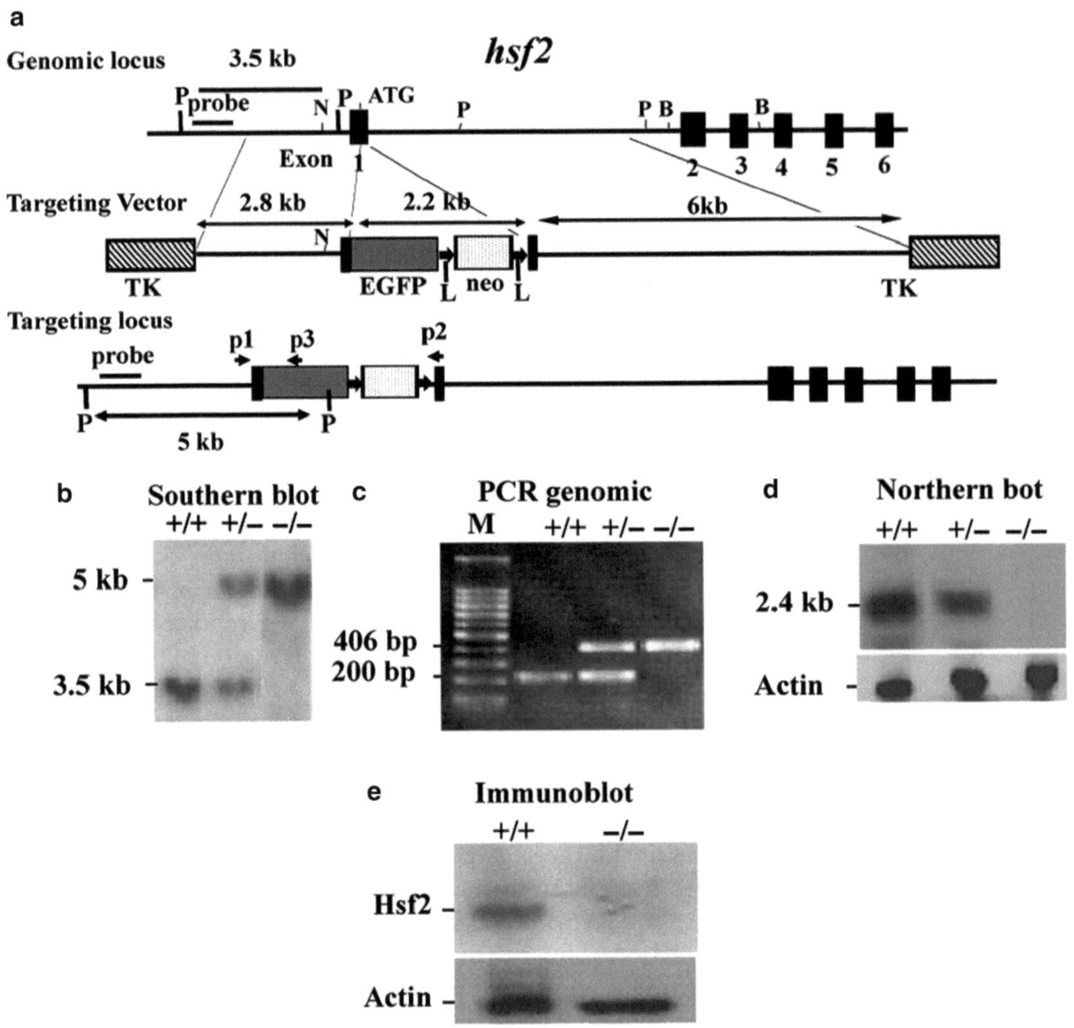

Fig. 4. Targeting strategy for the *hsf2* genomic locus and generation of *hsf2*-deficient mice. (**a**) Restriction map of the *hsf2* gene, showing the wild-type allele (*top*), targeting vector (*middle*), and the predicted targeted allele following homologous recombination (*bottom*). The ATG indicates the start codon (*top*). The position of the *EGFP-neo* and TK cassettes, probes for Southern blotting, and PCR primers P1, P2, and P3 are indicated. The vectors were designed so that the promoter of *hsf2* gene drives *EGFP* expression. Note that the Pvu II restriction enzyme site located upstream of exon 1 was destroyed in the targeting vector. The restriction enzymes are designated: P, PvuII; N, NheI; B, BamHI (13). (**b**) Southern blotting analysis of tail DNA derived from wild-type (+/+), heterozygous (+/–), or homozygous (–/–) *hsf2* mice. PvuII-digested genomic DNA was hybridized with an external probe to yield bands of 3.5 and 5 kb for the *hsf2* wild-type and targeted loci, respectively. (**c**) PCR-based genotyping assay amplifies wild-type and targeted *hsf2* locus fragments of 200 and 406 bp, respectively. (**d**) Northern blotting analysis. Total RNA extracted from the livers of 8-weeks-old mice of wild type (+/+), heterozygous (+/–), or homozygous (–/–) for the targeted *hsf2* allele adult mice was hybridized with a full-length murine *hsf2* cDNA probe. The expected 2.4 kb *hsf2* transcript was present in the wild-type and heterozygous mice, but absent in mice homozygous for the targeted *hsf2* allele. The level of actin mRNA is shown to indicate equal loading of RNA. (**e**) Western blot analysis. Equal amount of protein from cell extracts from the livers of 8-weeks-old mice of wild type (+/+) or homozygous (–/–) for the targeted *hsf2* allele adult mice was analyzed using SDS-PAGE followed by immunoblotting using antibody to Hsf2. The level of actin is shown to indicate equal loading of protein.

GCG GCG GCG G (included a BamHI site, underlined). This fragment lacks an unwanted PvuII site present 99 bp upstream of the start codon. This PCR fragment was digested by XhoI and BamHI and cloned into pBluescript II KS plasmid for amplification, and was sequenced to confirm the DNA sequence fidelity. The 2.8 kb proximal fragment was released by XhoI and BamHI digestion for subcloning into the final targeting vector.

A plasmid containing the 2.2 kb EGFP-neo cassette was digested with BamHI and ClaI to release the EGFP-neo fragment. The *EGFP* gene with the poly(A) signal was driven by the *hsf2* promoter (EGFP was from Clontech; neomycin was modified by the addition of two loxP sites by Dr. M. Capecchi's laboratory) (14, 15). Neomycin gene was driven by the TK promoter with a simian virus 40 poly(A) signal.

A 6 kb distal fragment, including the C-terminal 26 bp of exon 1 extending into the first intron, was PCR-amplified using the following primers: 5′-CCA TCG ATC CAA CGA GTT CAT CAC CTG GAG TC (included a ClaI site) and 5′-CTC ATA CTC GAG TTA ACT AAA CCA ATG CAT TCA ACTG-3′ (included XhoI site).

The 2.8 kb proximal fragment, 2.2 kb EGFP-neo fragment, and 6 kb distal fragment were ligated to phage DNA vector λDASHII-254-2TK at an XhoI site flanked by two TK genes (Fig. 1). The vector was then packaged into phage, and the positive phage clones were selected by PCR and restriction enzyme digestion. The positive phage DNA was digested by NotI to release the final targeting vector containing two TK genes. This final vector could be recircularized into a plasmid and could be amplified in bacteria. After amplification, the final vector was linearized with NotI and purified for transfection into ES cells.

ES cell electroporation: ES (D3; Incyte Genomics, St. Louis, MO) cells were cultured as described previously. ES cells were electroporated (BioRad Gene Plus, 250 V, 950 μF) with the linearized targeting vector and double selected by G418 (200 μg/ml) and ganciclovir (2 μM) (10, 13). Double-resistant ES cell clones were selected and expanded. Genomic DNA was isolated as described in Molecular Cloning (9).

Generation of mutant mice: Positive ES clones were selected by Southern blotting (Fig. 4b). The positive clones were microinjected into C57BL/6 blastocysts and germline-transmitting chimeric mice were obtained. The chimeric mice were then crossed with the wild-type mice to obtain *hsf2*$^{+/-}$ mice, which were crossed to generate *hsf2*$^{-/-}$ mice (13).

Genotyping of mutant mice: Southern blotting: ES clones, germline-transmitting chimeric mice, and first several litters of mice were genotyped by Southern blotting. Mouse genomic DNA was isolated (9) and digested with PvuII. Southern blotting was performed using an external probe (Fig. 4b) (13). This generates a 5 kb fragment for the targeted locus and a 3.5 kb fragment for the wild-type locus. PCR: When the *hsf2* mutant mouse line was established after Southern blotting, mice were genotyped routinely by PCR (Fig. 4c). Genomic DNA isolated from mouse tail was used as a template. PCR was performed using one common (i.e., recognized by both wild type and mutant) primer (P1; 5′-GTGG-TGTGCGTTCCCCGGAG-3′), a primer located in the wild-type locus (P2; 5′-TGACTCCAGGTGATGAACTC-3′) to identify the untargeted allele, and a primer (P3; 5′-CTTCGGGCATGGCGG-ACTTG-3′) located in the EGFP gene to identify the mutant gene. The locations of the primers P1, P2, and P3 are indicated by arrows in Fig. 4a. The expected PCR products for wild-type and targeted *hsf2* loci are fragments of 200 and 406 bp, respectively (Fig. 4c).

Figure 4d and e shows that cells deficient in Hsf2 do not express Hsf2 mRNA or protein.

3.3. Knockout of hsf4 Gene

Hsf4 gene structure: The mouse *hsf4* (*mhsf4*) gene is located on chromosome 8 and contains 13 exons (Fig. 5a) (16).

Design of the targeting vector to delete *hsf4*: To disrupt the *hsf4* gene, we designed a targeting vector by inserting the *EGFP-neo* cassette after the start codon. Ligation of the EGFP-neo fragment with the proximal fragment of the *hsf4* gene would result in an out-of-frame cDNA product for the *hsf4* gene by disruption of the gene at the ATG. The final targeting vector contains a 2.8 kb proximal fragment with homology to *hsf4*, a 2.2 kb *EGFP-neo* cassette (14, 15), and a 5.5 kb distal fragment with homology to *hsf4* (16).

Targeting vector: To generate the *hsf4* targeting vector, a 129/SvJ mouse genomic DNA phage library (Lambda FixII vector, Stratagene, La Jolla, CA) was used to identify clones containing the *hsf4* gene using a mouse *hsf4* cDNA probe. The isolated *hsf4* gene contained 13 exons within a 5.9 kb fragment, as well as several kb flanking sequences at both 5′ and 3′ regions.

The proximal 3.2 kb region was amplified by PCR using the following primers: 5′-TTCCCACGCGTCGACCCCTCCAGTCC CATTCTTTTGTTG-3′ and 5′-GAAGATCTGCCATGGCGCA-GTCTCGGCCGGCCGG-3′ (included BglII site, underlined). Because of the existence of an Sal I site within the amplified product, digestion with Sal I and Bgl II gives the final 2.8 kb proximal gene product.

A plasmid containing a 2.2 kb *EGFP-neo* cassette was digested with BamHI and ClaI to release the *EGFP-neo* fragment. The *EGFP* gene with the poly(A) signal was driven by the *hsf4* promoter, followed by the neomycin resistance gene, which was flanked by two loxP sites to allow its removal by Cre recombinase, and was driven by the TK promoter with a simian virus 40 poly(A) signal.

The distal 5.5 kb was PCR-amplified by using the following primers: 5′-CCATCGATGGCAGGAAGCGCCAGCTGCGCT-GCC-3′ (included ClaI site, underlined) and 5′-CCGCTCGAGC-GGGGCAGGGTCTTGTTGCATAGCCT-3′ (included Xho I site, underlined).

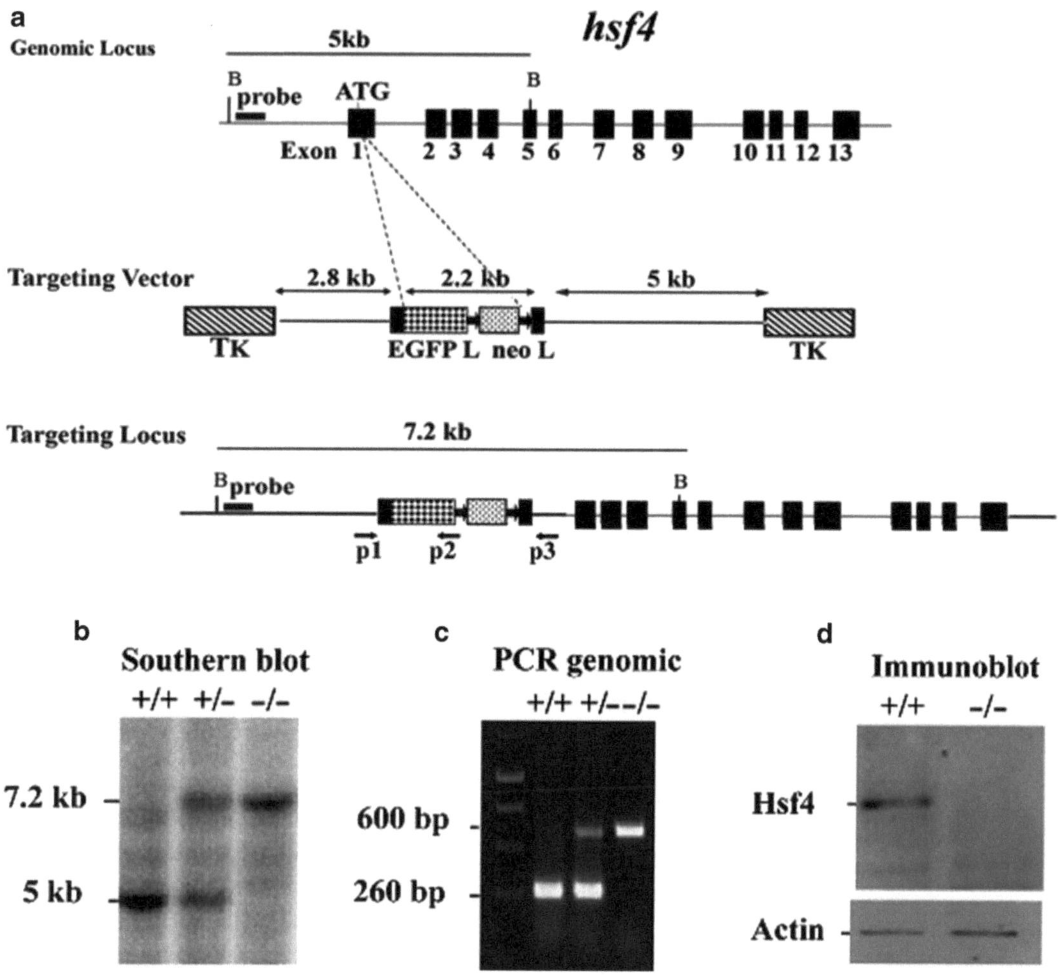

Fig. 5. Targeted disruption of the *hsf4* gene. (**a**) Wild-type *hsf4* locus, targeting vector, and the predicted targeted allele following homologous recombination are shown. Exons are presented by *black boxes*. The probe used for Southern blotting and PCR primers P1, P2, and P3 are indicated by *arrows* below the targeted allele (16). (**b**) Bgl II-digested tail DNA (10 μg) from wild-type (+/+), *hsf4*$^{+/-}$ (+/–), or *hsf4*$^{-/-}$ (–/–) mice was hybridized with an external probe to yield bands of 5 and 7.2 kb for the wild-type and targeted *hsf4* loci, correspondingly. (**c**) PCR-based genotyping assay amplifies fragments of 260 and 600 bp for wild-type and targeted *hsf4* allele, respectively. (**d**) 50 μg protein from lens extracts of wild-type (+/+) or *hsf4*$^{-/-}$ (–/–) mice at P28 was analyzed by Western blotting using antibody specific to Hsf4b. As a control for equal protein loading, the blot was probed for β-actin.

The 2.8 kb proximal fragment, 2.2 kb EGFP-neo fragment, and the 5.5 kb distal fragment were subcloned into phage DNA vector (λDASHII-254-2TK) at the XhoI sites to flank the targeting construct with two TK genes. The vector was then packaged into phage, and the positive phage clones were selected by PCR and restriction enzyme digestion. The positive phage DNA was digested by NotI to release the final targeting vector that contained the two TK genes. This vector could be circularized into a plasmid and amplified in bacteria. After amplification, the final vector was linearized by NotI digestion and used for electroporation of 129/SvJ ES cells (16).

ES cell electroporation: ES cells (D3; Incyte Genomics, St. Louis, MO) were electroporated with the linearized targeting vector. ES cells were then selected by G418 (200 μg/ml) and ganciclovir. Double-resistant ES cell clones were selected and expanded for screening by Southern blotting (9).

Genotyping of mutant mice: Southern blotting: Mouse genomic DNA from ES cells was isolated and digested with BglII. Restriction enzyme-digested genomic DNA was then hybridized with an external probe located upstream of the targeting vector, yielding 5 and 7.2 kb fragments for wild-type and *hsf4*-targeted alleles, respectively. From the 141 isolated ES clones that were analyzed by Southern blotting, 33 clones contained the correctly targeted allele (Fig. 2) (16). Two positive ES clones were injected into C57BL/6J blastocysts, and the resulting chimeric male mice were crossed with C57BL/6J females to generate germline transmission. Homozygous mice were obtained by interbreeding of F1 heterozygous mice (Fig. 5b).

Primers used for genotyping: For routine genotyping of mice, DNA extracted from tail was used for PCR analysis to verify a 260 bp wild type and a 600 bp targeted *hsf4* fragments using the following primers: P1: 5′-GCAAACGCAGCACTTTCGCG-3′; P2: 5′-CGGATCTTGAAGTTCACCTTGAT-3′; P3: 5′-TGGACAGGGGTGTTCACGACA-3′ (Fig. 5c).

Immunoblotting of lens extracts of *hsf4*-deficient mice showed no Hsf4 expression (Fig. 5d). Please also see Section 4. Notes for general considerations to knockout *hsf1*, *hsf2*, and *hsf4 genes.*

3.4. Physiological Function of Mice with a Targeted Disruption of the hsf1, hsf2, or hsf4 Gene

In this section, we briefly describe the major phenotypes of the *hsf1*-, *hsf2*-, and *hsf4*-deficient mice generated in our laboratory. Additionally, since some of our *hsf*-deficient mouse lines encode a reporter gene and we have also generated *hsp25*$^{-/-}$-lacZ and *hsp70.3*$^{-/-}$-lacZ (17, 18) reporter genes, we briefly describe the beneficial uses of knockout mice containing reporter genes in investigating the effects of hsf deletions on expression of their downstream target gene in vivo.

In addition to the *hsf* knockout mice that we have generated (11, 13, 16), there are two other *hsf1* (19, 20), two other *hsf2* (21, 22), and one other *hsf4* (23) mutant mouse lines that have been generated in other laboratories. The phenotypes of all mouse lines are almost comparable with each other. *Hsf1*-deficient mice generated in our laboratory exhibit complete female infertility (unpublished data). This phenotype was reported by Christians et al. (24), and the cause appears to be the inability of the *hsf1*$^{-/-}$ zygote to undergo zygotic gene activation following fertilization (24). Another major phenotype of *hsf1*$^{-/-}$ mice is their inability to mount a heat-shock response in every tissue that has been tested using immunoblotting or following crossing of *hsf1*$^{-/-}$ mice with *hsp70.3*$^{-/-}$-lacZ reporter mice (11). *Hsf1*$^{-/-}$ mice exhibit an age-dependent demyelinating disease (25), and cells deficient in *hsf1* exhibit accumulation of ubiquitinated proteins, including wild-type and mutant p53 (25, 26). *Hsf1*$^{-/-}$ mice in p53-deficient background (*hsf1*$^{-/-}$*p53*$^{-/-}$) exhibit a delay in development of lymphomas compared to *p53*$^{-/-}$ mice. There is a change in tumor spectrum that is observed in *hsf1*$^{-/-}$*p53*$^{-/-}$ mice compared to *p53*$^{-/-}$ mice, as double-knockout mice exhibit reduced lymphomas (7.9% in *hsf1*$^{-/-}$*p53*$^{-/-}$ mice versus 72.2% in *p53*$^{-/-}$ mice) while they exhibit increased levels of solid tumors (27). *Hsf2*$^{-/-}$ mice exhibit defects in spermatogenesis and males exhibit reduced fertility a few months after birth and in the background of *hsf1* deficiency, all males are infertile due to complete disruption in spermatogenesis (28). Hsf2 also expresses at high levels in the brain, and *hsf2*$^{-/-}$ mice exhibit developmental defects in the central nervous system (CNS) (13). Further studies on these mice are needed to reveal additional functions of Hsf2 in mammalian organisms. Before we generated *hsf4*$^{-/-}$ mice, there was no information on how Hsf4 becomes transcriptionally activated or in which tissues or cells it expresses. Previous reports indicated that the *hsf4* gene was mutated in humans, and humans who carry the mutation exhibit lamellar and marner cataracts (29). Interestingly, *hsf4*$^{-/-}$ mice exhibit developmental defects in fiber cell differentiation in the lens which leads to blindness in 100% of the progeny (16). Hsf4 activity was, for the first time, detected in the lens epithelial cells at 3 days postnatally (16). The activation of Hsf4 leads to expression of Hsp25, which was 1,000-fold lower in *hsf4*$^{-/-}$ lens (16).

Hsf4-EGFP is expressed in many tissues. Addition of a reporter gene (EGFP) under control of the *hsf2* and *hsf4* promoters has been a powerful means of revealing in which cell types these transcription factors express in vivo. As we already have reported, Hsf2-EGFP expression can be detected in the testis during spermatogenesis using flow cytometry (13). Flow cytometry and immunoblotting experiments also show that *hsf4-EGFP* (knockout/knock-in mice) is expressed in a number of adult tissues (Fig. 6). The expression of *hsf4-EGFP* was analyzed in spleen and found to be expressed in mature CD4+, CD8+, and CD3+ (T cell receptor) thymocytes, GR-1-positive granulocytes (not presented), neutrophils/macrophages (CD11b+), and dendritic cells (CD11c+) (Fig. 6). Interestingly, Hsf4 does not express in B cells or immature CD4^{-}CD8^{-} or CD4^{+}CD8^{+} T cells (Fig. 6, thymus), but it is expressed in spleen and peripheral lymph nodes (Fig. 6a and b). Interestingly, Hsf4-EGFP expression can be detected only once T cells leave the thymus and enter the periphery.

Hsf4 regulation of Hsp25 expression in vivo: Another unique method of analyzing knockout mice expressing a reporter gene is the use of intercrossing Hsfs with their downstream target genes that express LacZ (or EGFP when possible) under their endogenous promoters to determine the extent that they regulate each

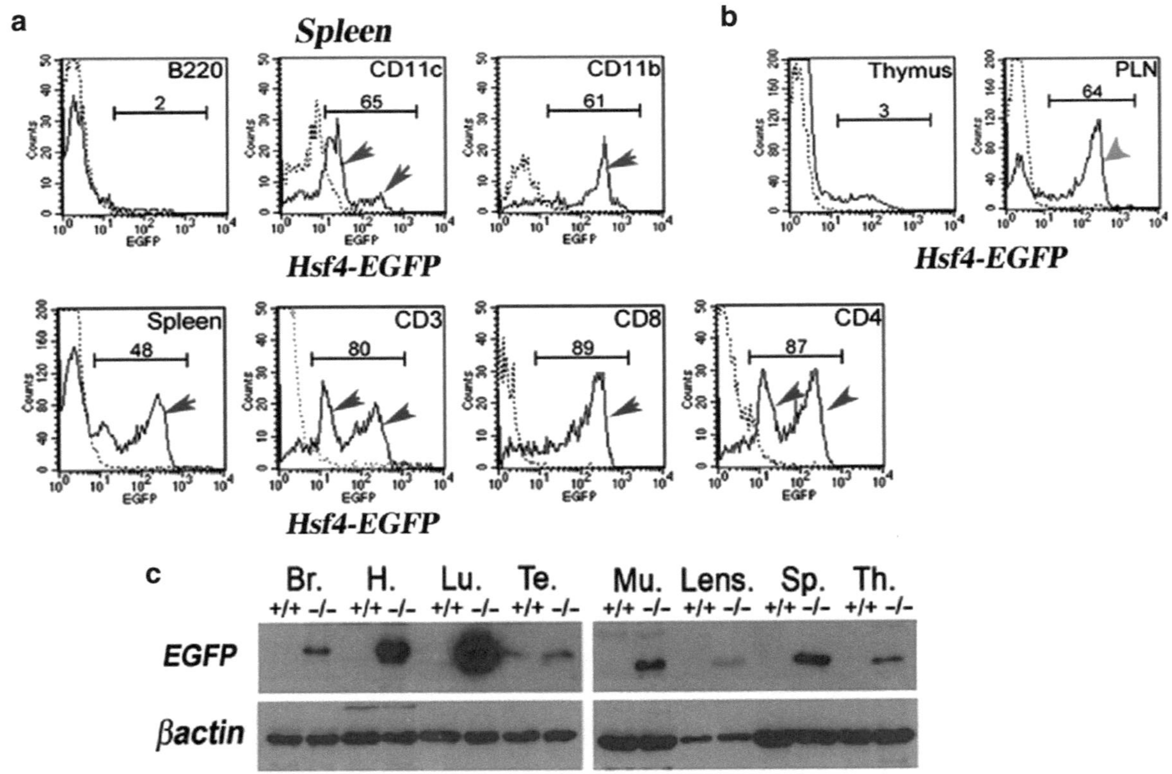

Fig. 6. Hsf4-EGFP is expressed in adult normal tissues. (**a**) Cells from *hsf4*$^{-/-}$ spleen were immunostained to detect a specific cell population expressing Hsf4-EGFP as indicated. B220 detects B cell population; CD11c detects dendritic cells; CD11b detects neutrophils/macrophages; CD3+, CD4+, or CD8+ are T cell-specific markers. *Dotted lines* are immunostained +/+ cells (no EGFP). (**b**) Cells from *hsf4*$^{-/-}$ thymus or peripheral lymph nodes (PLNs) were analyzed by flow cytometry for expression of EGFP. In (**a**) and (**b**), *numbers* represent percentages of cells expressing EGFP, and these populations are indicated by *green arrows*. (**c**) 30 μg of indicated adult tissue extracts from wild-type or *hsf4*$^{-/-}$ mice were used in immunoblotting experiments using antibody to EGFP. Hsf4-EGFP expression can be detected in the brain (Br), heart (H), lung (Lu), testis (Te), muscle (Mu), lens, spleen (Sp), and thymus tissue (Th). Wild type (+/+), *hsf4*$^{-/-}$(−/−). Note that EGFP expression is an indication of Hsf4 expression.

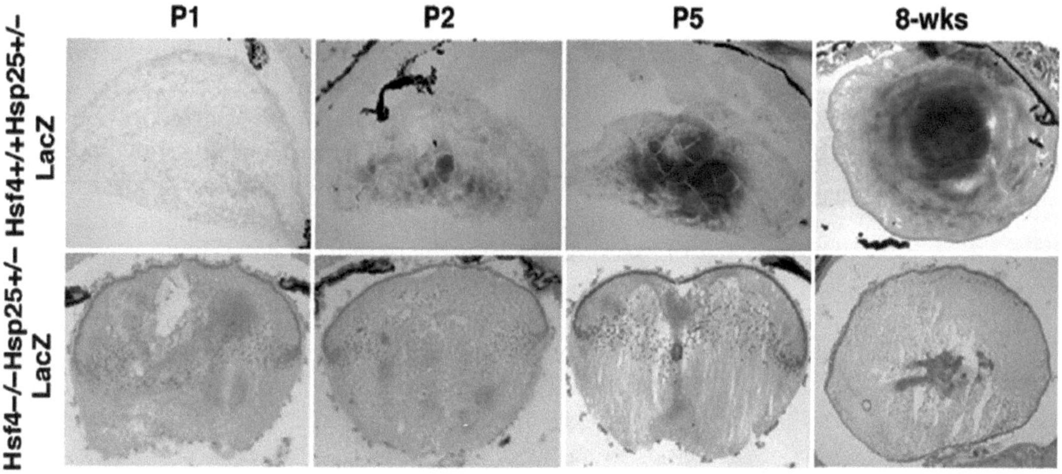

Fig. 7. Hsf4 controls the expression of Hsp25 in the lens during development. Histological analyses of *hsf4*$^{+/+}$*hsp25*$^{+/-}$*-LacZ* lens at P1 to 8-weeks to show the expression of β-gal in developing and mature lens. Note the positive X-gal staining at P2 *hsf4*$^{+/+}$*hsp25*$^{+/-}$*-lacZ* while no X-gal staining can be detected in *hsf4*$^{-/-}$*hsp25*$^{+/-}$*-lacZ*.

downstream target gene in vivo. In one study, we crossed *hsf4*-deficient mice with *hsp25*$^{-/-}$-lacZ reporter mice (18). We found that *hsp25-lacZ* is a downstream target of the *hsf4* gene in the lens. As we described earlier, Hsf4 DNA-binding and transcriptional activity was demonstrated in developing lens epithelial cells (16). Using gel mobility shift assays, Hsf4 DNA-binding activity could be detected between postnatal days P1–P5 lens extracts (data not presented, please see ref. 16). Using *hsp25*$^{+/-}$*-LacZ* knock-in mice, we were able to demonstrate that Hsp25-lacZ expression coincides with the onset of Hsf4 activity in the lens epithelium and fiber cells (Fig. 7, upper panel, X-gal staining). Crossing *hsf4*$^{-/-}$ mice with *hsp25*$^{-/-}$ mice, we were able to completely eliminate the Hsp25 promoter-driven β-galactosidase (β-gal) expression, suggesting that Hsp25 is a downstream target of the *hsf4* gene during lens epithelial cell differentiation.

4. Notes

General considerations to knock out *hsf1*, *hsf2*, and *hsf4* genes.

The following points are important considerations for successful gene targeting:

1. The genomic DNA source must be the same as the ES cells to be used. This facilitates homologous recombination.
2. Purchase a BAC clone containing the gene of interest (http://bacpac.chori.org).

3. For designing the targeting vector, the length of the 5′ and 3′ fragments is important in facilitating homologous recombination. Sizes between 0.8 and 5 kb have been successfully used. The total size of the fragment that can be inserted into the final vector needs to be considered before constructing the vector. The λDash II-254-2TK phage used by our laboratory can accommodate approximately 14 kb. Therefore, the sizes of the 5′ and 3′ fragments plus the neomycin gene that is required for positive selection of ES cells and a reporter gene (such as green fluorescent protein (GFP) or β-gal) should not exceed more than 14 kb pairs. Before attempting to construct the targeting vector, it is best to schematically draw the entire plan for the construct to be made. This should include all the restriction enzymes to be used.
4. It is best to insert the EGFP or LacZ genes at the ATG of the gene that is targeted. This design ensures that expression of EGFP or LacZ is under the direct control of the promoter of the gene to be targeted.
5. Although insertion of the EGFP or LacZ gene at the ATG interrupts the gene of interest, the design of the targeting construct could be such that a portion of the gene to be targeted is also deleted to ensure complete disruption of the gene.
6. Two probes need to be designed for detection of ES cell clones following electroporation of the targeting vector into ES cells. The outside probe is located outside of the 5′ and 3′ fragments. A restriction enzyme must be selected so that the outside probe can detect correct targeting into the intended locus. Sometimes, creating a unique restriction enzyme before the EGFP (or LacZ genes) or between the EGFP and neomycin genes (or after neomycin) is an option. Another consideration is that the fragment size created following restriction enzyme digest is not larger than 12–15 kb since large fragments are be more difficult to detect by Southern blotting. Once the outside probe (best size is 500 bp–1 kb) is selected, it is best to perform Southern blotting using the genomic DNA following digestion with one or two restriction enzymes to ensure that the probe generates a predicted band for the wild-type locus.
7. It is best that the neomycin gene is flanked by two loxP sites so that it can be removed following generation of the knockout mouse by crossing with transgenic female mouse expressing Cre recombinase (Splicer mice (30), Jackson laboratory).
8. The identity of all the fragments that have been amplified by PCR needs to be verified by DNA sequencing. The ligation sites of the final targeting vector also need to be confirmed by sequencing since deletions may occur during the ligation.

Acknowledgments

This work was supported by VA Award 1I01BX000161 and NIH grants CA062130 and CA132640 (NFM) and CA121951 and CA121951-07S2 (DM). For generation of *hsf* knockout mice, the microinjection of ES cells and generation of chimeras were conducted in the Medical College of Georgia Embryonic Stem Cell and Transgenic Core Facility.

References

1. Wu C. Heat shock transcription factors:structure and regulation. Ann Rev Cell Dev Biol 1995; 11:441–69.
2. Morimoto RI. Regulation of the heat shock transcriptional response: cross talk between a family of heat shock factors, molecular chaperones, and negative regulators. Genes Dev 1998;12:3788–96.
3. Nakai A, Tanabe M, Kawazoe Y, Inazawa J, Morimoto RI, Nagata K. HSF-4, a new member of the human heat shock factor family which lacks properties of a transcriptional activator. Mol Cell Biol 1997;17:469–81.
4. Hu Y, Mivechi NF. Association and regulation of heat shock transcription factor 4b with both extracellular signal-regulated kinase mitogen-activated protein kinase and dual-specificity tyrosine phosphatase DUSP26. Mol Cell Biol 2006;8:3282–94.
5. Muller U. Ten years of gene targeting:targeted mouse mutants, from vector design to phenootype analysis. Mechanisms of Development 1999;82:3–21.
6. Van Der Weyden L, Adams DJ, Bradley A. Tools for targeted manipulation of the mouse genome. Phsiol Genomics 2202;11:133–64.
7. Bockamp E, Sprengel R, Eshkind L, Lehmann T, Braun JM, Emmrich F, Hengstler JG. Conditional transgenic mouse models: from the basics to genome-wide sets of knockouts and current studies of tissue regeneration. Regen Med 2008;3:217–35.
8. Bockamp E, Maringer M, Spangenberg C, Fees S, Fraser S, Eshkind I, Oesch F, Zabel B. Of mice and models: improved animal models for biomedical research. Physiol Genomics 2002;11:115–32.
9. Sambrook J, Fritsch EF, Maniatis T. Molecular Cloning: A laboratroy Manual 1989; Second Edition.
10. Limaye A, Hall B, and Kulkarni AB. Manipulation of Mouse Embryonic Stem Cells for Knockout Mouse Production. Current Protocols in Cell Biology 2009;44 Unit 19. 13:1–24.
11. Zhang Y, Huang L, Zhang J, Moskophidis D, Mivechi NF. Targeted disruption of hsf1 leads to lack of thermotolerance and defines tissue-specific regulation for stress-inducible Hsp molecular chaperones. J Cell Biochem 2002; 86:376–93.
12. Zhang Y, Koushik S, Dai R, Mivechi NF. Structural organization and promoter analysis of murine heat shock transcription factor-1 gene. J Biol Chem 1998;273:32514–21.
13. Wang G, Zhang J, Moskophidis D, Mivechi NF. Targeted disruption of the heat shock transcription factor (hsf)-2 gene results in increased embryonic lethality, neuronal defects, and reduced spermatogenesis. Genesis 2003; 36:48–61.
14. Thomas KR, Capecchi MR. Site-directed mutagenesis by gene targeting in mouse embryo-derived stem cells. Cell 1987;51:503–12.
15. Godwin AR, Stadler HS, Nakamura K, Capecchi MR. Detection of targeted GFP-Hox gene fusions during mouse embryogenesis. Proc Natl Acad Sci USA 1998;95:13042–7.
16. Min J, Zhang Y, Moskophidis D, Mivechi NF. Unique contribution of heat shock transcription factor 4 in ocular lens development and fiber cell differentiation. Genesis 2004; 40:205–17.
17. Huang L, Mivechi NF, Moskophidis D. Insights into regulation and function of the major stress-induced hsp70 molecular chaperone in vivo: Analysis of mice with targeted gene disruption of the hsp70.1 or hsp70.3 genes. Mol Cell Biol 2001;21:8575–91.
18. Huang L, Min J, Maters S, Mivechi NF, Moskophidis DI. Insights into the function and regulation and of small hsp25 (HSPB1) in mouse model with targeted gene disruption. Genesis 2007;45:487–501.

19. Xiao X, Zuo X, Davis AA, McMillan DR, Curry BB, Richardson JA, Benjamin IJ. HSF1 is required for extra-embryonic development, postnatal growth and protection during inflammatory responses in mice. EMBO J 1999;18: 5943–52.
20. Sugahara K, Inouye S, Izu H, Katoh Y, Katsuki K, Takemoto T, Shimogori H, Yamashita, H, Nakai A. Heat shock transcription factor HSF1 is required for survival of sensory hair cells against acoustic overexposure. Hear Res 2003;182:88–96.
21. McMillan DR, Christians E, Forster M, Xiao X, Connell P, Plumier JC, Zuo X, Richardson J, Morgan S, Benjamin IJ. Heat shock transcription factor 2 is not essential for embryonic development, fertility, or adult cognitive and psychomotor function in mice. Mol Cell Biol 2002;22:8005–14.
22. Kallio M, Chang Y, Manuel M, Alastalo TP, Rallu M, Gitton Y, Pirkkala L, Loones, MT, Paslaru L, Larney S, Hiard S, Morange M, Sistonen L, Mezger V. Brain abnormalities, defective meiotic chromosome synapsis and female subfertility in HSF2 null mice. EMBO J 2002;21:2591–601.
23. Fujimoto M, Izu H, Seki K, Fukuda K, Nishida T, Shuichi Y, Kato K, Yonemura S, Inouye S, Nakai A. HSF4 is required for normal cell growth and differentiation during mouse lens developments. EMBO J 2004;23:4297–306.
24. Christians E, Davis AA, Thomas SD, Benjamin IJ. Maternal effect of Hsf1 on reproductive success. Nature 2000;407:693–4.
25. Homma S, Jin X, Wang G, et al. Demyelination, astrogliosis, and accumulation of ubiquitinated proteins, hallmarks of CNS disease in hsf1-deficient mice. J Neurosci 2007;27:7974–86.
26. Jin X, Moskophidis D, Hu Y, Phillips A, Mivechi NF. Heat shock factor 1 deficiency via its downstream target gene alphaB-crystallin (Hspb5) impairs p53 degradation. J Cell Biochem 2009;107:504–15.
27. Min J-N, Huang L, Zimonjic D, Moskophidis D, Mivechi NF. Selective suppression of lymphomas by functional loss of hsf1 in a p53-deficient mouse model of spontaneous tumors. Oncogene 2007;26:5086–97.
28. Wang G, Ying Z, Jin X, Tu N, Zhang Y, Phillips M, Moskophidis D, Mivechi NF. Essentail requirement for both *hsf1* and *hsf2* transcriptional activity activity in spermatogenesis and male fertility. Genesis 2004;38:66–80.
29. Bu L, Jin Y, Shi Y, et al. Mutant DNA-binding domain of HSF4 is associated with autosomal dominant lamellar and Marner cataract. Nat Genet 2002;31:276–8.
30. Koni PA, Joshi SK, Temann UA, Olson D, Burkly L, and Flavell RA. Conditional vascular cell adhesion molecule 1 deletion in mice: Impaired lymphocyte migration to bone marrow. J Exp Med 2001;193:741–54.

Chapter 2

The Role of Heat Shock Factors in Stress-Induced Transcription*

Yue Zhang, Shiuh-Dih Chou, Ayesha Murshid, Thomas L. Prince, Sheila Schreiner, Mary Ann Stevenson, and Stuart K. Calderwood

Abstract

Heat shock proteins (HSPs) are rapidly induced after stresses, such as heat shock, and accumulate at high concentrations in cells. HSP induction involves a family of heat shock transcription factors that bind the heat shock elements of the *HSP* genes and mediate transcription *in trans*. We discuss methods for the study of HSP binding to *HSP* promoters and the consequent increases in HSP gene expression in vitro and in vivo.

Key words: Heat, Shock, Factor, Binding, Transcription

1. Introduction

Heat shock factor (HSF) was first discovered in yeast as a sequence-specific transcription factor that binds to the promoters of heat shock protein (HSP) genes (1). HSF was shown to bind as a trimer to three inverted repeats of the sequence nGAAn at high affinity, an activity that was later shown in *Drosophila* HSF and human HSF1 (2–5). In more complex organisms, there at least four members of the family in avian and mammalian species and multiple members in higher plants (6–11). The current consensus in mammalian cells is that HSF1 is the most potent regulator of the heat shock response with the remaining factors playing supplementary roles in stress and perhaps more significant roles in

*Yue Zhang, Shiuh-Dih Chou, Ayesha Murshid, Thomas L. Prince, and Sheila Schreiner contributed equally to the article.

Stuart K. Calderwood and Thomas L. Prince (eds.), *Molecular Chaperones: Methods and Protocols*,
Methods in Molecular Biology, vol. 787, DOI 10.1007/978-1-61779-295-3_2, © Springer Science+Business Media, LLC 2011

development (6, 12, 13). The mechanisms by which HSF1 is triggered by stress are not entirely clear. HSF1 is thought to be repressed by its products – the HSPs through a feedback inhibition mechanism (14, 15). Activation is, thus, envisaged as a reversal of such inhibition as denatured protein sequester HSPs during heat shock and HSF1 liberated to bind to HSE elements in HSP genes. However, alternative hypotheses have been proposed involving stress-mediated HSF1 phosphorylation and binding to large noncoding RNA (16, 17). HSF1 and HSF2 are predicted to encode at least two splicing variants, with HSF2A and HSF2b showing differential expression during erythroid differentiation (6). HSF2A appears to be active in transcriptional regulation while HSF2B appears to be inactive (6).

Heat shock causes a rapid increase in HSF1 binding to HSP promoters and an acute elevation in the transcription of HSP genes (relative rates are indicated in Fig. 1). Transcription decays rapidly after initiation while HSF continues to bind to HSE for several hours after transcription is ceased (18, 19). HSP mRNA is then observed within 1 h of activation and is maintained at these high levels for up to 24 h due to enhanced stabilization after stress (20). After acute stress, HSP protein expression is delayed due to initial translational inhibition, but Hsp70, Hsp90, and Hsp110 are observed by 2–6 h after a 43°C heat shock and can persist in cells for up to 100 h (21) (Fig. 1). In this report, we are concerned with the early phase of HSP gene expression, involving HSF1 binding to HSP genes and activation of transcription.

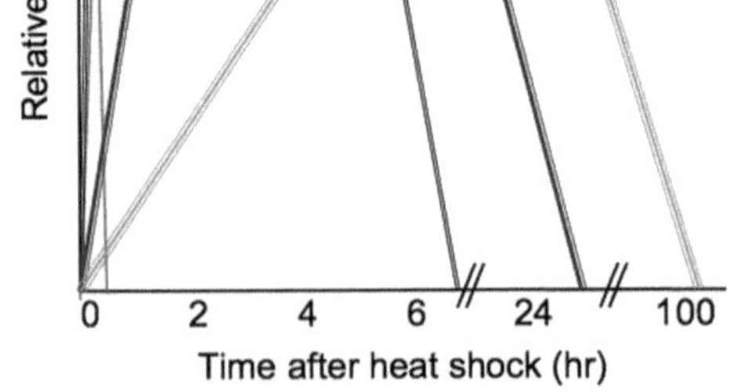

Fig. 1. Relative kinetics of *hsp* gene transcription, HSF binding, HSP mRNA expression, and heat shock protein expression after heat shock.

2. Materials

2.1. Purification of HSF2 and Electrophoretic Mobility Shift Assay

Primers

For human HSF2:

Forward primer was 5′–3′ GC[GAATCC]ATGAAGCAGAGT TCGA.

Reverse primer was 5′–3′ AAA[GTCGAC]TTCCTGGGGAT TTAGCTA.

For murine HSF2:

Forward primer was 5′–3′ GG[GAATCC]ATGAAGCAGAGT TCGAACG.

Reverse primer was 5′–3′ AGT[GTCGAC]TTGGGAGTTTA ACTATCT.

2.2. EMSA Oligonucleotides

Hsp70 HSE top strand: 5′-CACCTCGGCTGGAATATTCCCGA CCTGGCAGCCGA-3′.

2.3. Mutant Oligonucleotides

5′-CACCTCGGCTG**CAA**TAA**T**CCCGACCTGGCAGCCGA-3′.

Cells: BL21 (DE3) *Escherichia coli*, *Human HeLa*

2.4. Columns and Filters

20 ml glutathione-sepharose (Pierce Chemicals).

Mono-Q HR 5/5 (Pharmacia).

Centricon 10 ultrafilter.

2.5. Buffers

E. coli lysis buffer. 7M guanidine-HCl in 0.1M potassium phosphate buffer, pH 7.4, containing 50 mM DTT and 0.05% NP-40.

Dialysis buffer. 50 mM potassium phosphate buffer containing 0.1M KCl and 2 mM DTT.

BSA solution. 0.1 mg/ml bovine serum albumin.

Electrophoretic mobility shift assay (EMSA) lysis buffer. 10 mM HEPES, 10 mM NaCl, 0.1 mM EDTA, 1.0 mM dithiothreitol (DTT), 1.0 mM phenylmethylsulfonyl fluoride (PMSF), 2.0 mg/ml aprotinin, leupeptin, 20 mM NaF, and 2.0 mM Na_3VO_4 (pH 7.9).

HSF extraction buffer. Aprotinin, leupeptin, 20 mM NaF, and 2.0 mM Na_3VO_4 (pH 7.9) on ice. Cells are then lysed by addition of Nonidet P-40 to 0.6% and lysates clarified by spinning at 12,000 × *g*. Nuclear pellets.

EMSA incubation buffer. (12 μl) Contained 2.0 μl nuclear extract or recombinant protein, 2.0 mg/ml bovine serum albumin, 2.0 mg/ml poly dI–dC, 0.5–1.0 ng ^{32}P-labeled, double-stranded oligonucleotide probe, 12 mM HEPES, 12% glycerol, 0.12 mM

EDTA, 0.9 mM $MgCl_2$, 0.6 mM DTT, 0.6 mM PMSF, and 2.0 mg/ml aprotinin and leupeptin (pH 7.9).

2.6. ChIP Assay

Chromatin immunoprecipitation (ChIP) dilution buffer (1.1% Triton X-100, 1.2 mM EDTA, 16.7 mM Tris–HCl, pH 8.0, and 167 mM NaCl).

Protein A Agarose Slurry (Sigma Chemicals, St Louis, MO).

ChIP Washing buffers:

Washing buffer 1 (20 mM Tris–HCl, pH 8.0, 150 mM NaCl, 1% Triton X-100, 2 mM EDTA, and 0.1% SDS).

Washing buffer 2 (20 mM Tris–HCl, pH 8.0, 500 mM NaCl, 1% Triton X-100, 2 mM EDTA, and 0.1% SDS).

Washing buffer 3 (10 mM Tris–HCl, pH 8.0, 250 mM LiCl, 1% NP-40, 1% sodium deoxycholate, and 1 mM EDTA), TE (10 mM Tris–HCl, pH 7.5, 1 mM EDTA).

ChIP elution buffer (1% SDS, 0.1M $NaHCO_3$).

5M NaCl.

ChIP uncrosslinking buffer. 0.5M EDTA, 10 μl of 1M Tris–HCl, pH 6.5, and 2 μl proteinase K.

ChIP hsp70.1 primers. Exon region forward primers: {*hsp70.1* exn Forward: 5′ ggacatcagccagaacaagc 3′ *hsp70.1* exn Reverse: 5′ aagtcgatgccctca aac ag 3′ *hsp70.1* HSE Reverse: 5′ cggcttt tataagtcgtcgt 3′ *hsp70.1* HSE Forward: 5′ aggcgaaacccctg gaata 3′}.

2.7. Run-on Transcription

Run-on lysis buffer. 10 mM Tris–HCl (pH 7.4), 10 mM NaCl, 3 mM $MgCl_2$, and 0.5% nonidet-P40.

Run-on storage buffer. 50 mM Tris–HCl (pH 8.30), 40% glycerol, 5 mM $MgCl_2$, and 40 U of RNAsin (Roche Molecular Biochemicals).

Run-on reaction buffer. 10 mM Tris–HCl (pH 8.0), 5 mM $MgCl_2$, 0.3M KCl, 5 mM DTT, 1 mM ATP, 1 mM CTP, 1 mM GTP, and 50 μCi [α-^{32}P] UTP (3,000 Ci/mmol).

Hybridization solution. UltraHyb solution (*Ambion*).

Hybridization washing buffers. (1) 2× SSC, 0.1% SDS, (2) high-stringency solution (1× SSC, 0.1% SDS), (3) 2× SSC, 0.1% SDS with 10 μg RNase A.

3. Methods

3.1. Purification of Heat Shock Factors and In Vitro EMSA

In order to study the properties of HSF family members in vitro, we have prepared purified, recombinant HSF1 and HSF2.

3.1.1. Complementary DNA Cloning of Human and Mouse HSF2A and HSF2B

RNA was isolated from NIH-3T3 (mouse) or HeLa (human cells) and messenger RNA prepared by poly-T affinity chromatography (PolyAtract system, Promega, Madison, WI). cDNA was then prepared from the mRNA using the AMV reverse transcriptase system (Promega) and HSF2 cDNAs amplified using Taq polymerase and the polymerase chain reaction using primer sets mentioned above (6).

Forward primers contain Eco-R1 restriction site consensus sequences [marked in boxes], and the reverse primers contain Sal-1 sequences for subsequent cloning of amplified DNAs into the PGEX5 prokaryotic expression vector (Pharmacia). After transformation and growth of competent bacteria, colonies are screened for either total HSF2 using oligonucleotides (1764–1785; CAGGAGCAAGTTCACATAAATA and 1786–1807; GGCATAT CACTATCCAGAGGTG) predicted to detect all forms of HSF2 or for the larger form (HSF2A) using oligonucleotides predicted to hybridize specifically with this species (1420–1440; TTGTA TTATTGATGTAATCT and 1392–1412; CATCTGCACAGAAC TAG TGA). Oligonucleotides are then end-labeled with ^{32}P ATP and T4 polynucleotide kinase. Plasmids detected using these probes are isolated and screened for the presence of inserts and for the production of HSF2–glutathione transferase fusion proteins from representative cDNAs in bacteria exposed to the inducing agent IPTG (Pharmacia). After induction, bacterial lysates are prepared and screened by immunoblot with anti-GST antibodies (St Cruz Antibodies) and anti-HSF2 antibody Ab-3158 prepared in the Calderwood lab. Representative clones from human HSF2A and HSF2B and murine HSF2A and HSF2B are then further analyzed by dideoxynucleotide sequencing.

3.1.2. Purification of HSF2 Proteins

HSF2 variants are cloned into the pGEX-5 expression vector between the Eco RI and Sal I sites and the resulting plasmids are used to transform BL21 (DE3) *E. coli* bacteria. HSF2 was, thus, expressed as a fusion protein with glutathione-*S* transferase. All purification steps are carried out at 4°C. Briefly, IPTG-induced bacteria are pelleted and dissolved in *E. coli* lysis buffer and dialyzed against *E. coli* lysis buffer. The samples are centrifuged at 2,500 × *g* for 5 min and the supernatant loaded on a 20 ml volume glutathione sepharose column at a flow rate of 0.5 ml/min, washed extensively with dialysis buffer, and eluted with this buffer A containing 10 mM reduced glutathione. The eluate was loaded onto a Mono-Q HR 5/5 ion-exchange chromatography column at a flow rate of 0.8 ml/min and eluted with a 24 ml linear gradient from 0.1 to 1.0M final concentration of KCl in buffer A. Absorbance was monitored at 280 nm and the fractions corresponding to HSF2 assayed for binding to HSE, pooled, and concentrated with a Centricon 10 ultrafilter in the presence of 0.1 mg/ml bovine serum albumin. Relative concentrations of active HSF2 are estimated by quantitative EMSA (Fig. 1).

GST–HSF1 is purified using a similar protocol (22). Alternatively, we have described a detailed method for purifying recombinant HSF1 after expression in *E. coli* from the pET7.1 vector (23). Recombinant HSF1 without a GST tag is prepared by ammonium sulfate precipitation, heparin-agarose affinity, and ion-exchange chromatography in pure form as assessed by SDS-PAGE and reverse-phase HPLC (23).

Activity of purified GST–HSF2A, GST–HSF2B, or GST–HSF1 is estimated by EMSA. Proteins are incubated with ^{32}P-labeled HSE at a range of dilutions and then subjected to EMSA analysis as described below. GST–HSF2A and GST–HSF2B are serially diluted 1/2,200, 1/660, 1/220, 1/66, 1/22, and 3/22 prior to EMSA.

3.1.3. Nuclear Extraction from Tissue Culture Cells and EMSA

EMSA is carried out using purified recombinant HSF or after the extraction of intracellular HSF complexes from either whole cell or nuclear extracts from heat shocked cells and incubation of complexes with double-stranded oligonucleotides encoding heat shock elements (HSEs) in HSP genes. To prepare HSF from cells growing in vitro, nuclear extracts are prepared according to Schreiber (24) (see Note 1). In our standard assay, cells are incubated for 15 min in 200–800 μl of EMSA lysis buffer on ice. Cells are then lysed by addition of Nonidet P-40 to 0.6% and lysates clarified by spinning at 12,000×*g*. Nuclear pellets are then resuspended in 25 μl ice-cold EMSA extraction buffer. Extracts containing HSF are then aliquoted and stored at –80°C.

For incubation with oligonucleotide probe, each binding mixture (12 μl) contained 2.0 μl nuclear extract or recombinant protein, 2.0 mg/ml bovine serum albumin in EMSA incubation buffer. Samples are incubated at room temperature for 15 min, and then fractionated by electrophoresis on 4.0% polyacrylamide, 1× TBE gels. Oligonucleotide hHSE was synthesized, annealed, and labeled by end filling with ^{32}P-dCTP at 6,000 Ci/mmol (DuPont, NEN) to an activity of 100,000 cpm/ng. hHSE contains the HSE from the top strand of the human *HSP70.1* promoter (25). (We have found that double-stranded oligonucleotide end filling with Klenow fragment or end labeling of single-stranded oligonucleotides with T4 kinase to be equally effective.) The oligonucleotide shown in Subheading 2.2 (575 ng) and the complementary oligonucleotide (2,300 ng) (resulting in 1,150 ng double stranded oligo) are made up to 25 ng/μl in 46 μl of TE buffer, annealed by incubation at 100°C for 5 min, and cooled overnight. As a control, we carried out the EMSA procedure with a similar oligonucleotide containing mutations in the HSE elements, indicated in bold in the sequence shown in Subheading 2.2.

For experiments on cell extracts, a number of controls are used standardly. To determine specific binding of HSF1 to labeled HSE, we examine the ability to inhibit HSE–HSF association with a tenfold excess of unlabeled wild-type oligonucleotide included in the incubation. In addition, specific binding is further indicated by failure of a tenfold excess of the mutant HSE shown above to inhibit binding. The protein (HSF) in the HSF–HSE complex can be identified by the addition of specific anti-HSF1 or anti-HSF2 antibodies to the reaction mix. We used a 1:100 dilution of anti-HSF1 antibody 68-3 prepared in our laboratory to positively identify HSF1 in the complexes (26, 27). As a control, we use preimmune antiserum obtained from the same rabbit. For commercially obtained antibodies, a serial dilution approach was used to determine optimal antibody concentrations.

HSF1 from heat shocked cells is contained in large complexes of at least 600 kDa and is fractionated on 4% Tris–borate nondenaturing gels (28, 29). For most purposes, we found that the minigel (Bio-Rad, CA) format was quite adequate for separation, although for supershift assay and higher resolution a larger format was used (30).

3.2. Measuring HSF1 Binding to HSP Promoters In Vivo by the Chromatin Immunoprecipitation Assay

ChIP offers an attractive solution to transcription analysis by combining the specificity of immunoprecipitation and the sensitivity of PCR (see Note 2).

The method allows monitoring of the interactions between DNA and transcription factors and/or components of chromatin-remodeling complexes, but it is technically challenging due to the low abundance and/or only temporary interactions of these proteins (31).

Our ChIP assays were performed as described in previous publications with some modifications (32, 33). For each ChIP assay, heat shocked or control 10^6 HeLa cells were formaldehyde (1% final)-fixed for 10 min. (Note that the optimal cross-linking concentration and duration need be determined empirically with different tissue or cell type, state, and even intensity. Other protocols could rely either on other chemical reagents or UV-mediated physical cross-linking to preserve native nuclear structures for subsequent biochemical and molecular analysis.) After neutralization with 0.125M glycine, cell pellets were lysed in 200 μl *ChIP lysis buffer* with protease inhibitors. Samples were sonicated with DNA fragmentation centering around 300bp–1000bp and then diluted into 1,800 μl ChIP dilution buffer. ChIP was carried out with precipitating antibodies, either anti-HSF1 or anti-IgG as control from Stressgen (Vancouver, CA, USA), added to precleared chromatin with protein A agarose slurry at 4°C overnight. (Note that the optimal concentration of the primary antibody for ChIP must be determined empirically even with different lot of same name "antibody".) Then, four sequential washings were performed by adding the washing buffer 1, washing buffer 2, and washing buffer 3, finally elution was carried out in 500 μl elution buffer (1% SDS,

0.1M $NaHCO_3$). To reverse the cross-linking fixation, 20 μl of 5M NaCl were added and the mixture incubated at 65°C for 4 h. Afterward, we added 10 μl of uncrosslinking buffer and incubated at 45°C for 1 h, and then purified the DNAs. The immunoprecipitated DNA is now analyzed by PCR amplification using appropriate primer pairs for the HSE consensus region and the control region in the hsp70.1 exon. A total of 27–30 cycles of PCR were carried out with 2 μl of eluted DNA and primers to amplify the exon region (from +752 to +878) and the HSE containing region from −334 to −233. Amplified PCR products were analyzed by agarose gel/ethidium bromide. The input was used as positive control and anti-IgG mock ChIP as negative control. Alternatively, the PCR products were quantified by using ABI 7300 real-time PCR system, and $2^{-\Delta\Delta Ct}$ method for the fold increase in ChIP PCR products compared with the control (anti-IgG) was plotted for the respective region of hsp70.1. (Note that for some transcription factors, if a specific antibody is unavailable, a tagged construct could be made and transfect the cells and obtain its overexpression in cells.) Then, ChIP assay could be performed by using commercially available antibody, which is against such tag.

3.3. Measuring the Contribution of HSF1–HSE Binding to Transcription

3.3.1. Luciferase Reporter Assays for HSF Activity

To construct an intracellular reporter of HSF activity (pGL.hsp70B), we used 1.44 kb of the human *HSP70B* gene inserted into the pGL.Basic plasmid (*Promega*). The *HSP70B* gene is almost entirely silent at physiological temperatures, but powerfully activated by heat shock (34). pGL.hsp70B was constructed by digestion with Bgl*II* and Hind*III* and cloning into pGL.Basic. We have also used the human Hsp27 gene by a similar process, inserting the 730 kb Bgl*II* and Hind*III* digest of an HSP27 promoter fragment into pGL.Basic. For overexpression of HSF1, human HSF1 cDNA (35) is inserted into the pcDNA3.1 (−) expression vector (*Invitrogen*) at the *Xho*I and *Eco*RI sites (36). Human HSF2A was inserted into the pcDNA3.1(+) vector at the *Xho*I and *Eco*RI sites to produce pHSF2A (37).

To assay of HSF1 transcriptional activity in HeLa cells, the cells are maintained in HAM's F-12 (Mediatech) with 10% heat-inactivated fetal bovine serum (FBS). HeLa cells (2.5×10^5 cells/well) in 6-well plates are transfected with the pGL.hsp70B or pGL.hsp27 plasmids (38). pCMV-β-*lacZ* plasmid is cotransfected as an internal control for transfection efficiency. pHM6 empty vector is used as a blank plasmid to balance the amount of DNA transfected in transient transfection. Luciferase and β-galactosidase activity assays are performed after 24 h of transfection according to the Promega protocol. Luciferase activity is normalized to β-galactosidase activity. Results are expressed as relative luciferase (relative light units) activity of the appropriate control (see Note 3).

3.3.2. Nuclear Run-on Assay of Rate of Hsp70 Gene Transcription

To determine HSP70 gene transcriptional rate, cells are treated according to experiment and then quenched in ice-cold phosphate-buffered saline (PBS), pH 7.4, on ice. Cells are next washed in PBS and lysed in run-on lysis buffer. Nuclei are collected by centrifuge (500×*g*, 5 min) at 4°C and resuspended in storage buffer.

To assay rate of transcription, 100 μl of nuclei and 100 μl of run-on reaction buffer are added and samples incubated for 30 min at 30°C with shaking. RNA is then extracted from the reaction mix using Trizol (*invitrogen*) according to the manufacturer's protocol.

The hsp70 DNA containing cDNA probe (39) or control β-actin probe is linearized and purified by phenol/chloroform extraction and ethanol precipitation. Probes are then denatured and slot-blotted onto Hybond N+ membrane. (Membranes are first prehybridized with UltraHyb solution (*Ambion*) for 2 h at 42°C, before equivalent counts of newly transcribed RNA (10^6 cpm) are added to the solution.) Hybridization is then carried out for 24 h at 42°C. Membranes are then washed twice for 20 min at 42°C in low-stringency solution (2× SSC, 0.1% SDS), twice for 20 min in high-stringency solution (1× SSC, 0.1% SDS) and once for 30 min at 37°C in low-stringency solution containing 10 μg RNase A. Membranes are then rinsed in low-stringency solution and analyzed by incubation with X-ray film. We have successfully used this protocol for assay of transcription of the mouse *hsp70.1*, c-*fms*, IL-1β, and TNF-α genes (40) (see Note 4).

4. Notes

1. *The EMSA technique* has the advantages that it is rapid, sensitive, and straightforward to carry out. For assessing the significance of the transcription factor–response element interaction. The method it is however lacking in vivo signifcance in that response elements in chromatin are wound along nucelosomes and may not be available for binding. The EMSA reaction is carried out using naked DNA. In addition, as ChIP-on-CHIP and ChIP-seq studies begin to accumulate, it is evident that response elements for particular factors are more flexible than suspected from early studies.
2. Some of these problems can be avoided using the *ChIP assay* that measures HSF binding to chromosomal DNA in vivo. This technique is highly dependent on the availability of high-affinity and specific antibodies for transcription factors. This can be overcome by overexpression of the factor with a sequence tag and carrying out ChIP with anti-TAG antibody. However, this can introduce potential artifacts involved with protein overexpression.

3. To assess the results of HSF–DNA binding, we have used two approaches. We have used transfection of *reporter constructs* containing either HSP promoters or HSE coupled to reporter genes, CAT or luciferase. The assays have the advantages of being rapid and permitting accumulation of plentiful data. The promoter portion of the construct can be tailored to assess the activity of a single transcription factor, such as HSF1. The assay is, however, indirect and does not measure the transcription of the native, chromosomally embedded gene. There are other potential complications, as reporters require to be translated and yield enzymatically active proteins.
4. Transcriptional rate of HSP genes such can be assessed directly by run-on assay. This assay indicates joint activities of all response elements in the *HSP* gene promoters. Although genes, such as *HSP70B*, respond only to HSF1 or heat shock, others, such as *HSP70A*, have more complex promoters.

Acknowledgments

This work was supported by NIH research grants RO-1CA047407, R01CA119045, and RO-1CA094397.

References

1. Sorger PK, Pelham HRB. Purification and characterization of a heat shock element binding protein from yeast. EMBO Journal 1987;6:3035–41.
2. Sorger PK, Nelson HCM. Trimerization of a yeast transcriptional activator via a coiled-coil motif. Cell 1989;59:807–13.
3. Sorger PK, Pelham HRB. Yeast heat shock factor is an essential DNA-binding protein that exhibits temperature-dependent phosphorylation. Cell 1988;54:855–64.
4. Rabindran SK, Haroun RI, Clos J, Wisniewski J, Wu C. Regulation of heat shock factor trimer formation: role of a conserved leucine zipper. Science 1993;259:230–4.
5. Wu C. Heat shock transcription factors: structure and regulation. Ann Rev Cell Dev Biol 1995;11:441–69.
6. He H, Soncin F, Grammatikakis N, et al. Elevated expression of heat shock factor (HSF) 2A stimulates HSF1-induced transcription during stress. J Biol Chem 2003;278(37): 35465–75.
7. Fujimoto M, Hayashida N, Katoh T, et al. A Novel Mouse HSF3 Has the Potential to Activate Non-classical Heat Shock Genes during Heat Shock. Mol Biol Cell 2009.
8. Tanabe M, Kawazoe Y, Takeda S, Morimoto RI, Nagata K, Nakai A. Disruption of the HSF3 gene results in the severe reduction of heat shock gene expression and loss of thermotolerance. Embo J 1998;17(6):1750–8.
9. Tanabe M, Sasai N, Nagata K, et al. The mammalian HSF4 gene generates both an activator and a repressor of heat shock genes by alternative splicing. J Biol Chem 1999;274(39):27845–56.
10. Kumar M, Busch W, Birke H, Kemmerling B, Nurnberger T, Schoffl F. Heat Shock Factors HsfB1 and HsfB2b Are Involved in the Regulation of Pdf1.2 Expression and Pathogen Resistance in Arabidopsis. Mol Plant 2009;2(1):152–65.
11. Scharf KD, Rose S, Zott W, Schoffl F, Nover L. Three tomato genes code for heat stress transcription factors with a region of remarkable homology to the DNA-binding domain of the yeast HSF. Embo J 1990;9(13):4495–501.
12. McMillan DR, Xiao X, Shao L, Graves K, Benjamin IJ. Targeted disruption of heat shock transcription factor 1 abolishes thermotolerance

and protection against heat-inducible apoptosis. J Biol Chem 1998;273:7523–8.

13. Morange M. HSFs in development. Handb Exp Pharmacol 2006(172):153–69.
14. Abravaya K, Myers MP, Murphy SP, Morimoto RI. The human heat shock protein hsp70 interacts with HSF, the transcription factor that regulates heat shock protein expression. Genes Dev 1992;6:1153–64.
15. Zou J, Guo Y, Guettouche T, Smith DF, Voellmy R. Repression of heat shock transcription factor HSF1 activation by HSP90 (HSP90 complex) that forms a stress-sensitive complex with HSF1. Cell 1998;94(4):471–80.
16. Guettouche T, Boellmann F, Lane WS, Voellmy R. Analysis of phosphorylation of human heat shock factor 1 in cells experiencing a stress. BMC Biochem 2005;6(1):4.
17. Shamovsky I, Ivannikov M, Kandel ES, Gershon D, Nudler E. RNA-mediated response to heat shock in mammalian cells. Nature 2006;440(7083):556–60.
18. Mosser DD, Caron AW, Bourget L, Denis-Larose C, Massie B. Role of the human heat shock protein hsp70 in protection against stress- induced apoptosis. Mol Cell Biol 1997;17(9):5317–27.
19. Price BD, Calderwood SK. Heat-induced transcription from RNA polymerases II and III and HSF binding are co-ordinately regulated by the products of the heat shock genes. J Cell Physiol 1992;153:392–401.
20. Zhao M, Tang D, Lechpammer S, et al. Double-stranded RNA-dependent protein kinase (pkr) is essential for thermotolerance, accumulation of HSP70, and stabilization of ARE-containing HSP70 mRNA during stress. J Biol Chem 2002;277(46):44539–47.
21. Subjeck JR, Sciandra JJ, Johnson RJ. Heat shock proteins and thermotolerance; a comparison of induction kinetics. Br J Radiol 1982;55(656):579–84.
22. Wang X, Khaleque MA, Zhao MJ, Zhong R, Gaestel M, Calderwood SK. Phosphorylation of HSF1 by MAPK-activated protein kinase 2 on serine 121, inhibits transcriptional activity and promotes HSP90 binding. J Biol Chem 2006;281(2):782–91.
23. Soncin F, Prevelige R, Calderwood SK. Expression and purification of human heat-shock transcription factor 1. Protein Expr Purif 1997;9(1):27–32.
24. Schreiber E, Matthias P, Muller MM, Schaffner W. Rapid detection of octamer binding proteins with "mini-extracts" prepared from a small number of cells. Nucleic Acids Res 1989;17:6419.
25. Wu B, Hunt C, Morimoto RI. Structure and expression of the human gene encoding the major heat shock protein HSP70. Molecular and Cellular Biology 1985;5:330–41.
26. Bruce JL, Chen C, Xie Y, et al. Activation of heat shock transcription factor 1 to a DNA binding form during the G(1)phase of the cell cycle. Cell Stress Chaperones 1999;4(1):36–45.
27. Cahill CM, Waterman WR, Xie Y, Auron PE, Calderwood SK. Transcriptional repression of the prointerleukin 1beta gene by heat shock factor 1. J Biol Chem 1996;271(40):24874–9.
28. Nunes SL, Calderwood SK. Heat shock factor-1 and the heat shock cognate 70 protein associate in high molecular weight complexes in the cytoplasm of NIH-3 T3 cells. Biochem Biophys Res Commun 1995;213(1):1–6.
29. Westwood T, Wu C. Activation of *drosophila* heat shock factor: conformational changes associated with monomer-to-trimer transition. Mol Cell Biol 1993;13:3481–6.
30. Xie Y, Zhong R, Chen C, Calderwood SK. Heat shock factor 1 contains two functional domains that mediate transcriptional repression of the c-fos and c-fms genes. J Biol Chem 2003;278(7):4687–98.
31. Solomon MJ, Varshavsky A. Formaldehyde-mediated DNA-protein crosslinking: a probe for in vivo chromatin structures. Proc Natl Acad Sci USA 1985;82(19):6470–4.
32. Takacs-Vellai K, Vellai T, Chen EB, et al. Transcriptional control of Notch signaling by a HOX and a PBX/EXD protein during vulval development in C. elegans. Dev Biol 2007;302(2):661–9.
33. Khaleque MA, Bharti A, Gong J, et al. Heat shock factor 1 represses estrogen-dependent transcription through association with MTA1. Oncogene 2008;27(13):1886–93.
34. Tang D, Khaleque MA, Jones EL, et al. Expression of heat shock proteins and heat shock protein messenger ribonucleic acid in human prostate carcinoma in vitro and in tumors in vivo. Cell Stress Chaperones 2005;10(1):46–58.
35. Rabindran SK, Gioorgi G, Clos J, Wu C. Molecular Cloning and expression of a human heat shock factor, HSF1. Proceedings of the National Academy of Sciences (USA) 1991; 88:6906–10.
36. Oesterreich S, Hickey E, Weber L, Fuqua SA. Basal regulatory promoter elements in the hsp27 gene in human breast carcinoma cells. Biochem Biophys Res Commun 1996;222:155–63.
37. Chen C, Xie Y, Stevenson MA, Auron PE, Calderwood SK. Heat shock factor 1 represses Ras-induced transcriptional activation of the c-fos gene. J Biol Chem 1997;272(43):26803–6.

38. Wang XZ, Asea, A., Xie,Y., Kabingu,E., Stevenson, M.A., and Calderwood, S.K. RSK2 represses HSF1 activation during heat shock. Cell Stress & Chaperones 2000;5:432–7.
39. Hunt C, Calderwood SK. Characterization and sequence of a mouse HSP70 gene and its expression in mouse cell lines. Gene 1990;87:199–204.
40. Xie Y, Chen C, Stevenson MA, Auron PE, Calderwood SK. Heat shock factor 1 represses transcription of the IL-1beta gene through physical interaction with the nuclear factor of interleukin 6. J Biol Chem 2002;277(14):11802–10.

Chapter 3

Hsp90 and Client Protein Maturation*

Natalie Wayne, Parul Mishra, and Daniel N. Bolon

Abstract

Heat-shock protein 90 (Hsp90) is a molecular chaperone that assists in the maturation of a limited set of substrate proteins that are collectively referred to as clients. The majority of identified Hsp90 clients are involved in signal transduction, including many steroid hormone receptors and kinases. A handful of Hsp90 clients can be classified as nonsignal transduction proteins, including telomerase, cystic fibrosis transmembrane conductance regulator, and antigenic peptides destined for major histocompatibility complex. Because Hsp90 clients are causative agents in cancer and cystic fibrosis, research on Hsp90 has intensified in recent years. We review the historical path of Hsp90 research within each class of client (kinase, hormone receptor, and nonsignal transduction clients) and highlight current areas of active investigation.

Key words: Hsp90, Chaperone, ATPase, Kinase, Steroid hormone receptor, Signal transduction, Telomerase, CFTR, Antigen presentation

1. Introduction

Hsp90, as with most heat-shock proteins (Hsps), was initially identified from changes in transcriptional pattern (1) and relative translational rate (2) upon shifting cells to elevated temperature. Hsp90 is named for its induction in response to temperature stress as well as its molecular weight of about 90 kD. Elevated temperature challenges the ability of proteins to fold efficiently placing an elevated burden on cellular chaperones. In response to elevated temperature, and many other stressful conditions, eukaryotes induce expression of the heat-shock factor 1 (HSF1) transcription factor that in turn upregulates expression of chaperones, including Hsp90 (3). Many eukaryotes, including budding yeast and humans, have two genes encoding Hsp90 that encode nearly identical

*Natalie Wayne and Parul Mishra have contributed equally to this chapter.

Stuart K. Calderwood and Thomas L. Prince (eds.), *Molecular Chaperones: Methods and Protocols*,
Methods in Molecular Biology, vol. 787, DOI 10.1007/978-1-61779-295-3_3, © Springer Science+Business Media, LLC 2011

amino acid sequences, one of which is constitutively expressed while the other is HSF1 inducible (4). The constitutive expression level of Hsp90 under nonstress conditions is very high, and Hsp90 is one of the most abundant proteins in the cell, accumulating primarily in the cytosol (5, 6). The high expression level of Hsp90 in the absence of stress is consistent with its requirement for the activity of many critical signal transduction clients in eukaryotes. Hsp90 is essential for viability in eukaryotes (4). In bacteria, which lack the wealth of signal transduction proteins present in eukaryotes, Hsp90 knockouts are viable (7).

The Hsp90 chaperone is highly conserved in eukaryotes, both at the level of amino acid sequence and biochemical function. The *Saccharomyces cerevisiae* and human Hsp90 proteins are 59% identical in amino acid sequence alignments. The conservation of Hsp90 in eukaryotes extends to the functional level as *S. cerevisiae* with both endogenous Hsp90 genes knocked out can be rescued by expression of human Hsp90 (8). Hsp90 serves as a protein interaction hub, binding to about 10% of the yeast proteome (9) including many co-chaperones necessary for client maturation. The observation that human Hsp90 complements budding yeast knockouts indicates that essential elements of this hub are evolutionarily conserved. Consistent with the functional conservation of Hsp90, steroid hormone receptors that are not natural to budding yeast function in an Hsp90-dependent manner when introduced into yeast (8).

Hsp90 contains three domains: the N-terminal (N) domain, middle (M) domain, and the C-terminal (C) domain. The N-domain binds to ATP and together with the M-domain catalyzes ATP hydrolysis (10, 11). The C-domain forms a stable dimer (12). Dimerization of Hsp90 is required for client maturation (13). Recent tour de force structural studies have revealed that flexible hinges between all three domains enable Hsp90 to adopt dramatically different 3-dimensional conformations (14–16). Isolated Hsp90 is a slow ATPase, ranging from about 1 min^{-1} for yeast Hsp90 to about 0.1 min^{-1} for human Hsp90, and ATP binding and hydrolysis cause Hsp90 to preferentially shift between an open conformation with the N-domains separated and a closed conformation with the N-domains associated (17–19). These open and closed structural transitions are known to drive the binding of many co-chaperones (14, 20–22) and are believed to drive the maturation of clients (Fig. 1).

Hsp90 has a unique ATP binding site, and the discovery of small-molecule inhibitors that specifically inhibit Hsp90 has provided useful research tools that also show promise as anticancer therapeutic agents. Hsp90 inhibitors were originally uncovered from a natural product screen for compounds that reversed the transformed phenotype of the v-src oncogenic kinase.

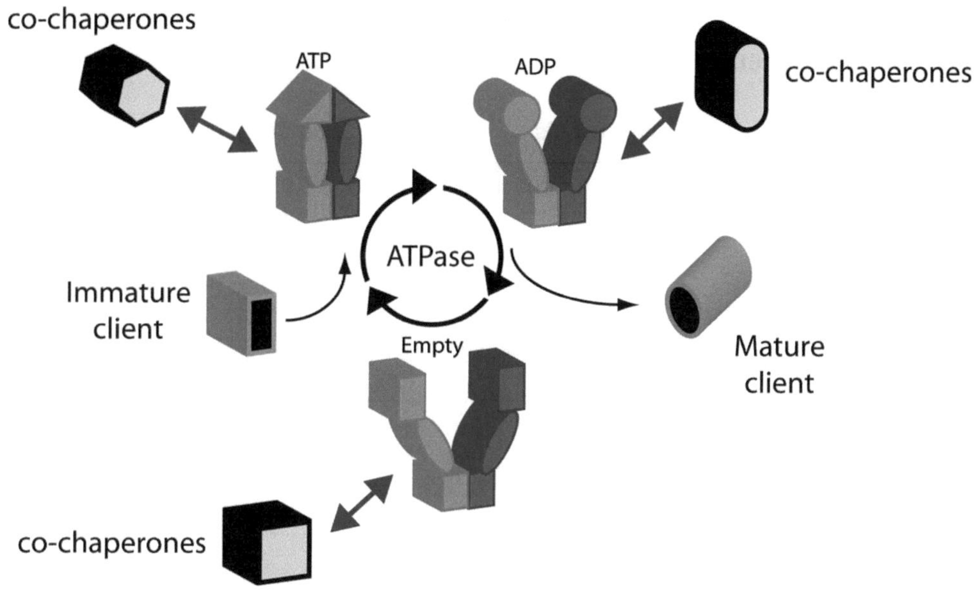

Fig. 1. Model of the ATPase-driven mechanism of Hsp90. ATP binding and hydrolysis along with cofactor binding are required for the maturation of most Hsp90 clients.

One of the natural products from this screen, geldanamycin (GA), was found to bind specifically to Hsp90 and inhibit its ability to chaperone v-src (23). While Hsp90 function is essential for the viability of healthy cells, cancer cells tend to have a higher level of Hsp90-dependent clients and a higher required level of Hsp90 activity. For many types of cancer, there appears to be a therapeutic window, where cancer cells can be preferentially impacted by Hsp90 inhibition compared to healthy cells. The medical interest in Hsp90 has spurred the synthesis of many small-molecule inhibitors as drug candidates (24).

Hsp90 is a specialized molecular chaperone that is required for the maturation of a limited set of clients. Many newly synthesized proteins require general molecular chaperones, including Hsp60 (GroEL in bacteria), Hsp70, and Hsp40, in order to fold properly, but only a small fraction of these rely upon Hsp90. The limited proteins that require Hsp90 to mature to their active state are referred to as Hsp90 clients. These Hsp90 clients fall into three main categories: protein kinases, steroid hormone receptors, and nonsignal transduction clients (Table 1). Within each greater client class, research has revealed important aspects of the mechanism of Hsp90-mediated client maturation, yet the molecular mechanism by which Hsp90 binds to and activates clients remains largely a mystery.

Table 1
Classes of Hsp90 clients

Client class	Example clients	Co-chaperones	References
Steroid hormone receptors	Progesterone and glucocorticoid receptors	FKBP51, FKBP52, Hop, p23	(38)
Kinase	v-src, B-Raf	Cdc37, Hop, p23	(23, 50)
Nonsignal transduction			
Telomerase	Telomerase	p23	(54)
Antigen presentation	MHC peptide loading		(59, 60)
CFTR	CFTR	Aha1	(61, 63, 64)

2. Steroid Hormone Receptor Hsp90 Clients

Steroid hormone receptors are a class of transcription factors that are activated by binding to steroid agonist and were some of the first proteins actively studied as Hsp90 clients. In 1984, Baulieu and colleagues reported that monoclonal antibodies raised against progesterone receptor (PR) recognized a 90 kD protein that tightly associated with apo-receptor, but not steroid-bound receptor (25), properties that indicated a regulatory role for the 90 kD protein. Soon after, Pratt and colleagues reported that monoclonal antibodies raised against glucocorticoid receptor (GR) were capable of immunoisolating the receptor as well as a 90 kD protein that cross-reacted with antibodies raised against Hsp90 (26), thus identifying Hsp90 as a steroid hormone receptor chaperone.

Hsp90 was found to bind strongly to the apo-form of many other steroid hormone receptors, including the estrogen and androgen receptors; however, some receptors, such as those for retinoic acid and thyroid hormone, showed little or no affinity for Hsp90 (27–30). Different DNA-binding properties in the absence of steroid agonist were noted for Hsp90-dependent and Hsp90-independent receptors in cell-free expression systems (27). In the absence of steroid agonist, Hsp90-independent receptors were found to bind to target DNA in the presence and absence of Hsp90. In contrast, Hsp90-dependent receptors without steroid agonist bound to DNA only when Hsp90 was absent. In the absence of steroid agonist, Hsp90-dependent receptors were found in the cytoplasm while Hsp90-independent receptors were found in the nucleus. At the time, these results led to the hypothesis that Hsp90 functions to repress steroid hormone receptors by sequestering them in the cytoplasm. However, it is now appreciated that Hsp90-dependent receptors require Hsp90 in order to bind to steroid.

Steroid binding controls the activity of all steroid receptors and for Hsp90-dependent receptors, this step requires Hsp90 activity.

Co-immunoprecipitation (co-IP) experiments revealed many components of the Hsp90 chaperone complexes involved in receptor maturation and paved the way for in vitro reconstitution of the chaperone reaction. Glutaraldehyde cross-linking studies revealed that the stoichiometry of receptor–Hsp90 complexes was 1:2 indicating that a receptor monomer binds to an Hsp90 dimer (31). From immunoisolation of steroid receptors expressed in cell-free expression systems, it was discovered that receptor–Hsp90 complexes included a number of co-chaperones, including Hsp70, p23, and Hop (32, 33). A combination of these three co-chaperones along with the general chaperones, Hsp40 and Hsp90, was found to be sufficient to develop a purified chaperone system for the maturation of progesterone receptor to the steroid-bound active form (34).

Structural and biochemical approaches have shown that steroid hormone receptors bind steroid ligand in an internal cavity and indicate that Hsp90 remodels this cavity to provide a solvent-accessible path for the hormone to enter. X-ray crystal (35, 36) structures clearly show that steroid ligands bind to receptors in an internal cavity that is completely inaccessible to solvent. Evidence for Hsp90 opening of this cleft comes from chemical modification studies that indicate modestly higher susceptibility of the steroid-binding cavity in the Hsp90-bound form of apo-GR (37). While the mechanism by which Hsp90 opens the steroid-binding cleft remains unknown, the data clearly indicate that isolated apo-GR and apo-PR are not able to bind to steroid agonist, but that when complexed with Hsp90 the binding to steroid becomes efficient (38).

While much is understood about Hsp90-mediated steroid receptor maturation, many important aspects of the molecular mechanism remain open questions. It is clear that ATP hydrolysis by Hsp90 is required for efficient PR maturation (39). However, it is not clear how the ATPase-driven conformational cycle of Hsp90 leads to opening of the steroid-binding cavity. What are the molecular interactions that cause apo-receptor to bind to Hsp90, Hsp90 to catalyze steroid binding, and steroid-bound receptor to dissociate from the chaperone complex? These questions are challenging to answer because of the dynamic nature of the Hsp90 conformational cycle, but may be addressed by structural and biochemical approaches that trap individual steps in the process.

3. Kinase Clients of Hsp90

The oncogenic v-src kinase was first shown to bind to Hsp90 through co-IP analysis (40), but it was not clear that Hsp90 was required for v-src kinase activity until it was reported that the natural compound GA prevents v-src activity indirectly by inhibiting Hsp90 (23).

When v-src is expressed in the presence of GA, it accumulates in an inactive form indicating that Hsp90 is required for the maturation of v-src to an active form. In addition, treatment of cells with GA causes an increase in the proteosome-mediated degradation rate of v-src and a decrease in the steady-state v-src level (41). Thus, Hsp90 is required for the maturation of v-src kinase to an active form and also protects v-src from degradation.

The maturation of kinases mediated by Hsp90 requires co-chaperones, including the kinase-specific co-chaperone Cdc37. Cdc37 was originally identified in a screen for genes required for cell cycle progression (42). Genetic evidence for a functional interaction between Cdc37 and Hsp90 came from studies showing that either mutations in Cdc37 or a reduction in the level of the yeast Hsp90 protein suppresses the lethality of v-src overexpression (43, 44). Mutations in Hsp90 and Cdc37 were also found to exacerbate a defect in the *Drosophila melanogaster* sevenless receptor tyrosine kinase (45). Thus, Hsp90 and Cdc37 are both involved in the maturation of these model kinases. A physical connection between Cdc37 and Hsp90 was first indicated by co-IP analysis with the cyclin-dependent kinase CDK4 (46). Both Hsp90 and Cdc37 were found to associate with CDK4 during co-IP analysis with an anti-CKD4 antibody suggesting that there may be a direct mechanistic link between Hsp90 and Cdc37 in chaperoning kinases.

The direct interaction between Cdc37 and Hsp90 has recently been confirmed and described in atomic detail through structural analyses. Comparison of full-length yeast Hsp90 to truncated Hsp90 constructs indicated that the isolated Hsp90 N-domain was primarily responsible for mediating binding to the C-terminal domain of human Cdc37 (22). Cocrystallization of these domains revealed the atomic details of this interaction (22). More recently, Pearl and colleagues have successfully purified a homogeneous complex of human Cdk4 and Cdc37 along with Hsp90 from the insect *Spodoptera frugiperda* expression system and analyzed the structure of this complex by negative-stained electron microscopy (16). This tour de force work indicates that the kinase-loaded complex consists of an Hsp90 dimer bound to single copies of Cdc37 and Cdk4. The observed electron density can be well-fit with a structural model, where Cdc37 binds to the Hsp90 N-domain and Cdk4 binds to the Hsp90 M-domain. Biochemical analyses of Cdk4–Cdc37 complexes in the absence of Hsp90 indicate a stoichiometry of one kinase to two Cdc37 molecules (16). The biochemical and structural data are consistent with an ordered model, where a dimer of Cdc37 initially binds to kinases and subsequent binding of this complex to Hsp90 results in the release of one Cdc37 molecule.

The association of human Cdc37 with yeast Hsp90 observed biochemically and structurally (22) is consistent with the high amino acid conservation of Hsp90 between these two species.

This conservation is also apparent at a functional level as metazoan Cdc37 (from flies) is capable of rescuing robust growth of *S. cerevisiae* made unhealthy by a point mutant in the yeast Cdc37 (45). Unlike Hsp90, the amino acid sequence of Cdc37 is divergent from yeast to humans with only 17% sequence identity after alignment. Of note, yeast Cdc37 binds to yeast Hsp90 with about 100-fold weaker affinity than human Cdc37 (47), indicating that while qualitative Cdc37 function may be similar across eukaryotes, the interaction details are quantitatively distinct.

The Hsp90 and Cdc37 chaperones stabilize the cellular accumulation of many kinases. Indeed, the observation that Hsp90 inhibition can lead to decreased cellular levels of client kinases has become a convenient screen to identify Hsp90-dependent kinases. In mammalian cells, inhibition of Hsp90 activity resulted in reduced levels of 80 out of 105 kinases analyzed (48). A recent study also found that more than 50% of the *S. cerevisiae* kinome was affected by the loss of functional Cdc37 (49) confirming the importance of a functional Hsp90:Cdc37 chaperone complex for the stability of cellular protein kinases.

Recent efforts have described the minimum chaperone requirements for the maturation of a kinase in vitro (50). Arlander et al. identified a five-protein system for chaperoning the Chk1 kinase in vitro that consisted of Hsp90, Hsp70, Hsp40, Cdc37, and CK2 (50). The in vitro chaperone systems for kinases and steroid hormone receptors differ in their co-chaperone requirements and sensitivities, indicating the influential role that co-chaperones play in the client maturation process (34, 51). As with steroid hormone receptors, the in vitro chaperone system for kinases provides a promising route to carefully analyze the client maturation process in detail.

4. Nonsignal Transduction Hsp90 Clients

Recent evidence indicates that Hsp90 is involved in the maturation of active telomerase. The catalytic core of telomerase consists of a protein component (TERT) and a template RNA (TR). Chaperones were originally implicated in the maturation of telomerase from the observation that reconstitution of active telomerase from isolated TERT and TR was stimulated by increasing concentration of chaperones present in reticulocyte lysates (52). A two-hybrid screen revealed that the Hsp90 co-chaperone p23 interacted with TERT, and co-IP analyses revealed that both Hsp90 and p23 are involved in complexes with telomerase in mammalian cells (52). More recent biochemical dissections of telomerase function indicate that the interplay between Hsp90 and p23 influences the dynamics of telomerase binding to telomeres (53). Hsp90 activity in vivo is important for the proper maintenance of telomeric DNA length that decreases when Hsp90 activity is impaired or when Hsp90 is overexpressed (54, 55).

Hsp90 can bind to antigenic peptides and has been implicated in the delivery of these peptides to major histocompatability complexes (MHCs). Hsp90 was first connected with antigen presentation from analyses of antigens in tumors that could confer immunity in syngenic recipients (56). Fractionation studies revealed that Hsp90 from the tumors was important in the development of immunity (57, 58). Cellular studies indicated that antigenic peptides complexed with Hsp90 are loaded on to MHC class I with much higher efficiency than free peptides (59), which may otherwise be prone to rapid degradation. In addition, the delivery of peptides to MHC class II proteins is reduced upon treatment of cells with Hsp90 inhibitors (60). The molecular details of Hsp90 involvement in antigen presentation are a current area of active investigation.

In addition to soluble proteins, Hsp90 has recently been implicated in the maturation of cystic fibrosis transmembrane conductance regulator (CFTR), an integral membrane protein. Hsp90 was originally implicated in CFTR maturation from the observation that treatment of mammalian cells with the Hsp90-specific inhibitors prevented newly synthesized wild-type CFTR from transitioning to the maturely glycosylated cell-surface form (61). Cystic fibrosis-causing mutations in CFTR, including the most commonly occurring deletion of a phenylalanine codon (ΔF508), cause defective folding and export of CFTR from the ER (62). Immunoprecipitation of wild-type and ΔF508 CFTR followed by mass spectrometry revealed that Hsp90 and many co-chaperones exhibited differential binding for wild-type and mutant CFTR (63). RNA interference screening of these co-chaperones demonstrated that silencing of the Hsp90 co-chaperone Aha1 rescued cell surface expression and chloride ion transport of ΔF508 CFTR (63). Recent biochemical analyses indicate that Aha1 associates with ΔF508 CFTR nearly twice as much as with wild-type CFTR (64). The observation that reducing the association of ΔF508 CFTR with Aha1 corrects the function of this anion channel provides a promising avenue for the development of therapeutics to treat cystic fibrosis.

Hsp90 can function as a promiscuous antiaggregation molecule in vitro for a large variety of hard-to-fold proteins, but the general antiaggregation properties of Hsp90 do not appear important in cells. Following the discovery that Hsp90 expression increased in response to conditions that stress protein folding, a number of studies examined the capacity of Hsp90 to assist in the folding of proteins that were difficult to fold on their own. The antiaggregation properties of Hsp90 for unfolded CFTR nucleotide-binding domain 1 (65), citrate synthase (66), and p53 (67) do not require the binding and hydrolysis of ATP by Hsp90. In contrast, both steroid hormone receptors and kinase clients of Hsp90 are dramatically impaired by mutations that destroy ATPase activity or competitive inhibition of the ATP-binding site with drugs (23, 39). Thus, the antiaggregation properties of Hsp90 do not seem to require the same ATPase-driven Hsp90 conformational cycle that is implicated in the maturation of kinases and steroid hormone receptors. Multiple chaperones in addition to Hsp90 are capable of providing antiaggregation properties in cells providing an explanation for the observation that Hsp90 is not required for the folding and solubility of most proteins in cells (68).

5. Current and Future Areas of Active Interest

Many important cellular roles of Hsp90 have been described, and much of the current and future research in Hsp90 involves analyses of molecular mechanism. The molecular mechanism by which Hsp90 catalyzes the maturation of any client is an open and active area of research. Current structural models indicate that Hsp90 undergoes dramatic rearrangements in response to ATP binding and hydrolysis (14, 15, 19). However, how these structural rearrangements lead to client maturation is not clear. From the engineering of a monomeric Hsp90 variant, it is clear that dimerization of Hsp90 is critical for efficient ATP hydrolysis as well as the activation of a kinase and a steroid hormone receptor client (13). How each subunit in the Hsp90 dimer binds to and hydrolyzes ATP during the maturation of a client remains unclear. In the coming years, research will help to reveal the molecular mechanism by which Hsp90 binds to client proteins and along with co-chaperones and ATP catalyzes the maturation of these clients to active states.

References

1. Ritossa, F. (1962) A new puffing pattern induced by temperature shock and DNP in drosophila. *Experientia* 18, 571–573.
2. McKenzie, S. L., Henikoff, S., and Meselson, M. (1975) Localization of RNA from heat-induced polysomes at puff sites in Drosophila melanogaster. *Proc Natl Acad Sci U S A* 72, 1117–21.
3. Lindquist, S. (1986) The heat-shock response. *Annu Rev Biochem* 55, 1151–91.
4. Borkovich, K. A., Farrelly, F. W., Finkelstein, D. B., Taulien, J., and Lindquist, S. (1989) hsp82 is an essential protein that is required in higher concentrations for growth of cells at higher temperatures. *Mol Cell Biol* 9, 3919–30.
5. Ghaemmaghami, S., Huh, W. K., Bower, K., Howson, R. W., Belle, A., Dephoure, N., O'Shea, E. K., and Weissman, J. S. (2003) Global analysis of protein expression in yeast. *Nature* 425, 737–41.
6. Huh, W. K., Falvo, J. V., Gerke, L. C., Carroll, A. S., Howson, R. W., Weissman, J. S., and O'Shea, E. K. (2003) Global analysis of protein localization in budding yeast. *Nature* 425, 686–91.
7. Bardwell, J. C., and Craig, E. A. (1988) Ancient heat shock gene is dispensable. *J Bacteriol* 170, 2977–83.
8. Picard, D., Khursheed, B., Garabedian, M. J., Fortin, M. G., Lindquist, S., and Yamamoto, K. R.

(1990) Reduced levels of hsp90 compromise steroid receptor action in vivo. *Nature* 348, 166–8.
9. Zhao, R., Davey, M., Hsu, Y. C., Kaplanek, P., Tong, A., Parsons, A. B., Krogan, N., Cagney, G., Mai, D., Greenblatt, J., Boone, C., Emili, A., and Houry, W. A. (2005) Navigating the chaperone network: an integrative map of physical and genetic interactions mediated by the hsp90 chaperone. *Cell* 120, 715–27.
10. Panaretou, B., Prodromou, C., Roe, S. M., O'Brien, R., Ladbury, J. E., Piper, P. W., and Pearl, L. H. (1998) ATP binding and hydrolysis are essential to the function of the Hsp90 molecular chaperone in vivo. *Embo J* 17, 4829–36.
11. Meyer, P., Prodromou, C., Hu, B., Vaughan, C., Roe, S. M., Panaretou, B., Piper, P. W., and Pearl, L. H. (2003) Structural and functional analysis of the middle segment of hsp90: implications for ATP hydrolysis and client protein and cochaperone interactions. *Mol Cell* 11, 647–58.
12. Minami, Y., Kimura, Y., Kawasaki, H., Suzuki, K., and Yahara, I. (1994) The carboxy-terminal region of mammalian HSP90 is required for its dimerization and function in vivo. *Mol Cell Biol* 14, 1459–64.
13. Wayne, N., and Bolon, D. N. (2007) Dimerization of Hsp90 is required for in vivo function. Design and analysis of monomers and dimers. *J Biol Chem* 282, 35386–95.
14. Ali, M. M., Roe, S. M., Vaughan, C. K., Meyer, P., Panaretou, B., Piper, P. W., Prodromou, C., and Pearl, L. H. (2006) Crystal structure of an Hsp90-nucleotide-p23/Sba1 closed chaperone complex. *Nature* 440, 1013–7.
15. Shiau, A. K., Harris, S. F., Southworth, D. R., and Agard, D. A. (2006) Structural Analysis of E. coli hsp90 reveals dramatic nucleotide-dependent conformational rearrangements. *Cell* 127, 329–40.
16. Vaughan, C. K., Gohlke, U., Sobott, F., Good, V. M., Ali, M. M., Prodromou, C., Robinson, C. V., Saibil, H. R., and Pearl, L. H. (2006) Structure of an Hsp90-Cdc37-Cdk4 complex. *Mol Cell* 23, 697–707.
17. Hessling, M., Richter, K., and Buchner, J. (2009) Dissection of the ATP-induced conformational cycle of the molecular chaperone Hsp90. *Nat Struct Mol Biol* 16, 287–93.
18. Mickler, M., Hessling, M., Ratzke, C., Buchner, J., and Hugel, T. (2009) The large conformational changes of Hsp90 are only weakly coupled to ATP hydrolysis. *Nat Struct Mol Biol* 16, 281–6.
19. Southworth, D. R., and Agard, D. A. (2008) Species-dependent ensembles of conserved conformational states define the Hsp90 chaperone ATPase cycle. *Mol Cell* 32, 631–40.
20. McLaughlin, S. H., Sobott, F., Yao, Z. P., Zhang, W., Nielsen, P. R., Grossmann, J. G., Laue, E. D., Robinson, C. V., and Jackson, S. E. (2006) The co-chaperone p23 arrests the Hsp90 ATPase cycle to trap client proteins. *J Mol Biol* 356, 746–58.
21. Panaretou, B., Siligardi, G., Meyer, P., Maloney, A., Sullivan, J. K., Singh, S., Millson, S. H., Clarke, P. A., Naaby-Hansen, S., Stein, R., Cramer, R., Mollapour, M., Workman, P., Piper, P. W., Pearl, L. H., and Prodromou, C. (2002) Activation of the ATPase activity of hsp90 by the stress-regulated cochaperone aha1. *Mol Cell* 10, 1307–18.
22. Roe, S. M., Ali, M. M., Meyer, P., Vaughan, C. K., Panaretou, B., Piper, P. W., Prodromou, C., and Pearl, L. H. (2004) The Mechanism of Hsp90 regulation by the protein kinase-specific cochaperone p50(cdc37). *Cell* 116, 87–98.
23. Whitesell, L., Mimnaugh, E. G., De Costa, B., Myers, C. E., and Neckers, L. M. (1994) Inhibition of heat shock protein HSP90-pp60v-src heteroprotein complex formation by benzoquinone ansamycins: essential role for stress proteins in oncogenic transformation. *Proc Natl Acad Sci U S A* 91, 8324–8.
24. Whitesell, L., and Lindquist, S. L. (2005) HSP90 and the chaperoning of cancer. *Nat Rev Cancer* 5, 761–72.
25. Joab, I., Radanyi, C., Renoir, M., Buchou, T., Catelli, M. G., Binart, N., Mester, J., and Baulieu, E. E. (1984) Common non-hormone binding component in non-transformed chick oviduct receptors of four steroid hormones. *Nature* 308, 850–3.
26. Sanchez, E. R., Toft, D. O., Schlesinger, M. J., and Pratt, W. B. (1985) Evidence that the 90-kDa phosphoprotein associated with the untransformed L-cell glucocorticoid receptor is a murine heat shock protein. *J Biol Chem* 260, 12398–401.
27. Dalman, F. C., Koenig, R. J., Perdew, G. H., Massa, E., and Pratt, W. B. (1990) In contrast to the glucocorticoid receptor, the thyroid hormone receptor is translated in the DNA binding state and is not associated with hsp90. *J Biol Chem* 265, 3615–8.
28. Dalman, F. C., Sturzenbecker, L. J., Levin, A. A., Lucas, D. A., Perdew, G. H., Petkovitch, M., Chambon, P., Grippo, J. F., and Pratt, W. B. (1991) Retinoic acid receptor belongs to a subclass of nuclear receptors that do not form "docking" complexes with hsp90. *Biochemistry* 30, 5605–8.
29. Redeuilh, G., Moncharmont, B., Secco, C., and Baulieu, E. E. (1987) Subunit composition of the molybdate-stabilized "8-9 S" nontransformed estradiol receptor purified from calf uterus. *J Biol Chem* 262, 6969–75.

30. Veldscholte, J., Berrevoets, C. A., Zegers, N. D., van der Kwast, T. H., Grootegoed, J. A., and Mulder, E. (1992) Hormone-induced dissociation of the androgen receptor-heat-shock protein complex: use of a new monoclonal antibody to distinguish transformed from nontransformed receptors. *Biochemistry* 31, 7422–30.
31. Denis, M., Wikstrom, A. C., and Gustafsson, J. A. (1987) The molybdate-stabilized nonactivated glucocorticoid receptor contains a dimer of Mr 90,000 non-hormone-binding protein. *J Biol Chem* 262, 11803–6.
32. Smith, D. F., Schowalter, D. B., Kost, S. L., and Toft, D. O. (1990) Reconstitution of progesterone receptor with heat shock proteins. *Mol Endocrinol* 4, 1704–11.
33. Smith, D. F., and Toft, D. O. (1992) Composition, assembly and activation of the avian progesterone receptor. *J Steroid Biochem Mol Biol* 41, 201–7.
34. Kosano, H., Stensgard, B., Charlesworth, M. C., McMahon, N., and Toft, D. (1998) The assembly of progesterone receptor-hsp90 complexes using purified proteins. *J Biol Chem* 273, 32973–9.
35. Wagner, R. L., Apriletti, J. W., McGrath, M. E., West, B. L., Baxter, J. D., and Fletterick, R. J. (1995) A structural role for hormone in the thyroid hormone receptor. *Nature* 378, 690–7.
36. Williams, S. P., and Sigler, P. B. (1998) Atomic structure of progesterone complexed with its receptor. *Nature* 393, 392–6.
37. Stancato, L. F., Silverstein, A. M., Gitler, C., Groner, B., and Pratt, W. B. (1996) Use of the thiol-specific derivatizing agent N-iodoacetyl-3-[125I]iodotyrosine to demonstrate conformational differences between the unbound and hsp90-bound glucocorticoid receptor hormone binding domain. *J Biol Chem* 271, 8831–6.
38. Pratt, W. B., and Toft, D. O. (2003) Regulation of signaling protein function and trafficking by the hsp90/hsp70-based chaperone machinery. *Exp Biol Med (Maywood)* 228, 111–33.
39. Grenert, J. P., Johnson, B. D., and Toft, D. O. (1999) The importance of ATP binding and hydrolysis by hsp90 in formation and function of protein heterocomplexes. *J Biol Chem* 274, 17525–33.
40. Brugge, J. S., Erikson, E., and Erikson, R. L. (1981) The specific interaction of the Rous sarcoma virus transforming protein, pp60src, with two cellular proteins. *Cell* 25, 363–72.
41. An, W. G., Schulte, T. W., and Neckers, L. M. (2000) The heat shock protein 90 antagonist geldanamycin alters chaperone association with p210bcr-abl and v-src proteins before their degradation by the proteasome. *Cell Growth Differ* 11, 355–60.
42. Reed, S. I. (1980) The selection of S. cerevisiae mutants defective in the start event of cell division. *Genetics* 95, 561–77.
43. Dey, B., Lightbody, J. J., and Boschelli, F. (1996) CDC37 is required for p60v-src activity in yeast. *Mol Biol Cell* 7, 1405–17.
44. Xu, Y., and Lindquist, S. (1993) Heat-shock protein hsp90 governs the activity of pp60v-src kinase. *Proc Natl Acad Sci U S A* 90, 7074–8.
45. Cutforth, T., and Rubin, G. M. (1994) Mutations in Hsp83 and cdc37 impair signaling by the sevenless receptor tyrosine kinase in Drosophila. *Cell* 77, 1027–36.
46. Dai, K., Kobayashi, R., and Beach, D. (1996) Physical interaction of mammalian CDC37 with CDK4. *J Biol Chem* 271, 22030–4.
47. Siligardi, G., Panaretou, B., Meyer, P., Singh, S., Woolfson, D. N., Piper, P. W., Pearl, L. H., and Prodromou, C. (2002) Regulation of Hsp90 ATPase activity by the co-chaperone Cdc37p/p50cdc37. *J Biol Chem* 277, 20151–9.
48. Citri, A., Harari, D., Shohat, G., Ramakrishnan, P., Gan, J., Lavi, S., Eisenstein, M., Kimchi, A., Wallach, D., Pietrokovski, S., and Yarden, Y. (2006) Hsp90 recognizes a common surface on client kinases. *J Biol Chem* 281, 14361–9.
49. Mandal, A. K., Lee, P., Chen, J. A., Nillegoda, N., Heller, A., DiStasio, S., Oen, H., Victor, J., Nair, D. M., Brodsky, J. L., and Caplan, A. J. (2007) Cdc37 has distinct roles in protein kinase quality control that protect nascent chains from degradation and promote posttranslational maturation. *J Cell Biol* 176, 319–28.
50. Arlander, S. J., Felts, S. J., Wagner, J. M., Stensgard, B., Toft, D. O., and Karnitz, L. M. (2006) Chaperoning checkpoint kinase 1 (Chk1), an Hsp90 client, with purified chaperones. *J Biol Chem* 281, 2989–98.
51. Dittmar, K. D., Banach, M., Galigniana, M. D., and Pratt, W. B. (1998) The role of DnaJ-like proteins in glucocorticoid receptor.hsp90 heterocomplex assembly by the reconstituted hsp90.p60.hsp70 foldosome complex. *J Biol Chem* 273, 7358–66.
52. Holt, S. E., Aisner, D. L., Baur, J., Tesmer, V. M., Dy, M., Ouellette, M., Trager, J. B., Morin, G. B., Toft, D. O., Shay, J. W., Wright, W. E., and White, M. A. (1999) Functional requirement of p23 and Hsp90 in telomerase complexes. *Genes Dev* 13, 817–26.
53. DeZwaan, D. C., Toogun, O. A., Echtenkamp, F. J., and Freeman, B. C. (2009) The Hsp82 molecular chaperone promotes a switch between unextendable and extendable telomere states. *Nat Struct Mol Biol* 16, 711–6.
54. Grandin, N., and Charbonneau, M. (2001) Hsp90 levels affect telomere length in yeast. *Mol Genet Genomics* 265, 126–34.
55. Toogun, O. A., Dezwaan, D. C., and Freeman, B. C. (2008) The hsp90 molecular chaperone

modulates multiple telomerase activities. *Mol Cell Biol* 28, 457–67.

56. Prehn, R. T., and Main, J. M. (1957) Immunity to methylcholanthrene-induced sarcomas. *J Natl Cancer Inst* 18, 769–78.
57. DuBois, G. C., Law, L. W., and Appella, E. (1982) Purification and biochemical properties of tumor-associated transplantation antigens from methylcholanthrene-induced murine sarcomas. *Proc Natl Acad Sci U S A* 79, 7669–73.
58. Srivastava, P. K., DeLeo, A. B., and Old, L. J. (1986) Tumor rejection antigens of chemically induced sarcomas of inbred mice. *Proc Natl Acad Sci U S A* 83, 3407–11.
59. Binder, R. J., Blachere, N. E., and Srivastava, P. K. (2001) Heat shock protein-chaperoned peptides but not free peptides introduced into the cytosol are presented efficiently by major histocompatibility complex I molecules. *J Biol Chem* 276, 17163–71.
60. Rajagopal, D., Bal, V., Mayor, S., George, A., and Rath, S. (2006) A role for the Hsp90 molecular chaperone family in antigen presentation to T lymphocytes via major histocompatibility complex class II molecules. *Eur J Immunol* 36, 828–41.
61. Loo, M. A., Jensen, T. J., Cui, L., Hou, Y., Chang, X. B., and Riordan, J. R. (1998) Perturbation of Hsp90 interaction with nascent CFTR prevents its maturation and accelerates its degradation by the proteasome. *Embo J* 17, 6879–87.
62. Riordan, J. R. (2005) Assembly of functional CFTR chloride channels. *Annu Rev Physiol* 67, 701–18.
63. Wang, X., Venable, J., LaPointe, P., Hutt, D. M., Koulov, A. V., Coppinger, J., Gurkan, C., Kellner, W., Matteson, J., Plutner, H., Riordan, J. R., Kelly, J. W., Yates, J. R., 3rd, and Balch, W. E. (2006) Hsp90 cochaperone Aha1 downregulation rescues misfolding of CFTR in cystic fibrosis. *Cell* 127, 803–15.
64. Sun, F., Mi, Z., Condliffe, S. B., Bertrand, C. A., Gong, X., Lu, X., Zhang, R., Latoche, J. D., Pilewski, J. M., Robbins, P. D., and Frizzell, R. A. (2008) Chaperone displacement from mutant cystic fibrosis transmembrane conductance regulator restores its function in human airway epithelia. *Faseb J* 22, 3255–63.
65. Youker, R. T., Walsh, P., Beilharz, T., Lithgow, T., and Brodsky, J. L. (2004) Distinct roles for the Hsp40 and Hsp90 molecular chaperones during cystic fibrosis transmembrane conductance regulator degradation in yeast. *Mol Biol Cell* 15, 4787–97.
66. Jakob, U., Lilie, H., Meyer, I., and Buchner, J. (1995) Transient interaction of Hsp90 with early unfolding intermediates of citrate synthase. Implications for heat shock in vivo. *J Biol Chem* 270, 7288–94.
67. Muller, L., Schaupp, A., Walerych, D., Wegele, H., and Buchner, J. (2004) Hsp90 regulates the activity of wild type p53 under physiological and elevated temperatures. *J Biol Chem* 279, 48846–54.
68. Nathan, D. F., Vos, M. H., and Lindquist, S. (1997) In vivo functions of the Saccharomyces cerevisiae Hsp90 chaperone. *Proc Natl Acad Sci U S A* 94, 12949–56.

Chapter 4

The Role of p23, Hop, Immunophilins, and Other Co-chaperones in Regulating Hsp90 Function

Marc B. Cox and Jill L. Johnson

Abstract

Molecular chaperones are a diverse group of highly conserved proteins that transiently interact with partially folded polypeptide chains during normal cellular processes, such as protein translation, translocation, and disassembly of protein complexes (1). Prior to folding or after denaturation, hydrophobic residues that are normally sequestered within a folded protein are exposed to the aqueous environment and are prone to aggregation or misfolding. Multiple classes of molecular chaperones, such as Hsp70s and Hsp40s, recognize and transiently bind polypeptides with exposed hydrophobic stretches in order to prevent misfolding. Other types of chaperones, such as Hsp90, have more specialized functions in that they appear to interact with only a subset of cellular proteins. This chapter focuses on the role of Hsp90 and partner co-chaperones in promoting the folding and activation of a diverse group of proteins with critical roles in cellular signaling and function.

Key words: Hsp90, p23, Co-chaperone, Immunophilin, Protein folding

1. Hsp90 Molecular Chaperone

The highly conserved and abundant molecular chaperone Hsp90 (heat-shock protein, 90 kDa) interacts with a wide array of cellular proteins or clients in a dynamic ATP-dependent cycle. As shown in Table 1, there are three main classes of Hsp90 clients: transcription factors, such as nuclear receptors for steroid hormones; a wide variety of protein kinases; and a third class of dissimilar proteins, including telomerase, matrix metalloproteinase 2, and a variety of viral proteins. There is no obvious sequence or structural homology between these diverse clients. Hsp90 interaction promotes client folding, stability, transport, and/or assembly into multiprotein complexes (2, 3). Clients interact with Hsp90 while in a nearly

Stuart K. Calderwood and Thomas L. Prince (eds.), *Molecular Chaperones: Methods and Protocols*, Methods in Molecular Biology, vol. 787, DOI 10.1007/978-1-61779-295-3_4, © Springer Science+Business Media, LLC 2011

Table 1
Representative list of Hsp90 client proteins[a]

Transcription factors	References
Nuclear hormone receptors	
Progesterone, glucocorticoid, estrogen receptors	(4)
Heme-responsive transcription factor, Hap1	(120)
Aryl hydrocarbon receptor	(121)
Mutant p53	(122)
Heat-shock factor, Hsf1	(121)
HIF1a	(123)
Protein kinases	
v-src	(124)
Raf, Ste11	(125)
PKR, dsRNA-dependent kinase	(126)
Wee1 kinase	(127)
ErbB2	(128)
Fes tyrosine kinase	(121)
AKT	(129)
Other	
Telomerase	(4)
Hepatitise B reverse transcriptase	(4)
Matrix metalloproteinase 2 (MMP2)	(130)

[a]A complete list of client proteins is maintained at http://www.picard.ch/downloads/Hsp90interactors.pdf

completely folded conformation, stabilizing client conformations in such a way to allow them to quickly respond to an activation signal, such as ligand binding or phosphorylation (4). A classic example is the interaction of Hsp90 with the progesterone (PR) and glucocorticoid receptors (GR). Hsp90 interacts with these receptors in the absence of hormone, repressing transcriptional activity. Hsp90 also maintains the receptor in a conformation capable of high-affinity hormone binding. Once hormone binds, Hsp90 and associated co-chaperones are released and the receptors become active as transcription factors.

Up to 10% of all proteins in *Saccharomyces cerevisiae* are directly or indirectly dependent on Hsp90 for function, indicating that Hsp90 functions as a global cellular regulator (5, 6). Two isoforms of Hsp90 are expressed in yeast cytosol and expression of at least one form is essential for viability. Two isoforms of cytosolic Hsp90 are present in mammalian cells, as well as two related proteins, mitochondrial Trap-1 and Grp94 of the endoplasmic reticulum. These proteins also have critical cellular functions: Trap-1 is critical for regulating mitochondrial integrity and Grp94, which is essential in mouse and has roles in ER-associated degradation and protein quality control pathways. Somewhat surprisingly, the *Escherichia coli* homolog, HtpG, although it shares a similar

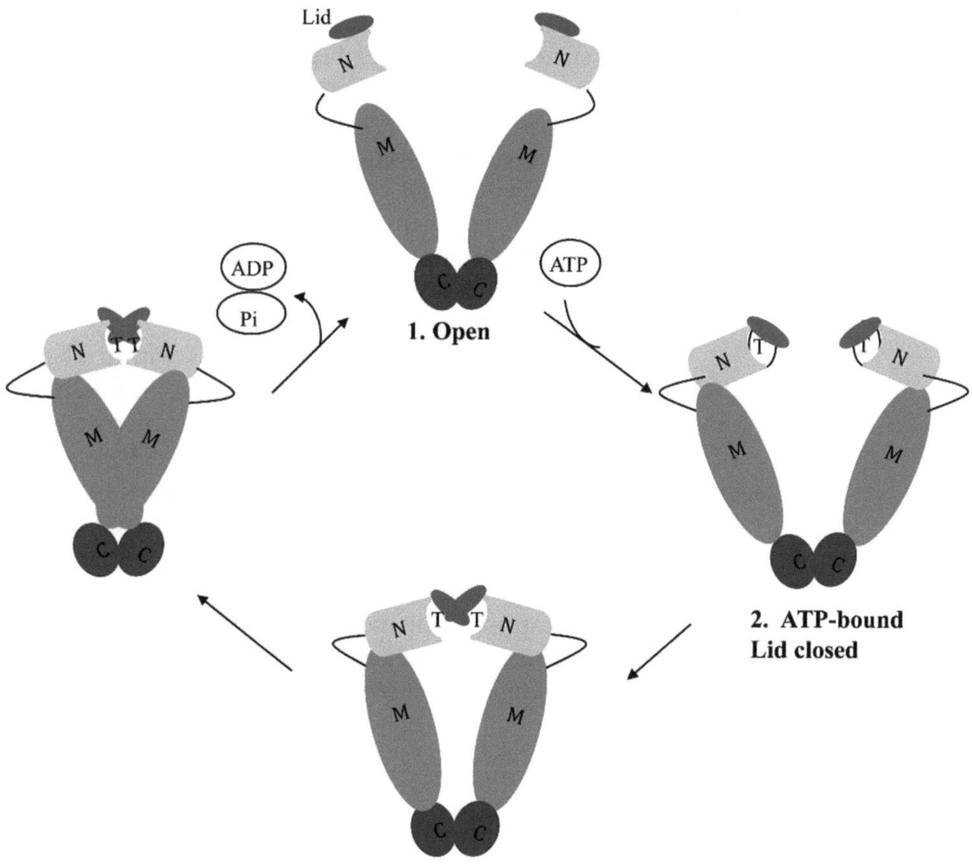

Fig. 1. Hsp90 ATPase cycle. In the absence of ATP (1), Hsp90 is dimerized at its C-terminus in the "open" conformation. Upon ATP binding (2), the N-terminal domains undergo subsequent conformational changes that result in closing of a "lid" over the bound nucleotide. Formation of a second dimerization interface between the amino-termini results in the "closed" conformation (3). Continued rearrangements of the closed conformation allow interaction of the N-terminal and middle domains, resulting in the "closed and twisted" conformation which is able to hydrolyze ATP (4). After ATP hydrolysis, the lid opens and the N-terminal domains release from one another returning to the open conformation.

structure and mechanism, is not essential for viability and its in vivo functions remain unclear (7).

Hsp90 contains three conserved domains: an N-terminal ATP-binding domain, a middle domain, and a carboxy-terminal domain. The ability of Hsp90 to bind and hydrolyze ATP is essential for its function. Recent structural evidence supports a unified model for the conformational changes of cytosolic Hsp90, Trap-1, Grp94, and HtpG (reviewed in refs. 3, 8). A simplified model is presented in Fig. 1. In the absence of nucleotide, Hsp90 is dimerized at the carboxy-terminus, resulting in an open conformation. Nucleotide binding induces the closing of a lid over bound nucleotide. The N-terminal domains then associate to form the "closed" conformation, in which portions of the amino-terminal domain "cross over" to associate with the other protomer. The Hsp90 dimer then forms a compact, twisted structure that results in association of a flexible loop from the middle domain with the nucleotide-binding pocket. This structure is capable of ATP hydrolysis. After hydrolysis, Hsp90 returns to the open conformation. Recent studies indicate that some forms of Hsp90 adopt the "closed" conformation even in the absence of nucleotide, suggesting that nucleotide shifts the equilibrium between distinct conformations rather than triggering specific conformational changes (9). As discussed in more detail below, cytosolic Hsp90 function is dependent on a battery of interacting co-chaperones, many of which regulate the ATPase activity and/or bind specific conformations of Hsp90 (3, 7).

The location of the client-binding site(s) within Hsp90 and the conformation of Hsp90 that is active in protein folding remain unclear. However, the single-particle electron microscopy structure of an Hsp90–Cdc37–Cdk4 kinase complex (10) suggests that clients bind the amino-terminal and/or middle domains on the "outside" of the dimer. Given the diversity of Hsp90 clients, it is possible that clients have different binding sites and/or that the client binding site changes as Hsp90 undergoes conformational changes. Clarification of the mechanism by which Hsp90 interacts with diverse clients and facilitates protein folding remains one of the biggest challenges in understanding Hsp90 function.

2. Cytosolic Hsp90 Function is Dependent on Co-chaperone Proteins

Hsp90 client protein complexes contain additional proteins that function as part of an ATP-dependent, ordered pathway in which clients are transferred from Hsp70 to Hsp90. For the purpose of this review, co-chaperones are proteins that have physical interaction with Hsp90 but cooperate with Hsp90 to fold clients rather than being clients themselves. Some, but not all, co-chaperones have molecular chaperone activity and can prevent aggregation of model substrates. Some also may have functions not directly related to Hsp90 client activity. Co-chaperones are variously characterized by the domains of Hsp90 with which they interact, their own biochemical functions, or in the order in which they appear to interact with particular client proteins. In this review, the co-chaperones are grouped according to Hsp90-interacting domain and discussed in terms of what is known about their structure and function followed by what is known about how they interact with client proteins. Co-chaperones exhibit specificity for certain clients (11), but the basis of this selectivity remains largely unknown. Table 2 sorts co-chaperones by their interaction domains and also lists their functions as well as some known client interactions. Importantly, co-chaperones frequently exhibit mutually exclusive binding, either by competing for the same binding site or because they stabilize conformations of Hsp90 that preclude interactions with other co-chaperones (4, 12, 13).

Table 2
Hsp90 co-chaperones and their functions

Interaction domain	Biochemical functions	Known client protein–Hsp90 complex interactions
Amino-terminus		
p23 (Sba1)	Chaperone activity (20); inhibits Hsp90 ATPase activity (14, 16)	Steroid hormone receptors, telomerase, hepatitis B virus reverse transcriptase (4)
Sgt1 (Sgt1)	Adaptor protein (24)	Kinetochore complex, NLR immune sensors; Polo kinase (27)
Cdc37	Adaptor, binds protein kinases (28); inhibits Hsp90 ATPase activity (131)	Protein kinases (28)
GCUNC-45	Regulates Hsp90 (30)	Progesterone receptor (30)
Middle domain		
Aha1	Activates Hsp90 ATPase activity (132)	Mutant CFTR (32)
Carboxy-terminus		
Hop (Sti1)	Inhibits Hsp90 ATPase activity; stimulates Hsp70 ATPase activity (35, 41)	Steroid hormone receptors (47), kinases (38)
Cyp40 (Cpr6/7)	Chaperone activity; PPIase activity (20, 42)	Steroid hormone receptors (47)
FKBP51/52	Chaperone activity; PPIase activity (40)	Steroid hormone receptors (47)
XAP2	Immunophilin-like domain (4)	Dioxin receptor (4)
AIPL1	Immunophilin-like domain (48)	Nub1 (48)
PP5 (Ppt1)	Dephosphorylates Hsp90 (50)	Steroid hormone receptors (47), kinases (49)
Tah1	Weakly stimulates Hsp90 ATPase activity (52)	DNA helicases Rvb1/2; ribonucleoprotein biogenesis (5, 53)
TTC4 (Cns1)	Activates Hsp70 ATPase activity (55)	DNA polymerase alpha, replication protein Cdc6 (4, 56)
Tom70	Mitochondrial import receptor (57)	Mitochondrial precursor proteins (57)
CHIP	E3 ligase (133)	Glucocorticoid receptor, phosphorylated tau (4, 58)
Tpr2	Stimulates Hsp70 ATPase activity (59)	Steroid hormone receptors (59, 60)
αSgt	Inhibits Hsp70 ATPase activity (64), putative adaptor/client transfer protein (74)	Steroid hormone receptors (74)

2.1. Co-chaperones that Bind the Amino-Terminal ATPase Domain of Hsp90 p23

p23 and its yeast homolog, Sba1, preferentially bind Hsp90 in the presence of ATP or nonhydrolyzable analogs. p23 inhibits the ATPase activity, stabilizing the closed conformation, which results in prolonged interaction of the client with Hsp90 (14–18). p23 has molecular chaperone activity and is able to suppress the aggregation of citrate synthase (19, 20). p23 is found in complex with Hsp90 and a range of client proteins, including steroid hormone receptors, telomerase, and duck hepatitis B reverse transcriptase (4).

As evidence for a critical role for p23 in viral function, hepatitis C expresses a homolog of p23 that may help to recruit Hsp90 to the site of viral replication (21). Although not essential in yeast, Sba1 is required for optimal growth and activity of Hsp90 clients (22). However, p23 is essential in mice and mice lacking p23 die in an early embryonic stage (23).

2.2. Sgt1

Sgt1 interacts with Hsp90 through a CHORD (CS) domain that shares homology with p23. However, Sgt1 does not modulate the ATPase activity of Hsp90 and binds preferentially to Hsp90 in the absence of ATP (24). Like many other co-chaperones, Sgt1 contains a tetratricopeptide repeat (TPR) domain, which typically mediates interaction with the carboxy-terminus of Hsp90 (4). However, the TPR domain of Sgt1 is not required for Hsp90 interaction and is absent from Sgt1 orthologs in some species. *SGT1* is essential in yeast, and Sgt1 appears to act early in the Hsp90 folding cycle, serving as an adaptor protein that helps to recruit client proteins to Hsp90 (24). Although the structures of p23 and the CS domain of Sgt1 are similar, they exhibit distinct interactions with Hsp90 (25, 26). Sgt1 interacts with the CBF3 kinetochore complex, Polo kinase, plant R (resistance) proteins, and related Nod-like receptors (reviewed in ref. 27).

2.3. Cdc37

The middle domain of Cdc37 (also known as $p50^{cdc37}$) binds directly to the amino-terminal domain of Hsp90. Cdc37 inhibits ATPase activity, apparently by preventing amino-terminal dimerization and contacting a residue required for ATP hydrolysis (2). The amino-terminal domain of Cdc37 is able to bind protein kinases directly, suggesting a role for Cdc37 in promoting Hsp90–kinase interaction (reviewed in ref. 28). Cdc37 is essential in yeast and is required for the folding and stabilization of multiple mammalian kinases and a significant proportion of protein kinases in *S. cerevisiae* (29).

2.4. GCUNC-45

GCUNC-45 is a member of the UCS (UNC-45/Cro2/She4p) family of proteins that bind myosin. There is no apparent yeast homolog. Like many co-chaperones discussed below, GCUNC-45 interacts with Hsp90 through TPR domains. However, in contrast to other TPR-containing co-chaperones, GCUNC-45 binds the ATP-binding domain of Hsp90. GCUNC-45 did not affect the ATPase activity of Hsp90, but did disrupt the ability of Aha1 to stimulate the ATPase activity of Hsp90 (30).

2.5. Co-chaperones that Bind the Middle Domain of Hsp90: Aha1

Aha1 (activator of Hsp90 ATPase) binds the middle domain of Hsp90. Aha1 and Hch1, a related protein found only in certain fungi, dramatically stimulate the ATPase activity of Hsp90. Hsp90 has a split ATPase domain, which requires association of the amino-terminal and middle domains for hydrolysis to occur.

Aha1 stimulates ATP hydrolysis by promoting a conformational change that allows a catalytic loop of the middle domain to associate with the amino-terminus (reviewed in ref. 2). Surprisingly, Aha1 can promote these structural rearrangements even in the absence of ATP (31). The in vivo functions of Aha1 are poorly understood, and only slight defects are observed upon combination of *AHA1* and *HCH1* deletion in yeast. However, a recent study of the mutant cystic fibrosis transmembrane conductance regulator (CFTR) provided new insight into Aha1 functions (32). The stability of the mutant CFTR was specifically enhanced in the absence of Aha1, but not p23, suggesting that Aha1 plays a critical role in the quality control process that leads to degradation of misfolded Hsp90 clients.

2.6. Co-chaperones that Bind the Carboxy-Terminal Domain of Hsp90

2.6.1. TPR-Containing Co-chaperones

A number of co-chaperones contain TPR domains and dynamically compete for a common binding site at the carboxy-terminus of Hsp90, which contains the terminal sequence EEVD (4, 33). A slightly different collection of these TPR proteins copurify with different steroid hormone receptors, indicating client specificity, but the basis for this selectivity is poorly understood (11).

2.6.2. Hop

Hop (Hsp70–Hsp90 organizing protein) interacts with a wide range of Hsp90 client proteins. Separate TPR domains of Hop bind the carboxy-terminus of Hsp70 and Hsp90, both of which end in the sequence EEVD. Simultaneous binding of Hop to Hsp70 and Hsp90 helps mediate transfer of client protein from Hsp70 to Hsp90 (34). Sti1, the yeast homolog, preferentially binds Hsp90 in the absence of ATP and inhibits dimerization of the amino-terminal domain, resulting in reduced ATPase activity (2, 4). Sti1 was also able to stimulate the ATPase activity of the Hsp70 Ssa1 (35). In addition to three distinct TPR domains, one with specificity for Hsp70, one with specificity for Hsp90, and one that appears to weakly interact with both chaperones, Hop/Sti1 also contains domains rich in aspartic acid and proline residues that are important for Sti1 function (33, 36, 37). Deletion of *STI1* from yeast results in slow growth at high and low temperatures, as well as defects in Hsp90 clients, such as heterologous steroid hormone receptors and kinases (38). To date, no functional effect on Hsp90 client proteins in mammalian cells has been observed upon Hop overexpression or knockdown. Given the clear evidence of a role for yeast Sti1 in Hsp90 client protein function, it is likely that an unidentified protein is present within higher vertebrates that can complement the loss of Hop.

2.6.3. Immunophilins

Immunophilins are a group of proteins that bind immunosuppressant drugs, such as FK506, rapamycin, or cyclosporin A (39). Although they vary in size and domain structure, all members of the immunophilin family have peptidyl-prolyl isomerase (PPIase) activity.

The immunosuppressant drugs bind the catalytic pocket and inhibit PPIase activity. Two types of immunophilins, Cyp40 (which binds cyclosporin A) and FKBP51/FKBP52 (which binds FK506) also contain TPR domains that mediate interaction with Hsp90 (4). These two classes of immunophilins have dissimilar sequence and structures, yet both types of co-chaperones mediate the activity of known Hsp90 client proteins and have molecular chaperone activity that is independent of their PPIase activity (20, 40).

2.6.4. Cyp40

Cyp40 associates with nuclear steroid hormone receptors, including the progesterone, estrogen, and glucocorticoid receptors (4). Cyp40 contains an amino-terminal PPIase domain and a TPR domain separated by a small linker region. In *S. cerevisiae*, there are two homologs of Cyp40, Cpr6, and Cpr7. Purified Cpr6 had no effect on the ATPase activity of Hsp90 (41). Cells lacking *CPR6* exhibit few discernible phenotypes while loss of *CPR7* results in slow growth and defects in Hsp90 client protein activity. Cpr6 and Cpr7 exhibit different PPIase and molecular chaperone activities, but it is unclear how these activities affect Hsp90 client activity (42, 43). The role of the PPIase domain remains unclear, but the TPR domain and/or the hydrophobic linker that separates the PPIase and TPR domains appear to be critical for molecular chaperone and Hsp90 client activity (44, 45).

2.6.5. FKBP51 and FKBP52

FKBP51 and FKBP52 are two related proteins that are found in complex with steroid hormone receptors. There are no FKBP homologs that contain TPR domains in yeast. Similar to CyP40, FKBP51 and FKBP52 contain a PPIase domain and a TPR domain separated by a short linker. Despite their sequence similarities, FKBP52 and FKBP51 have distinct effects on steroid receptor activity. The presence of FKBP52 in steroid hormone receptor complexes results in higher hormone-binding affinity. FKBP51 offsets the positive effects of FKBP52. The specific ability of FKBP52 to promote steroid hormone receptor activity is dependent on noncatalytic residues in the PPIase domain that may directly interact with the ligand-binding domain (LBD) (46). Consistent with a positive role in steroid hormone receptor function, mice lacking FKBP52 exhibit androgen and progesterone receptor-related reproductive defects. Similar effects were not observed in mice lacking FKBP51. However, the importance of FKBP51 in regulating glucocorticoid receptor is highlighted in New World primates. These primates overexpress FKBP51 relative to Old World primates, resulting in reduced GR hormone-binding affinity and glucocorticoid resistance. The overexpression of FKBP51 may be an adaptation to higher circulating levels of glucocorticoids in New World primates (47). These studies emphasize that the type of TPR-containing co-chaperone bound to Hsp90 may have dramatic effects on client protein activity.

2.6.6. XAP2 and AIPL1

XAP2 contains TPR domains and was originally identified as the hepatitis B virus X-associated protein. It was subsequently found to interact with the dioxin or aryl hydrocarbon receptor, along with Hsp90 and p23. Although XAP2 contains an FKBP-related immunophilin-like domain, it does not bind FK506 (4). AIPL1 is homologous to XAP2 and interacts with Hsp70 and Hsp90 in a TPR-dependent manner. Mutations in AIPL1 cause Leber congenital amaurosis, a form of inherited blindness. AIPL1 may interact with the proteasome-associated protein, NUB1, and have other functions in photoreceptors (48).

2.7. Other TPR-Containing Co-chaperones

2.7.1. PP5

PP5 is a serine–threonine phosphatase that interacts with Hsp90 through its TPR domains. PP5 interacts with a wide range of proteins involved in signaling pathways, including ATM/ATR and the Rac GTPase (reviewed in ref. 49). Deletion of the yeast homolog, *PPT1*, results in Hsp90 client-specific defects (50). Intriguingly, Ppt1 appears to be able to dephosphorylate both Hsp90 and Cdc37 (51), indicating that Ppt1 may have a general role in regulating the activity of the Hsp90 molecular chaperone machine.

2.7.2. Tah1

Tah1 binds Hsp90 through a single TPR domain that contains two TPR motifs, and is able to weakly stimulate ATPase activity (52). Deletion of *TAH1* has little effect on the growth of *S. cerevisiae.* However, Tah1 forms complexes with Hsp90 and the essential DNA helicases, Rvb1 and Rvb2, and has a conserved role in ribonucleoprotein biogenesis (5, 53).

2.7.3. TTC4

The most well-studied homologs of TTC4 (TPR domain 4) are Cns1 of *S. cerevisiae* and Dpit47 of *Drosophila melanogaster. CNS1* is one of the few co-chaperones that is essential in *S. cerevisiae* (54). Cns1 was able to activate the ATPase activity of the Hsp70 Ssa1, but did not affect the ATPase activity of Hsp90 (55). TTC4 and homologs are nucleoplasmic and interact with DNA polymerase alpha and the replication protein Cdc6 (4, 56).

2.7.4. Tom70

Tom70 is protein of the mitochondrial outer membrane with TPR domains exposed to the cytoplasm. Tom70 is required for the post-translational import of mitochondrial preproteins. In mammals, Hsp70 and Hsp90 precursors are targeted to the mitochondria by Hsp70 and Hsp90. *TOM70* is not essential in yeast, but Hsp70 and Tom70 maintain a critical role in preprotein import (57).

2.7.5. CHIP

CHIP is an E3 ubiquitin ligase that interacts with either Hsp70 or Hsp90 through TPR domains. Incorporation of CHIP into Hsp90 complexes leads to client ubiquitination and degradation. CHIP plays a key role in regulating chaperone-mediated protein folding versus degradation. Some specific clients known to be affected by CHIP are the glucocorticoid receptor and phosphorylated tau (4, 58).

2.7.6. Tpr2

Tpr2 contains both TPR domains and a J-domain characteristic of *E. coli* DnaJ and Hsp40 molecular chaperones. Tpr2 was able to interact with both Hsp70 and Hsp90 through its TPR domains. In addition, Tpr2 was able to stimulate the ATPase activity of Hsp70. Tpr2 was required for steroid receptor activity in vivo and was also able to mediate assembly of Hsp90 complexes with the progesterone receptor and Chk1 kinase in vitro (59, 60).

2.7.7. Small Glutamine-Rich TPR-Containing Protein α

Small glutamine-rich TPR-containing protein α (αSGT) was originally identified for the ability to interact with the NS1 protein of parvovirus (61). αSGT has also been referred to as U-binding protein (Ubp) due to its interaction with the human immune-deficiency virus Vpu protein (62, 63). αSGT is reported to interact with a variety of cellular proteins, to have a role in mitotic progression, and to be a pro-apoptotic factor (64–73). Recently, αSGT was characterized as a co-chaperone that functionally regulates the androgen receptor (74). Although originally described as an AR-specific co-chaperone because of its lack of effect on ER, αSGT is also a functional affecter of GR and PR (Cox unpublished observation). αSGT contains an N-terminal dimerization domain, a central TPR domain that has a high degree of similarity to the Hop TPR2A domain, and a glutamine-rich carboxy-terminal domain of unknown function (63). αSGT binds to both Hsp70 and Hsp90 presumably by interaction with the carboxy-terminal EEVD motif and, in the case of Hsp70, has been shown to regulate chaperone activity (64, 73, 75). In addition, αSGT is predicted to bind to an EEVD-like motif just upstream of the androgen receptor hormone-binding domain, serving as a molecular rheostat for AR activity. Knockdown of αSGT protein in human prostate cancer cells results in an increase in both hormone-dependent and hormone-independent functions, which can be restored upon exogenous αSGT expression (74). αSGT is evolutionarily conserved from yeast to humans (76). Deletion of the *S. cerevisiae* homolog, *SGT2*, results in a cell wall defect and abnormal cell cycle progression (77, 78). Synthetic lethality between *SGT2* and *YDJ1* has been observed in stressed cells and Sgt2 interacts indirectly with Ydj1 by way of its interaction with Mdy2 (79).

2.8. Other Co-chaperones that Do Not Appear to Bind Hsp90 Directly but Are Known to Affect Hsp90 Client Protein Activity

In addition to the proteins listed above, which directly bind Hsp90, a number of other proteins are known to be critical for Hsp90 function. One particular class of Hsp70s, represented by the Ssa family in *S. cerevisiae*, interacts with many client proteins prior to Hsp90 interaction (4). Accordingly, proteins that modulate Hsp70 ATPase activity are also required for Hsp90 client activity. Hsp40s stimulate the ATPase activity of Hsp70s (1). One class of Hsp40s, DNAJA2/Hdj2 (Ydj1 in yeast), is well-known to be required for Hsp90 client activity (80). Members of the Hsp110 family function as nucleotide exchange factors for Hsp70 (81). Sse1, a member of the Hsp110 family of proteins, is a known Hsp90 co-chaperone (82). Finally, Hip stabilizes the ADP-bound form of Hsp70 (83) and is found in complexes with the progesterone receptor (84).

3. Assembly of Hsp90 Client Protein Complexes

Analysis of the interaction of Hsp90 with steroid hormone receptors separated assembly into early, intermediate, and late stages (4) (Fig. 2). In this model, newly synthesized or misfolded receptor first interacts with Hsp40 and Hsp70. Subsequent interaction of Hop/Sti1, which simultaneously binds Hsp70 and Hsp90 through separate TPR domains, promotes Hsp90 interaction. Hop/Sti1 is then displaced by other TPR-containing co-chaperones, such as the immunophilins, which compete for the same binding site at the carboxy-terminus of Hsp90. Nucleotide binding results in formation of the ATP-bound closed form of Hsp90 and p23 interaction. Client dissociation results after ATP hydrolysis and/or p23 release (15). Studies with purified progesterone receptor indicate that Hsp70, Hsp40, Hop, Hsp90, and p23 are sufficient to restore high-affinity hormone binding to receptors and indicate that Hsp70 interaction precedes Hsp90 interaction (85). Studies with purified Chk1 kinase indicate a similar ordered pathway of assembly, but the required co-chaperones vary. Hsp40, Hsp70, Cdc37,

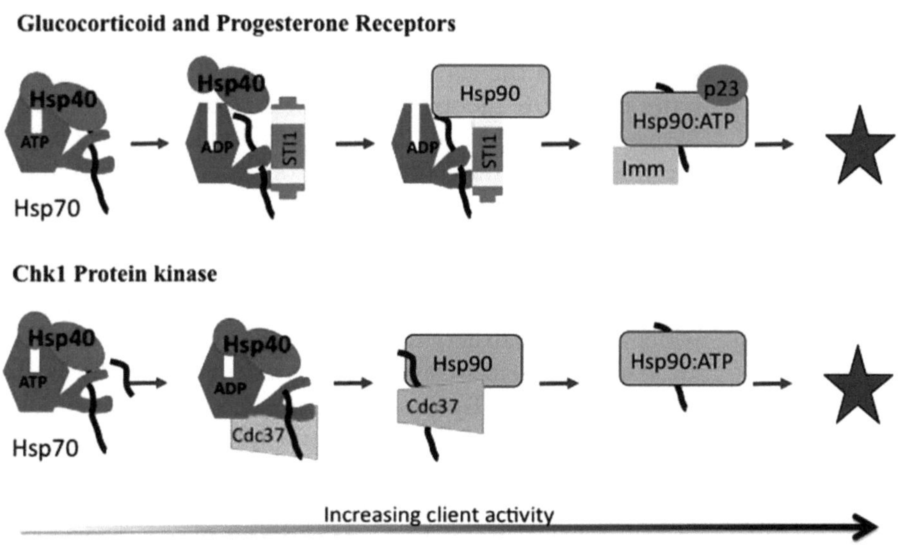

Fig. 2. Client proteins require Hsp90 plus distinct sets of co-chaperones to become active in a purified system. *Top*: Assembly of Hsp90 with steroid hormone receptors (85). Immunophilins are shown to illustrate the late complex, but were not required for activity. *Bottom*: Assembly of Hsp90 with the Chk1 kinase. Hop (not shown) enhanced kinase activation, but was not required for activity. Also not shown is the CK2 protein kinase, which is required for Cdc37 activity (86).

and Hsp90 were sufficient to restore Chk1 kinase activity in a purified system (86). In both cases, low levels of client activity were present even during the early stages and were maximal during the late stage, suggesting that client folding progresses throughout the cycle. The following sections go into more detail about the stages of Hsp90 client assembly and discuss client-specific differences.

3.1. Early Steps in the Assembly of Hsp90 Complexes: Interaction with Hsp70 and Hsp40

Hsp70s of the abundant, essential Ssa class in *S. cerevisiae* were among the first proteins identified in Hsp90 complexes with steroid receptors (4, 87). A role for the Hsp40 Ydj1, which regulates the ATPase activity of Ssa1, in Hsp90 client maturation was first identified in yeast (80). Ydj1 is required for the in vivo maturation of heterologous Hsp90 client proteins, such as steroid receptors, (80, 88) as well as endogenous yeast client proteins, such as kinases and the heme-regulated transcription factor, Hap1 (89, 90). The cytosol of *S. cerevisiae* has 13 different cytosolic Hsp40s (91), but only Ydj1 appears to function with Hsp90. Ydj1 and Ssa1 cooperate in many cellular processes, including a specific role early in the folding of Hsp90 clients. It is unclear whether they also function at later stages of the Hsp90–client interaction cycle.

Ydj1 and/or the human homolog, Hdj2/DNAJA2, enhance in vitro assembly of Hsp90 complexes with the progesterone and glucocorticoid receptors as well as the Chk1 kinase (85, 86, 92). Ydj1 interacts directly with the progesterone receptor and is required to promote Hsp70 interaction. In contrast, Hsp70 appears to bind glucocorticoid receptor prior to Ydj1 (93, 94). Alteration of different domains of Ydj1 had similar effects on the interaction of Ydj1 with the glucocorticoid receptor and the Ste11 MAP kinase in vivo (95). Thus, cooperative Hsp70 and Hsp40 interaction is a shared step in the early folding of representatives of two different types of Hsp90 clients, transcription factors, and kinases (Fig. 2). Many of the differences in client protein folding may occur later in the folding pathway and result from the use of adaptor proteins that promote client interaction with Hsp90.

3.2. Intermediate Stage: Adaptor Proteins Recruit Hsp90 to Client Complexes

Although it is widely accepted that both Hsp70 and Hsp90 act sequentially during maturation of Hsp90 clients, the process by which Hsp70 transfers clients to Hsp90 is still unclear. The idea of an adaptor protein arose largely from studies with PR. It was known that Hsp70 and Hsp90 were both present in PR complexes, but purified Hsp70 and Hsp90 did not interact directly. However, an Hsp70–Hop–Hsp90 ternary complex formed readily. Thus, Hop was proposed to serve as an adaptor to allow transfer of clients from Hsp70 to Hsp90, likely without direct Hop–client contact (34). A second type of adaptor is Cdc37, which is able to simultaneously bind both protein kinases and Hsp90 through separate domains. Sgt1 also acts as this type of adaptor, interacting with clients through its TPR domains and Hsp90 through its p23-like CS domain (24). A common feature of all three of these adaptor proteins is that they preferentially bind Hsp90 in the nucleotide-free or ADP-bound form, which may facilitate "loading" of the client onto Hsp90 (2, 3). Additional co-chaperones may also serve an adaptor role. αSGT may act as an adaptor protein to promote transfer of androgen receptor from Hsp70 to Hsp90 (74). An alternate adaptor in promoting retrograde transfer of substrates from Hsp90 to Hsp70 was proposed for Tpr2 (59). This may provide an additional level of regulation of the Hsp90 molecular chaperone machine.

In a model of PR assembly, Hop first binds Hsp70–client complexes and then targets them to Hsp90 to form the intermediate complex containing Hsp70, Hop, and Hsp90. Hsp70 and Hop then dissociate to form the late or mature complexes characterized by p23 and immunophilins (Fig. 2). This model is supported by time-course experiments monitoring the interaction of chaperones with PR (96). The intermediate client complex becomes predominant in the presence of the Hsp90 inhibitor geldanamycin, which competes for ATP binding to Hsp90 (4). Although the mechanism remains unclear, in vitro studies demonstrated that Hop was able to promote transfer of a model-denatured client protein from Hsp70 to Hsp90 (4, 97), consistent with the idea that it helps to load clients onto Hsp90. One of the unanswered questions about Hop function arises from studies of Sti1 in *S. cerevisiae*. Surprisingly, loss of *STI1* causes only mild growth defects, and Hsp90 complexes isolated from cells with or without Sti1 contain similar levels of Hsp70 (38). However, as discussed below, the most likely explanation for this result is that other proteins also act as adaptor proteins to promote transfer of clients from Hsp70 to Hsp90.

Cdc37 binds Hsp90 and kinases through distinct domains and helps target protein kinases to Hsp90. Cdc37 also binds a few non-kinase clients, such as the androgen receptor (28). As mentioned above (Fig. 2), purified Chk1 kinase required Hsp40, Hsp70, Cdc37, and Hsp90 to become active as a kinase. Although not required for kinase activity, the presence of Hop further enhanced activity (86). Genetic evidence also suggests that the functions of Hop and Cdc37 at least partially overlap (98). The functions of Hop/Sti1 may also overlap with the adaptor protein Sgt1, since ternary complexes among Sgt1, Sti1, and Hsp90 have been observed (24). Sgt1 appears to enable Hsp90 to interact with a wide range of client proteins involved in the innate immune responses of animals as well as those required for kinetochore function in yeast (27). Future studies will be required to establish adaptor protein:client pairs and the degree of functional overlap between the different adaptor proteins.

3.3. Late Stage: Interaction with p23 and Immunophilins

After Hsp70 release and ATP-associated conformational changes in Hsp90, other co-chaperones, such as p23 and immunophilins, bind and stabilize the closed conformation of Hsp90. This "late" complex is actually a collection of complexes distinguished by which TPR-containing proteins are bound at the carboxy-terminus of Hsp90. Studies with different steroid hormone receptors showed that different levels of Cyp40, PP5, FKBP51, and FKBP52 bind the receptors (11). Thus, the composition of the late complex varies in a client-specific manner. However, there is not yet enough evidence about the composition of other "late" client complexes to know the extent of this variation. Upon activation, ATP hydrolysis occurs and Hsp90 and co-chaperones are released from the client protein. In the case of the steroid hormone receptors, chaperone interactions are restricted to the LBDs. Thus, it is generally thought that the late-stage chaperones assist the LBD in achieving and maintaining the high-affinity hormone-binding conformation through the classic "folding and holding" chaperone role. However, it is also clear that some of the chaperone components have a direct regulatory role in the receptor-signaling pathways. The Hsp90 complex silences steroid hormone receptor function until hormone binding occurs by blocking access to the receptor nuclear localization signal, blocking receptor dimerization, and blocking the association of transcriptional coactivators (reviewed in ref. 99). Additionally, the immunophilin FKBP52 can directly influence receptor hormone binding (100–102) and receptor translocation to the nucleus (103, 104).

Based on studies of the steroid hormone receptors, it is generally assumed that the late-stage complexes are relatively short-lived. In the case of PR, the mature conformation of the receptor lasts only several minutes (96), which correlates with the slow ATPase cycle of Hsp90 (reviewed in ref. 3, 8). After several minutes, the complex dissociates and the receptor is cycled back into the chaperoning pathway. Thus, the receptor-folding pathway is a dynamic, ordered process that involves many transient interactions. In the absence of Hsp90, the receptors are unstable and degraded rapidly through the proteosomal degradation pathway (7).

Upon hormone binding, the steroid hormone receptors are translocated to the nucleus, if not already within that compartment, where they dimerize, associate with hormone-responsive enhancer elements on the DNA, and mediate gene transcription. The role of the late-stage chaperones in these processes remains unclear. There is evidence that chaperones are involved in the nuclear translocation of GR, including chaperone and co-chaperone interactions with dynein motor proteins and interactions with nuclear pore proteins (105, 106). These findings suggest that the late-stage complex is required at least until GR reaches the nuclear pore. However, whether or not the Hsp90 chaperone complex dissociates in the cytosol or travels into the nucleus with GR is debated. Given that some of the receptors (ER and PR) already exist largely within the nucleus (107, 108), one would assume that late-stage complexes with these receptors are present in the nucleus. Given their presence in the nucleus, it is plausible that the late-stage chaperones could have a role in all steps of the steroid hormone receptor-signaling pathway, including DNA binding and the regulation of gene transcription. Some evidence exists suggesting that Hsp90 and p23 can regulate the receptors at the level of gene transcription, but the exact role of the late-stage chaperones on the DNA remains unclear (109).

3.4. Evidence for Direct Interaction of Co-chaperones with Client Proteins

When assessing the regulation of Hsp90 client proteins by co-chaperones, one question that arises is whether the functional effects observed are the result of a direct interaction between client protein and the co-chaperone or indirect interaction with the Hsp90:client protein complex. It is also possible that a co-chaperone does not interact directly with client protein alone, but co-chaperone association with the Hsp90 complex could bring the co-chaperone into direct contact with the client protein. Identifying and characterizing direct client protein:co-chaperone interactions has traditionally been a challenging task. Hsp90 is one of the most abundant proteins in the cell. Thus, any system in which Hsp90 is present cannot determine direct interactions without the use of additional tools as one cannot be sure that the client:co-chaperone interaction observed is simply through both proteins associating with the Hsp90 complex. The yeast two-hybrid system is a classic example of a system in which Hsp90 is present and likely mediates many of the client protein:co-chaperone interactions observed. Purification of recombinant proteins and assessing direct interactions in vitro in the absence of any other factors, including Hsp90, is the logical experimental approach. However, even this approach is made difficult by the fact that many client proteins are extremely unstable in the absence of Hsp90. The steroid hormone receptors are an example of a client protein that cannot be purified in full-length form without the presence of Hsp90. Expression and purification of the steroid hormone receptors in a bacterial system are not possible. The receptors can be stably expressed in insect cells and yeast cells solely because of the presence of Hsp90 and associated co-chaperones. Again, in this scenario, Hsp90 is present and interferes with the assessment of direct interactions. Despite these challenges, some evidence for direct client protein:co-chaperone interactions have been obtained.

One approach that has been used is to immunodeplete Hsp90 from cell extracts prior to immunoprecipitation or immunoprecipitating client protein and stripping off the chaperones with high salt prior to the addition of the interacting co-chaperone. The latter approach was used to demonstrate a direct interaction between FKBP52 and GR (110). The functional significance of this interaction remains unclear given the fact that FKBP52 regulation of GR requires FKBP52 interaction with Hsp90 (101). Introducing mutations that abrogate co-chaperone interaction with Hsp90 could also be used to assess direct interactions. An FKBP52 mutant containing a single-point mutation in the TPR domain that abrogates binding to Hsp90 exists (101), but the use of this mutant to assess direct interactions has not been attempted and/or published to date. Multiple groups have demonstrated an interaction between kinase clients and a mutant Cdc37 protein that is deficient in Hsp90 binding (28). In addition, a time course of assembly in a purified system suggested that Cdc37 binding to Hsp90 preceded binding to Chkl (86). The Hsp90 inhibitor, geldanamycin, has also been used as a tool to assess direct interactions. The presence of geldanamycin disrupted the interaction of a reverse transcriptase client with Hsp90, but the interaction with p23 was not affected suggesting a direct interaction between p23 and the reverse transcriptase (111). In addition, geldanamycin reduced the interaction of αSGT with Hsp90, but the interaction between αSGT and AR was unaffected, suggesting a direct interaction of αSGT with AR (74). Finally, although clients, such as the steroid hormone receptors, are difficult to purify, some clients can be purified and assessed for binding in vitro. Using affinity purification and pull-down assays in vitro, the co-chaperone Sgtl was shown to directly interact with the yeast kinetochore protein Skpl independent of Hsp82 (24).

4. Co-chaperones as Drug Targets

Hsp90 has attracted considerable attention as a target for cancer chemotherapy because Hsp90 is required for the functions of proteins that influence multiple steps in cancer progression (7). Small molecule inhibitors that bind Hsp90, destabilize clients, and induce apoptosis have been used in clinical trials to treat a range of tumor types. The most promising results with Hsp90 inhibitors in clinical trials are as part of combination therapies for breast cancer, lung cancer, and some leukemias (112, 113). Most Hsp90 inhibitors bind the ATPase domain of Hsp90, and thus inhibit the wide range of Hsp90 functions. Similar to Hsp90, some Hsp90 co-chaperones are overexpressed in tumor cells (114). Thus, inhibitors that target co-chaperones may also have clinical significance. For example, an inhibitor of Cdc37 might selectively disrupt the activity of oncogenic kinases and an inhibitor of Ahal might be used to alleviate the effects of cystic fibrosis. In addition, the immunosuppressive drug, FK506, effectively reduces androgen-dependent prostate cancer cell growth presumably through disruption of FKBP52 (115). Thus, compounds that target FKBP52, but lack immunosuppressive effects, might be used for the treatment of prostate cancer. Interestingly, cells may have already devised mechanisms to disrupt co-chaperone function. Cleavage of Hop and p23 was observed upon apoptosis and certain compounds appear to induce cleavage of p23 (116–118), although the biological significance of these events remains unclear.

5. Remaining Questions About Co-chaperone Function

One of the main questions about co-chaperone function is why cytosolic Hsp90 requires co-chaperones, whereas other forms of Hsp90 (Trap-1, Grp94, and HtpG) do not. In addition, the co-chaperones encoded by diverse eukaryotic species vary, suggesting that each organism has a slightly different component of co-chaperones (119). More work is required to determine whether the requirement for specific co-chaperones is due to how it regulates Hsp90 versus how co-chaperones directly affect client function. Many additional questions remain about how specific co-chaperones direct the function of Hsp90 to properly fold and activate the diverse array of client proteins.

Acknowledgments

J.L.J. is funded by a grant from the National Science Foundation MCB-0744522. M.B.C. is supported in part by NIH/NIGMS grant No. 1SC1GM084863 and Grant Number 5G12RR008124 (to the Border Biomedical Research Center/University of Texas at El Paso) from the National Center for Research Resources (NCRR/NIH). The contents of this publication are solely the responsibility of the authors and do not necessarily represent the official views of NCRR or NIH.

References

1. Mayer MP, Bukau B. Hsp70 chaperones: cellular functions and molecular mechanism. Cell Mol Life Sci 2005;62:670–84.
2. Pearl LH, Prodromou C. Structure and mechanism of the hsp90 molecular chaperone machinery. Annu Rev Biochem 2006;75:271–94.
3. Wandinger SK, Richter K, Buchner J. The Hsp90 chaperone machinery. J Biol Chem 2008;283:18473–7.
4. Pratt WB, Toft DO. Regulation of signaling protein function and trafficking by the hsp90/hsp70-based chaperone machinery. Exp Biol Med (Maywood) 2003;228:111–33.
5. Zhao R, Davey M, Hsu YC, et al. Navigating the chaperone network: an integrative map of physical and genetic interactions mediated by the hsp90 chaperone. Cell 2005;120:715–27.
6. McClellan AJ, Xia Y, Deutschbauer AM, Davis RW, Gerstein M, Frydman J. Diverse cellular functions of the hsp90 molecular chaperone uncovered using systems approaches. Cell 2007;131:121–35.
7. Pearl LH, Prodromou C, Workman P. The Hsp90 molecular chaperone: an open and shut case for treatment. Biochem J 2008;410:439–53.
8. Neckers L, Mollapour M, Tsutsumi S. The complex dance of the molecular chaperone Hsp90. Trends Biochem Sci 2009;34:223–6.
9. Southworth DR, Agard DA. Species-dependent ensembles of conserved conformational states define the Hsp90 chaperone ATPase cycle. Mol Cell 2008;32:631–40.
10. Vaughan CK, Gohlke U, Sobott F, et al. Structure of an Hsp90-Cdc37-Cdk4 complex. Mol Cell 2006;23:697–707.
11. Riggs D, Cox M, Cheung-Flynn J, Prapapanich V, Carrigan P, Smith D. Functional specificity of co-chaperone interactions with Hsp90 client proteins. Crit Rev Biochem Mol Biol 2004;39:279–95.

12. Harst A, Lin H, Obermann WM. Aha1 competes with Hop, p50 and p23 for binding to the molecular chaperone Hsp90 and contributes to kinase and hormone receptor activation. Biochem J 2005;387:789–96.
13. Siligardi G, Hu B, Panaretou B, Piper PW, Pearl LH, Prodromou C. Co-chaperone regulation of conformational switching in the Hsp90 ATPase cycle. J Biol Chem 2004;279: 51989–98.
14. McLaughlin SH, Sobott F, Yao ZP, et al. The co-chaperone p23 arrests the Hsp90 ATPase cycle to trap client proteins. J Mol Biol 2006;356:746–58.
15. Young JC, Hartl FU. Polypeptide release by Hsp90 involves ATP hydrolysis and is enhanced by the co-chaperone p23. Embo J 2000;19: 5930–40.
16. Richter K, Walter S, Buchner J. The Co-chaperone Sba1 connects the ATPase reaction of Hsp90 to the progression of the chaperone cycle. J Mol Biol 2004;342:1403–13.
17. Ali MM, Roe SM, Vaughan CK, et al. Crystal structure of an Hsp90-nucleotide-p23/Sba1 closed chaperone complex. Nature 2006;440: 1013–7.
18. Cox MB, Miller CA, 3rd. Cooperation of heat shock protein 90 and p23 in aryl hydrocarbon receptor signaling. Cell Stress Chaperones 2004;9:4–20.
19. Weikl T, Abelmann K, Buchner J. An unstructured C-terminal region of the Hsp90 co-chaperone p23 is important for its chaperone function. J Mol Biol 1999;293:685–91.
20. Freeman BC, Toft DO, Morimoto RI. Molecular chaperone machines: chaperone activities of the cyclophilin Cyp-40 and the steroid aporeceptor-associated protein p23. Science 1996;274:1718–20.
21. Taguwa S, Kambara H, Omori H, et al. Co-chaperone activity of human butyrate-induced transcript 1 facilitates hepatitis C virus replication through an Hsp90-dependent pathway. J Virol 2009.
22. Fang Y, Fliss AE, Rao J, Caplan AJ. SBA1 encodes a yeast hsp90 cochaperone that is homologous to vertebrate p23 proteins. Mol Cell Biol 1998;18:3727–34.
23. Grad I, McKee TA, Ludwig SM, et al. The Hsp90 cochaperone p23 is essential for perinatal survival. Mol Cell Biol 2006;26:8976–83.
24. Catlett MG, Kaplan KB. Sgt1p is a unique co-chaperone that acts as a client-adaptor to link Hsp90 to Skp1p. J Biol Chem 2006; 281:33739–48.
25. Zhang M, Boter M, Li K, et al. Structural and functional coupling of Hsp90- and Sgt1-centred multi-protein complexes. Embo J 2008;27:2789–98.
26. Kadota Y, Amigues B, Ducassou L, et al. Structural and functional analysis of SGT1-HSP90 core complex required for innate immunity in plants. EMBO Rep 2008;9:1209–15.
27. Picard D. A stress protein interface of innate immunity. EMBO Rep 2008;9:1193–5.
28. Caplan AJ, Mandal AK, Theodoraki MA. Molecular chaperones and protein kinase quality control. Trends Cell Biol 2007;17: 87–92.
29. Mandal AK, Lee P, Chen JA, et al. Cdc37 has distinct roles in protein kinase quality control that protect nascent chains from degradation and promote posttranslational maturation. J Cell Biol 2007;176:319–28.
30. Chadli A, Graham JD, Abel MG, et al. GCUNC-45 is a novel regulator for the progesterone receptor/hsp90 chaperoning pathway. Mol Cell Biol 2006;26:1722–30.
31. Hessling M, Richter K, Buchner J. Dissection of the ATP-induced conformational cycle of the molecular chaperone Hsp90. Nat Struct Mol Biol 2009;16:287–93.
32. Wang X, Venable J, LaPointe P, et al. Hsp90 cochaperone Aha1 downregulation rescues misfolding of CFTR in cystic fibrosis. Cell 2006;127:803–15.
33. Scheufler C, Brinker A, Bourenkov G, et al. Structure of TPR domain-peptide complexes: critical elements in the assembly of the Hsp70-Hsp90 multichaperone machine. Cell 2000; 101:199–210.
34. Chen S, Smith DF. Hop as an adaptor in the heat shock protein 70 (Hsp70) and hsp90 chaperone machinery. J Biol Chem 1998;273: 35194–200.
35. Wegele H, Haslbeck M, Reinstein J, Buchner J. Sti1 is a novel activator of the Ssa proteins. J Biol Chem 2003;278:25970–6.
36. Flom G, Behal RH, Rosen L, Cole DG, Johnson JL. Definition of the minimal fragments of Sti1 required for dimerization, interaction with Hsp70 and Hsp90 and in vivo functions. Biochem J 2007;404:159–67.
37. Carrigan PE, Riggs DL, Chinkers M, Smith DF. Functional comparison of human and Drosophila Hop reveals novel role in steroid receptor maturation. J Biol Chem 2005;280: 8906–11.
38. Chang HC, Nathan DF, Lindquist S. In vivo analysis of the Hsp90 cochaperone Sti1 (p60). Mol Cell Biol 1997;17:318–25.
39. Barik S. Immunophilins: for the love of proteins. Cell Mol Life Sci 2006;63:2889–900.

40. Bose S, Weikl T, Bugl H, Buchner J. Chaperone function of Hsp90-associated proteins. Science 1996;274:1715–7.
41. Prodromou C, Siligardi G, O'Brien R, et al. Regulation of Hsp90 ATPase activity by tetratricopeptide repeat (TPR)-domain co-chaperones. Embo J 1999;18:754–62.
42. Mayr C, Richter K, Lilie H, Buchner J. Cpr6 and Cpr7, two closely related Hsp90-associated immunophilins from Saccharomyces cerevisiae, differ in their functional properties. J Biol Chem 2000;275:34140–6.
43. Duina AA, Chang HC, Marsh JA, Lindquist S, Gaber RF. A cyclophilin function in Hsp90-dependent signal transduction. Science 1996;274:1713–5.
44. Duina AA, Marsh JA, Kurtz RB, Chang HC, Lindquist S, Gaber RF. The peptidyl-prolyl isomerase domain of the CyP-40 cyclophilin homolog Cpr7 is not required to support growth or glucocorticoid receptor activity in Saccharomyces cerevisiae. J Biol Chem 1998;273:10819–22.
45. Mok D, Allan RK, Carrello A, Wangoo K, Walkinshaw MD, Ratajczak T. The chaperone function of cyclophilin 40 maps to a cleft between the prolyl isomerase and tetratricopeptide repeat domains. FEBS Lett 2006;580:2761–8.
46. Riggs DL, Cox MB, Tardif HL, Hessling M, Buchner J, Smith DF. Noncatalytic role of the FKBP52 peptidyl-prolyl isomerase domain in the regulation of steroid hormone signaling. Mol Cell Biol 2007;27:8658–69.
47. Smith DF, Toft DO. The Intersection of Steroid Receptors with Molecular Chaperones: Observations and Questions. Mol Endocrinol 2008;22:2229–40.
48. Hidalgo-de-Quintana J, Evans RJ, Cheetham ME, van der Spuy J. The Leber congenital amaurosis protein AIPL1 functions as part of a chaperone heterocomplex. Invest Ophthalmol Vis Sci 2008;49:2878–87.
49. Hinds TD, Jr., Sanchez ER. Protein phosphatase 5. Int J Biochem Cell Biol 2007.
50. Wandinger SK, Suhre MH, Wegele H, Buchner J. The phosphatase Ppt1 is a dedicated regulator of the molecular chaperone Hsp90. Embo J 2006;25:367–76.
51. Vaughan CK, Mollapour M, Smith JR, et al. Hsp90-dependent activation of protein kinases is regulated by chaperone-targeted dephosphorylation of Cdc37. Mol Cell 2008;31:886–95.
52. Millson SH, Vaughan CK, Zhai C, et al. Chaperone ligand-discrimination by the TPR-domain protein Tah1. Biochem J 2008;413:261–8.
53. Boulon S, Marmier-Gourrier N, Pradet-Balade B, et al. The Hsp90 chaperone controls the biogenesis of L7Ae RNPs through conserved machinery. J Cell Biol 2008;180:579–95.
54. Tesic M, Marsh JA, Cullinan SB, Gaber RF. Functional interactions between Hsp90 and the Co-chaperones Cns1 and Cpr7 in saccharomyces cerevisiae. J Biol Chem 2003;278:32692–701.
55. Hainzl O, Wegele H, Richter K, Buchner J. Cns1 is an activator of the Ssa1 ATPase activity. J Biol Chem 2004;279:23267–73.
56. Crevel G, Bennett D, Cotterill S. The Human TPR Protein TTC4 Is a Putative Hsp90 Co-Chaperone Which Interacts with CDC6 and Shows Alterations in Transformed Cells. PLoS ONE 2008;3:e0001737.
57. Young JC, Hoogenraad NJ, Hartl FU. Molecular chaperones Hsp90 and Hsp70 deliver preproteins to the mitochondrial import receptor Tom70. Cell 2003;112:41–50.
58. Dickey CA, Kamal A, Lundgren K, et al. The high-affinity HSP90-CHIP complex recognizes and selectively degrades phosphorylated tau client proteins. J Clin Invest 2007;117:648–58.
59. Brychzy A, Rein T, Winklhofer KF, Hartl FU, Young JC, Obermann WM. Cofactor Tpr2 combines two TPR domains and a J domain to regulate the Hsp70/Hsp90 chaperone system. EMBO J 2003;22:3613–23.
60. Moffatt NS, Bruinsma E, Uhl C, Obermann WM, Toft D. Role of the cochaperone Tpr2 in Hsp90 chaperoning. Biochemistry 2008;47:8203–13.
61. Cziepluch C, Kordes E, Poirey R, Grewenig A, Rommelaere J, Jauniaux JC. Identification of a novel cellular TPR-containing protein, SGT, that interacts with the nonstructural protein NS1 of parvovirus H-1. J Virol 1998;72:4149–56.
62. Callahan MA, Handley MA, Lee YH, Talbot KJ, Harper JW, Panganiban AT. Functional interaction of human immunodeficiency virus type 1 Vpu and Gag with a novel member of the tetratricopeptide repeat protein family. J Virol 1998;72:5189–97.
63. Dutta S, Tan YJ. Structural and functional characterization of human SGT and its interaction with Vpu of the human immunodeficiency virus type 1. Biochemistry 2008;47:10123–31.
64. Angeletti PC, Walker D, Panganiban AT. Small glutamine-rich protein/viral protein U-binding protein is a novel cochaperone that affects heat shock protein 70 activity. Cell Stress Chaperones 2002;7:258–68.

65. Fielding BC, Gunalan V, Tan TH, et al. Severe acute respiratory syndrome coronavirus protein 7a interacts with hSGT. Biochem Biophys Res Commun 2006;343:1201–8.
66. Fonte V, Kapulkin V, Taft A, Fluet A, Friedman D, Link CD. Interaction of intracellular beta amyloid peptide with chaperone proteins. Proc Natl Acad Sci USA 2002;99:9439–44.
67. Liou ST, Wang C. Small glutamine-rich tetratricopeptide repeat-containing protein is composed of three structural units with distinct functions. Arch Biochem Biophys 2005;435:253–63.
68. Natochin M, Campbell TN, Barren B, et al. Characterization of the G alpha(s) regulator cysteine string protein. J Biol Chem 2005;280: 30236–41.
69. Schantl JA, Roza M, De Jong AP, Strous GJ. Small glutamine-rich tetratricopeptide repeat-containing protein (SGT) interacts with the ubiquitin-dependent endocytosis (UbE) motif of the growth hormone receptor. Biochem J 2003;373:855–63.
70. Wang H, Shen H, Wang Y, et al. Overexpression of small glutamine-rich TPR-containing protein promotes apoptosis in 7721 cells. FEBS Lett 2005;579:1279–84.
71. Wang H, Zhang Q, Zhu D. hSGT interacts with the N-terminal region of myostatin. Biochem Biophys Res Commun 2003;311: 877–83.
72. Winnefeld M, Rommelaere J, Cziepluch C. The human small glutamine-rich TPR-containing protein is required for progress through cell division. Exp Cell Res 2004;293: 43–57.
73. Yin H, Wang H, Zong H, et al. SGT, a Hsp90beta binding partner, is accumulated in the nucleus during cell apoptosis. Biochem Biophys Res Commun 2006;343:1153–8.
74. Buchanan G, Ricciardelli C, Harris JM, et al. Control of androgen receptor signaling in prostate cancer by the cochaperone small glutamine rich tetratricopeptide repeat containing protein alpha. Cancer Res 2007;67:10087–96.
75. Wu SJ, Liu FH, Hu SM, Wang C. Different combinations of the heat-shock cognate protein 70 (hsc70) C-terminal functional groups are utilized to interact with distinct tetratricopeptide repeat-containing proteins. Biochem J 2001;359:419–26.
76. Kordes E, Savelyeva L, Schwab M, Rommelaere J, Jauniaux JC, Cziepluch C. Isolation and characterization of human SGT and identification of homologues in Saccharomyces cerevisiae and Caenorhabditis elegans. Genomics 1998;52:90–4.
77. de Groot PW, Ruiz C, Vazquez de Aldana CR, et al. A genomic approach for the identification and classification of genes involved in cell wall formation and its regulation in Saccharomyces cerevisiae. Comp Funct Genomics 2001;2:124–42.
78. Niu W, Li Z, Zhan W, Iyer VR, Marcotte EM. Mechanisms of cell cycle control revealed by a systematic and quantitative overexpression screen in S. cerevisiae. PLoS Genet 2008;4: e1000120.
79. Liou ST, Cheng MY, Wang C. SGT2 and MDY2 interact with molecular chaperone YDJ1 in Saccharomyces cerevisiae. Cell Stress Chaperones 2007;12:59–70.
80. Kimura Y, Yahara I, Lindquist S. Role of the protein chaperone YDJ1 in establishing Hsp90-mediated signal transduction pathways. Science 1995;268:1362–5.
81. Cyr DM. Swapping nucleotides, tuning Hsp70. Cell 2008;133:945–7.
82. Liu XD, Morano KA, Thiele DJ. The yeast Hsp110 family member, Sse1, is an Hsp90 cochaperone. J Biol Chem 1999;274:26654–60.
83. Hohfeld J, Minami Y, Hartl FU. Hip, a novel cochaperone involved in the eukaryotic Hsc70/Hsp40 reaction cycle. Cell 1995;83: 589–98.
84. Prapapanich V, Chen S, Toran EJ, Rimerman RA, Smith DF. Mutational analysis of the hsp70-interacting protein Hip. Mol Cell Biol 1996;16:6200–7.
85. Kosano H, Stensgard B, Charlesworth MC, McMahon N, Toft D. The assembly of progesterone receptor-hsp90 complexes using purified proteins. J Biol Chem 1998;273:32973–9.
86. Arlander SJ, Felts SJ, Wagner JM, Stensgard B, Toft DO, Karnitz LM. Chaperoning checkpoint kinase 1 (Chk1), an Hsp90 client, with purified chaperones. J Biol Chem 2006;281: 2989–98.
87. Chang HC, Lindquist S. Conservation of Hsp90 macromolecular complexes in Saccharomyces cerevisiae. J Biol Chem 1994; 269:24983–8.
88. Johnson JL, Craig EA. A role for the Hsp40 Ydj1 in repression of basal steroid receptor activity in yeast. Mol Cell Biol 2000;20:3027–36.
89. Mandal AK, Nillegoda N, Chen JA, Caplan AJ. Ydj1 protects nascent protein kinases from degradation and controls the rate of their maturation. Mol Cell Biol 2008;28:4434–44.
90. Hon T, Lee HC, Hach A, et al. The Hsp70-Ydj1 molecular chaperone represses the activity of the heme activator protein Hap1 in the absence of heme. Mol Cell Biol 2001; 21:7923–32.

91. Sahi C, Craig EA. Network of general and specialty J protein chaperones of the yeast cytosol. Proc Natl Acad Sci USA 2007;104:7163–8.
92. Dittmar KD, Banach M, Galigniana MD, Pratt WB. The role of DnaJ-like proteins in glucocorticoid receptor.hsp90 heterocomplex assembly by the reconstituted hsp90.p60. hsp70 foldosome complex. J Biol Chem 1998;273:7358–66.
93. Hernandez MP, Chadli A, Toft DO. HSP40 binding is the first step in the HSP90 chaperoning pathway for the progesterone receptor. J Biol Chem 2002;277:11873–81.
94. Murphy PJ, Morishima Y, Chen H, et al. Visualization and mechanism of assembly of a glucocorticoid receptor.Hsp70 complex that is primed for subsequent Hsp90-dependent opening of the steroid binding cleft. J Biol Chem 2003;278:34764–73.
95. Flom GA, Lemieszek M, Fortunato EA, Johnson JL. Farnesylation of Ydj1 is required for in vivo interaction with Hsp90 client proteins. Mol Biol Cell 2008;19:5249–58.
96. Smith DF. Dynamics of heat shock protein 90- progesterone receptor binding and the disactivation loop model for steroid receptor complexes. Molecular Endocrinology 1993;7: 1418–29.
97. Wegele H, Wandinger SK, Schmid AB, Reinstein J, Buchner J. Substrate transfer from the chaperone Hsp70 to Hsp90. J Mol Biol 2006;356:802–11.
98. Lee P, Shabbir A, Cardozo C, Caplan AJ. Sti1 and Cdc37 can stabilize Hsp90 in chaperone complexes with a protein kinase. Mol Biol Cell 2004;15:1785–92.
99. Cheung J, Smith DF. Molecular chaperone interactions with steroid receptors: an update. Mol Endocrinol 2000;14:939–46.
100. Davies TH, Ning YM, Sanchez ER. Differential control of glucocorticoid receptor hormone-binding function by tetratricopeptide repeat (TPR) proteins and the immunosuppressive ligand FK506. Biochemistry 2005;44:2030–8.
101. Riggs DL, Roberts PJ, Chirillo SC, et al. The Hsp90-binding peptidylprolyl isomerase FKBP52 potentiates glucocorticoid signaling in vivo. Embo J 2003;22:1158–67.
102. Tranguch S, Cheung-Flynn J, Daikoku T, et al. Cochaperone immunophilin FKBP52 is critical to uterine receptivity for embryo implantation. Proc Natl Acad Sci USA 2005;102:14326–31.
103. Galigniana MD, Harrell JM, Murphy PJ, et al. Binding of hsp90-associated immunophilins to cytoplasmic dynein: direct binding and in vivo evidence that the peptidylprolyl isomerase domain is a dynein interaction domain. Biochemistry 2002;41: 13602–10.
104. Galigniana MD, Radanyi C, Renoir JM, Housley PR, Pratt WB. Evidence that the peptidylprolyl isomerase domain of the hsp90-binding immunophilin FKBP52 is involved in both dynein interaction and glucocorticoid receptor movement to the nucleus. J Biol Chem 2001;276:14884–9.
105. Echeverria PC, Mazaira G, Erlejman A, Gomez-Sanchez C, Piwien Pilipuk G, Galigniana MD. Nuclear import of the glucocorticoid receptor-hsp90 complex through the nuclear pore complex is mediated by its interaction with Nup62 and importin beta. Mol Cell Biol 2009;29:4788–97.
106. Pratt WB, Galigniana MD, Harrell JM, DeFranco DB. Role of hsp90 and the hsp90-binding immunophilins in signalling protein movement. Cell Signal 2004;16: 857–72.
107. Htun H, Holth LT, Walker D, Davie JR, Hager GL. Direct visualization of the human estrogen receptor alpha reveals a role for ligand in the nuclear distribution of the receptor. Mol Biol Cell 1999;10:471–86.
108. Lim CS, Baumann CT, Htun H, et al. Differential localization and activity of the A- and B-forms of the human progesterone receptor using green fluorescent protein chimeras. Mol Endocrinol 1999;13: 366–75.
109. Freeman BC, Yamamoto KR. Disassembly of transcriptional regulatory complexes by molecular chaperones. Science 2002;296: 2232–5.
110. Silverstein AM, Galigniana MD, Kanelakis KC, Radanyi C, Renoir JM, Pratt WB. Different regions of the immunophilin FKBP52 determine its association with the glucocorticoid receptor, hsp90, and cytoplasmic dynein. J Biol Chem 1999;274: 36980–6.
111. Hu J, Toft DO, Seeger C. Hepadnavirus assembly and reverse transcription require a multi-component chaperone complex which is incorporated into nucleocapsids. Embo J 1997;16:59–68.
112. Koga F, Kihara K, Neckers L. Inhibition of cancer invasion and metastasis by targeting the molecular chaperone heat-shock protein 90. Anticancer Res 2009;29:797–807.
113. Sawai A, Chandarlapaty S, Greulich H, et al. Inhibition of Hsp90 down-regulates mutant epidermal growth factor receptor (EGFR) expression and sensitizes EGFR mutant

tumors to paclitaxel. Cancer Res 2008;68: 589–96.

114. McDowell CL, Bryan Sutton R, Obermann WM. Expression of Hsp90 chaperome proteins in human tumor tissue. Int J Biol Macromol 2009;45:310–4.

115. Periyasamy S, Warrier M, Tillekeratne MP, Shou W, Sanchez ER. The immunophilin ligands cyclosporin A and FK506 suppress prostate cancer cell growth by androgen receptor-dependent and -independent mechanisms. Endocrinology 2007;148:4716–26.

116. Bredemeyer AJ, Carrigan PE, Fehniger TA, Smith DF, Ley TJ. Hop cleavage and function in granzyme B-induced apoptosis. J Biol Chem 2006;281:37130–41.

117. Radanyi C, Le Bras G, Bouclier C, et al. Tosylcyclonovobiocic acids promote cleavage of the hsp90-associated cochaperone p23. Biochem Biophys Res Commun 2009;379: 514–8.

118. Mollerup J, Berchtold MW. The co-chaperone p23 is degraded by caspases and the proteasome during apoptosis. FEBS Lett 2005; 579:4187–92.

119. Johnson JL, Brown C. Plasticity of the Hsp90 chaperone machine in divergent eukaryotic organisms. Cell Stress Chaperones 2009;14: 83–94.

120. Zhang L, Hach A, Wang C. Molecular mechanism governing heme signaling in yeast: a higher-order complex mediates heme regulation of the transcriptional activator HAP1. Mol Cell Biol 1998;18:3819–28.

121. Nair SC, Toran EJ, Rimerman RA, Hjermstad S, Smithgall TE, Smith DF. A pathway of multi-chaperone interactions common to diverse regulatory proteins: estrogen receptor, Fes tyrosine kinase, heat shock transcription factor Hsf1, and the aryl hydrocarbon receptor. Cell Stress Chaperones 1996;1: 237–50.

122. Blagosklonny MV, Toretsky J, Bohen S, Neckers L. Mutant conformation of p53 translated in vitro or in vivo requires functional HSP90. Proc Natl Acad Sci USA 1996;93:8379–83.

123. Minet E, Ernest I, Michel G, et al. HIF1A gene transcription is dependent on a core promoter sequence encompassing activating and inhibiting sequences located upstream from the transcription initiation site and cis elements located within the 5′UTR. Biochem Biophys Res Commun 1999;261:534–40.

124. Xu Y, Lindquist S. Heat-shock protein hsp90 governs the activity of pp60v-src kinase. Proc Natl Acad Sci 1993;90:7074–8.

125. Louvion JF, Abbas-Terki T, Picard D. Hsp90 is required for pheromone signaling in yeast. Mol Biol Cell 1998;9:3071–83.

126. Donze O, Abbas-Terki T, Picard D. The Hsp90 chaperone complex is both a facilitator and a repressor of the dsRNA-dependent kinase PKR. Embo J 2001;20:3771–80.

127. Aligue R, Akhavan-Niak H, Russell P. A role for Hsp90 in cell cycle control: Wee1 tyrosine kinase activity requires interaction with Hsp90. EMBO J 1994;13:6099–106.

128. Xu W, Mimnaugh E, Rosser MF, et al. Sensitivity of mature Erbb2 to geldanamycin is conferred by its kinase domain and is mediated by the chaperone protein Hsp90. The Journal of biological chemistry 2001; 276(5):3702–8.

129. Sato S, Fujita N, Tsuruo T. Modulation of Akt kinase activity by binding to Hsp90. Proc Natl Acad Sci USA 2000;97:10832–7.

130. Eustace BK, Sakurai T, Stewart JK, et al. Functional proteomic screens reveal an essential extracellular role for hsp90 alpha in cancer cell invasiveness. Nat Cell Biol 2004;6: 507–14.

131. Siligardi G, Panaretou B, Meyer P, et al. Regulation of Hsp90 ATPase activity by the co-chaperone Cdc37p/p50cdc37. J Biol Chem 2002;277:20151–9.

132. Panaretou B, Siligardi G, Meyer P, et al. Activation of the ATPase activity of hsp90 by the stress-regulated cochaperone aha1. Mol Cell 2002;10:1307–18.

133. Murata S, Minami Y, Minami M, Chiba T, Tanaka K. CHIP is a chaperone-dependent E3 ligase that ubiquitylates unfolded protein. EMBO Rep 2001;2:1133–8.

Chapter 5

Detecting *HSP90* Phosphorylation

Mehdi Mollapour and Len Neckers

Abstract

Heat-shock protein 90 (*HSP90*) is an essential molecular chaperone in eukaryotes. It is important for chaperoning proteins that are important determinants of multistep carcinogenesis. *HSP90*'s ATPase activity is associated with its chaperone function. Co-chaperones as well as posttranslational modifications (phosphorylation, acetylation, and *S*-nitrosylation) are important for regulating its ATPase activity. Yeast can be used to express and purify *HSP90* and also detect its phosphorylation by pan-phosphoserine or phosphothreonine antibodies.

Key words: *HSP90*, Molecular chaperones, Posttranslational modification, Phosphorylation

1. Introduction

Heat-shock protein 90 (*HSP90*) is an essential molecular chaperone in eukaryotes (1–3). Its cellular functions have been most clearly identified in mammalian cells, *Drosophila* and baker's yeast *Saccharomyces cerevisiae* (4–7). *HSP90* creates and maintains the functional conformation of a subset of proteins (8). These targets (or "clients") are key components of signal transduction pathways and numerous transcription factors. *HSP90* and a discrete set of co-chaperone proteins "hold" these clients in a state from which they can respond to activating signals.

HSP90 chaperone activity depends on ATP binding and hydrolysis (9–11) which is coupled to a conformational cycle involving the opening and closing of a dimeric "molecular clamp" via transient association of *HSP90*'s N-terminal domain (12, 13), which also binds the antitumor antibiotics geldanamycin and radicicol (14–17).

ATPase activity of *HSP90* is also regulated by co-chaperones. For example, Hop^{Sti1} (18–21), $p50^{Cdc37}$ (22–25), and $p23^{Sba1}$ (26–28)

Stuart K. Calderwood and Thomas L. Prince (eds.), *Molecular Chaperones: Methods and Protocols*,
Methods in Molecular Biology, vol. 787, DOI 10.1007/978-1-61779-295-3_5, © Springer Science+Business Media, LLC 2011

have an inhibitory effect on the ATPase cycle of *HSP90* while Aha1 (29–31) and Cpr6 (32, 33) have an activating effect.

HSP90 is a phosphoprotein (34–42). Cells treated with the serine/threonine phosphatase inhibitor, okadaic acid, demonstrate *HSP90* hyperphosphorylation and decreased association of the chaperone with $pp60^{v\text{-}Src}$, suggesting a link between *HSP90* phosphorylation and chaperoning of its target proteins (35, 43). Recent work has shown that c-Src directly phosphorylates Tyr-300 of *HSP90* under basal conditions, reducing its ability to chaperone client proteins (41). Therefore, *HSP90* phosphorylation provides an additional means of fine-tuning its chaperone activity. Lack of phospho-specific antibodies has made it difficult to study *HSP90* phosphorylation in mammalian cells. Also *HSP90* gene knockouts are lethal in mammalian systems, so any mutant *HSP90* must be investigated in a background of highly expressed native mammalian *HSP90* proteins.

Simple baker's yeast, *S. cerevisiae*, is a well-established and valuable tool for studying various aspects of conserved protein chaperone machinery. The yeast system has provided us with a powerful tool to study *HSP90* phosphorylation, since it readily allows plasmid exchange whereby any introduced *HSP90* gene – provided it is partially functional – can provide 100% of the *HSP90* of the cell (Fig. 1). Such genetic modifications are simply not

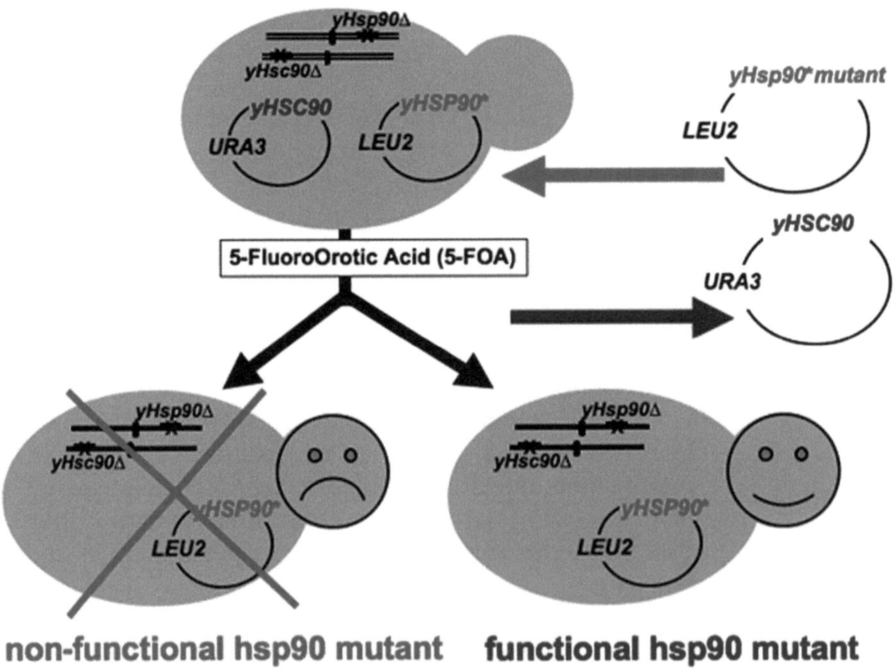

Fig. 1. With plasmid shuffling, a mutant *HSP90* gene can be made to provide all the *HSP90* of the yeast cell (y*HSP90* = Hsp82 and yHsc90 = Hsc82). This involves introducing the mutation into y*HSP90* on Leu2 plasmid and then introducing it into haploid yeast cells (y*HSP90*Δ, y*HSP90*Δ). Growth of these cells on 5-fluoroorotic acid (5-FOA) "cures" the yeast cells of the wild-type yHsc90, therefore creating *HSP90* mutant.

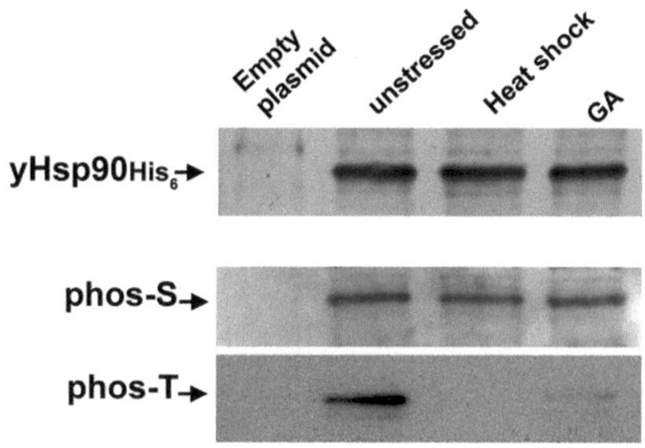

Fig. 2. Yeast *HSP90* phosphorylation on serine (phos-S) and threonine (phos-T) residues. y*HSP90*-His$_6$ was purified from yeast cells that were heat shocked at 39°C for 40 min or treated with 100 μM GA for 1 h. Wild-type cells containing the empty plasmid were used as negative control.

achievable in cultured mammalian cells. This plasmid exchange (Fig. 1) was used to isolate temperature-sensitive (ts) *HSP90* mutants.

This chapter describes the isolation and identification of yeast *HSP90* phosphorylation using immunoblotting procedures. Using the yeast system, it is possible to show that *HSP90* is constitutively phosphorylated on serine and threonine residues. However, *HSP90* threonine phosphorylation is lost upon either heat-shock stress or treatment with the *HSP90* inhibitor GA (Fig. 2).

2. Materials

1. YPD (2% (w/v) Bacto peptone, 1% (w/v) yeast extract, 2% (w/v) glucose, 20 mg/l adenine).
2. Yeast protein extraction buffer (EB): 50 mM Tris–HCI, pH 6.8, 100 mM NaCl, 50 mM $MgCl_2$. One tablet of complete EDTA-free protease inhibitor cocktail (Roche) and one tablet of PhosphoSTOP (Roche) are added to 50 ml EB.
3. Bio-Rad Protein Assay solution (Bio-Rad).
4. 425–600 μm glass beads (acid washed), (Sigma).
5. SDS-PAGE sample buffer (2×): 125 mM Tris–HCI, pH 6.8, 20% glycerol, 2% SDS, 10% 2-mercaptoethanol, 0.01% bromophenol blue, stable at –20°C. Aliquot and avoid freeze–thaw cycles.
6. Ponceau S solution (Sigma).
7. Tris-buffered saline (TBS): 150 mM NaCl, 25 mM Tris base. Adjust pH to 7.4 using HCI. Sterile filter and incubate at 4°C.

8. Albumin, bovine serum (minimum purity 98%).
9. Dried skimmed milk.
10. Protran BA85, 0.45 μm Nitrocellulose membrane (Whatman).
11. ECL plus Western Blotting Detection System (GE Healthcare).
12. Ni-NTA agarose (Qiagen).
13. Imidazole (Sigma).
14. Phospho-serine Q5 antibody (Qiagen).
15. Phospho-threonine Q7 antibody (Qiagen).
16. Tetra-His antibody (Qiagen).
17. Anti-secondary mouse antibody; ECL™ antimouse IgG, Horseradish Persoxidase linked whole antibody (GE Healthcare).
18. X-ray film, X-ray cassette, and X-ray film-developing machine.

3. Methods

The extraction of total yeast protein:

1. Grow PP30 cells (9) expressing His_6 linked at the N-domain of Hsp82 (y*HSP90*) on 150 ml YPD overnight at 28°C.
2. Harvest and wash cells two to three times in ice-cold deionized water (dH_2O).
3. Transfer the cell pellet into a screw cap 2-ml tube.
4. Pellet the cells and remove the supernatant (see Note 1).
5. Add equal volume of cell pellets, ice-cold glass beads.
6. Add half the volume of pellet/glass beads, EB.
7. Bead beat the cells using the mini-beadbeater (BioSpec Products, Inc.) for 30 s.
8. Incubate the cells on ice for 30 s.
9. Repeat (steps 7 and 8) 10–12 times.
10. Centrifuge the tubes at (10,000 × *g*) for 10 min at 4°C (see Note 2).
11. Transfer the supernatants into fresh 1.5-ml microcentrifuge tubes.
12. Centrifuge the tubes at (10,000 × *g*) for 10 min at 4°C (see Note 2).
13. Transfer the supernatants (soluble protein) into fresh 1.5-ml microcentrifuge tubes.
14. Protein concentrations were determined using Bio-Rad Protein Assay solution (Bio-Rad).

15. Transfer 40 μl of Ni-NTA agarose slurry into a 1.5-ml microcentrifuge tube. (see Note 3).
16. Add 1.0 ml of EB to the Ni-NTA agarose and spin at 10,000 × *g* for 1 min at 4°C.
17. Remove supernatant and add 1.0 ml of EB to the Ni-NTA agarose (see Note 4).
18. Repeat step (steps 16 and 17) four times.
19. Resuspend the Ni-NTA agarose in 30 μM imidazole in EB and incubate at 4°C for 30 min (see Note 5).
20. Repeat steps 16 and 17 twice and remove supernatant.
21. Add 1 mg of total protein to the Ni-NTA agarose in a total volume of 600 μl.
22. Incubate the total proteins/Ni-NTA agarose at 4°C for 2 h (see Note 6).
23. Centrifuge the tubes at (1,000 × *g*) for 1 min at 4°C.
24. Gently remove the supernatant (see Note 7).
25. Add 1 ml of EB to the Ni-NTA agarose.
26. Repeat (steps 23–25) five times.
27. Wash the Ni-NTA agarose with 30-μl EB.
28. Wash the Ni-NTA agarose with EB once.
29. Centrifuge the microcentrifuge tube at 15,000 × *g* for 1 min at 4°C.
30. Remove as much supernatant as possible.
31. Add 40 μl of the protein sample buffer.
32. Boil the samples for 3–5 min.
33. Centrifuge the samples at 1,000 × *g* and load the supernatant on to a 7.5% SDS-PAGE Tris–HCI gel (see Note 8).
34. Transfer the proteins from SDS-PAGE gel on to ProtranBA85, 0.45 μm nitrocellulose membrane (Whatman).
35. Examine the quality and efficiency of the transfer by staining the membrane with Ponceau S solution (Sigma) for 2 min (see Note 9).
36. Wash the membrane with dH_2O.
37. Incubate the membrane in 5% milk–TBS-T for 15–20 min at room temperature.
38. Wash the membrane with 1× TBS-T for 5 min at room temperature.
39. Repeat (step 38) three times.
40. Incubate the membrane with 1:500–1,000 dilution of either phospho-serine (Q5) or phospho-threonine (Q7) antibodies, (Qiagen), in 2% BSA–TBS-T overnight at 4°C (see Note 10).

41. Wash the membrane three times with 1× TBS-T for 5 min at room temperature.
42. Incubate the membrane with 1:2,000 dilution of anti-secondary mouse antibody in 5% milk–TBS-T for 1 h (see Note 11) at room temperature.
43. Wash the membrane three times with 1× TBS-T for 5 min at room temperature.
44. Remove 1× TBS-T and then apply ECL plus (GE Healthcare) to nitrocellulose membrane for 2–3 min.
45. Drain nitrocellulose membrane of excess developing solution (do not let dry).
46. Wrap the blot in saran wrap.
47. Place the blot in the X-ray film cassette (see Note 12).
48. Expose the blots to X-ray films by placing X-ray film directly against the western blot for different lengths of time.

4. Notes

1. The cell pellet must be kept on ice.
2. At this stage, Bio-Rad Protein Assay solution (Bio-Rad) should be prepared.
3. Ni-NTA agarose is precharged with Ni^{2+} ions and appears blue in color. It is provided as a 50% slurry in 30% ethanol.
4. Do not disturb the Ni-NTA agarose pellet.
5. Imidazole at low concentrations is commonly used in the binding and wash buffer to minimize binding of unwanted host cell proteins.
6. Use Eppendorf Thermomixer R to gently mix total proteins/Ni-NTA agarose solution.
7. Avoid disturbing the Ni-NTA agarose.
8. Criterion precast gels from Bio-Rad are suitable for this purpose.
9. Prepare 5% dry milk (LabScientific Inc.) in 1× TBS-T (0.1% Tween-20, Sigma) buffer before examining the membrane.
10. Phospho-antibodies stock concentration is 0.1 μg/μl.
11. 1:2,000 dilution of anti-secondary mouse antibody; ECL™ antimouse IgG horseradish peroxidase-inked whole antibody (GE Healthcare).
12. This procedure must be performed in the dark.

References

1. Pearl, L. H., and Prodromou, C. (2006) Structure and mechanism of the Hsp90 molecular chaperone machinery, *Annu Rev Biochem* 75, 271–294.
2. Rutherford, S., Knapp, J. R., and Csermely, P. (2007) Hsp90 and developmental networks, *Adv Exp Med Biol* 594, 190–197.
3. Wandinger, S. K., Richter, K., and Buchner, J. (2008) The Hsp90 chaperone machinery, *J Biol Chem* 283, 18473–18477.
4. Neckers, L. (2007) Heat shock protein 90: the cancer chaperone, *J Biosci* 32, 517–530.
5. Rutherford, S., Hirate, Y., and Swalla, B. J. (2007) The Hsp90 capacitor, developmental remodeling, and evolution: the robustness of gene networks and the curious evolvability of metamorphosis, *Crit Rev Biochem Mol Biol* 42, 355–372.
6. Picard, D. (2006) Chaperoning steroid hormone action, *Trends Endocrinol Metab* 17, 229–235.
7. Truman, A. W., Millson, S. H., Nuttall, J. M., King, V., Mollapour, M., Prodromou, C., Pearl, L. H., and Piper, P. W. (2006) Expressed in the yeast Saccharomyces cerevisiae, human ERK5 is a client of the Hsp90 chaperone that complements loss of the Slt2p (Mpk1p) cell integrity stress-activated protein kinase, *Eukaryot Cell* 5, 1914–1924.
8. Prodromou, C., and Pearl, L. H. (2003) Structure and functional relationships of Hsp90, *Curr Cancer Drug Targets* 3, 301–323.
9. Panaretou, B., Prodromou, C., Roe, S. M., O'Brien, R., Ladbury, J. E., Piper, P. W., and Pearl, L. H. (1998) ATP binding and hydrolysis are essential to the function of the Hsp90 molecular chaperone in vivo, *EMBO J* 17, 4829–4836.
10. Obermann, W. M., Sondermann, H., Russo, A. A., Pavletich, N. P., and Hartl, F. U. (1998) In vivo function of Hsp90 is dependent on ATP binding and ATP hydrolysis, *J Cell Biol* 143, 901–910.
11. Grenert, J. P., Johnson, B. D., and Toft, D. O. (1999) The importance of ATP binding and hydrolysis by hsp90 in formation and function of protein heterocomplexes, *J Biol Chem* 274, 17525–17533.
12. Prodromou, C., Roe, S. M., O'Brien, R., Ladbury, J. E., Piper, P. W., and Pearl, L. H. (1997) Identification and structural characterization of the ATP/ADP-binding site in the Hsp90 molecular chaperone, *Cell* 90, 65–75.
13. Grenert, J. P., Sullivan, W. P., Fadden, P., Haystead, T. A., Clark, J., Mimnaugh, E., Krutzsch, H., Ochel, H. J., Schulte, T. W., Sausville, E., Neckers, L. M., and Toft, D. O. (1997) The amino-terminal domain of heat shock protein 90 (hsp90) that binds geldanamycin is an ATP/ADP switch domain that regulates hsp90 conformation, *J Biol Chem* 272, 23843–23850.
14. Stebbins, C. E., Russo, A. A., Schneider, C., Rosen, N., Hartl, F. U., and Pavletich, N. P. (1997) Crystal structure of an Hsp90-geldanamycin complex: targeting of a protein chaperone by an antitumor agent, *Cell* 89, 239–250.
15. Roe, S. M., Prodromou, C., O'Brien, R., Ladbury, J. E., Piper, P. W., and Pearl, L. H. (1999) Structural basis for inhibition of the Hsp90 molecular chaperone by the antitumor antibiotics radicicol and geldanamycin, *J Med Chem* 42, 260–266.
16. Workman, P., Burrows, F., Neckers, L., and Rosen, N. (2007) Drugging the cancer chaperone HSP90: combinatorial therapeutic exploitation of oncogene addiction and tumor stress, *Ann N Y Acad Sci* 1113, 202–216.
17. Neckers, L. (2006) Chaperoning oncogenes: Hsp90 as a target of geldanamycin, *Handb Exp Pharmacol*, 259–277.
18. Abbas-Terki, T., Briand, P. A., Donze, O., and Picard, D. (2002) The Hsp90 co-chaperones Cdc37 and Sti1 interact physically and genetically, *Biol Chem* 383, 1335–1342.
19. Chang, H. C., Nathan, D. F., and Lindquist, S. (1997) In vivo analysis of the Hsp90 cochaperone Sti1 (p60), *Mol Cell Biol* 17, 318–325.
20. Richter, K., Muschler, P., Hainzl, O., Reinstein, J., and Buchner, J. (2003) Sti1 is a non-competitive inhibitor of the Hsp90 ATPase. Binding prevents the N-terminal dimerization reaction during the atpase cycle, *J Biol Chem* 278, 10328–10333.
21. Song, Y., and Masison, D. C. (2005) Independent regulation of Hsp70 and Hsp90 chaperones by Hsp70/Hsp90-organizing protein Sti1 (Hop1), *J Biol Chem* 280, 34178–34185.
22. Lee, P., Shabbir, A., Cardozo, C., and Caplan, A. J. (2004) Sti1 and Cdc37 can stabilize Hsp90 in chaperone complexes with a protein kinase, *Mol Biol Cell* 15, 1785–1792.
23. MacLean, M., and Picard, D. (2003) Cdc37 goes beyond Hsp90 and kinases, *Cell Stress Chaperones* 8, 114–119.

24. Siligardi, G., Panaretou, B., Meyer, P., Singh, S., Woolfson, D. N., Piper, P. W., Pearl, L. H., and Prodromou, C. (2002) Regulation of Hsp90 ATPase activity by the co-chaperone Cdc37p/p50cdc37, *J Biol Chem* 277, 20151–20159.
25. Vaughan, C. K., Mollapour, M., Smith, J. R., Truman, A., Hu, B., Good, V. M., Panaretou, B., Neckers, L., Clarke, P. A., Workman, P., Piper, P. W., Prodromou, C., and Pearl, L. H. (2008) Hsp90-dependent activation of protein kinases is regulated by chaperone-targeted dephosphorylation of Cdc37, *Mol Cell* 31, 886–895.
26. McLaughlin, S. H., Sobott, F., Yao, Z. P., Zhang, W., Nielsen, P. R., Grossmann, J. G., Laue, E. D., Robinson, C. V., and Jackson, S. E. (2006) The co-chaperone p23 arrests the Hsp90 ATPase cycle to trap client proteins, *J Mol Biol* 356, 746–758.
27. Picard, D. (2006) Intracellular dynamics of the Hsp90 co-chaperone p23 is dictated by Hsp90, *Exp Cell Res* 312, 198–204.
28. Sullivan, W. P., Owen, B. A., and Toft, D. O. (2002) The influence of ATP and p23 on the conformation of hsp90, *J Biol Chem* 277, 45942–45948.
29. Lotz, G. P., Lin, H., Harst, A., and Obermann, W. M. (2003) Aha1 binds to the middle domain of Hsp90, contributes to client protein activation, and stimulates the ATPase activity of the molecular chaperone, *J Biol Chem* 278, 17228–17235.
30. Meyer, P., Prodromou, C., Liao, C., Hu, B., Mark Roe, S., Vaughan, C. K., Vlasic, I., Panaretou, B., Piper, P. W., and Pearl, L. H. (2004) Structural basis for recruitment of the ATPase activator Aha1 to the Hsp90 chaperone machinery, *EMBO J* 23, 511–519.
31. Panaretou, B., Siligardi, G., Meyer, P., Maloney, A., Sullivan, J. K., Singh, S., Millson, S. H., Clarke, P. A., Naaby-Hansen, S., Stein, R., Cramer, R., Mollapour, M., Workman, P., Piper, P. W., Pearl, L. H., and Prodromou, C. (2002) Activation of the ATPase activity of hsp90 by the stress-regulated cochaperone aha1, *Mol Cell* 10, 1307–1318.
32. Johnson, J. L., Halas, A., and Flom, G. (2007) Nucleotide-dependent interaction of Saccharomyces cerevisiae Hsp90 with the cochaperone proteins Sti1, Cpr6, and Sba1, *Mol Cell Biol* 27, 768–776.
33. Mayr, C., Richter, K., Lilie, H., and Buchner, J. (2000) Cpr6 and Cpr7, two closely related Hsp90-associated immunophilins from Saccharomyces cerevisiae, differ in their functional properties, *J Biol Chem* 275, 34140–34146.
34. Scroggins, B. T., and Neckers, L. (2007) Post-translational modification of heat shock protein 90: impact on chaperone function, *Expert Opin. Drug Discov* 2, 1403–1414.
35. Mimnaugh, E. G., Worland, P. J., Whitesell, L., and Neckers, L. M. (1995) Possible role for serine/threonine phosphorylation in the regulation of the heteroprotein complex between the hsp90 stress protein and the pp60v-src tyrosine kinase, *J Biol Chem* 270, 28654–28659.
36. Garnier, C., Lafitte, D., Jorgensen, T. J., Jensen, O. N., Briand, C., and Peyrot, V. (2001) Phosphorylation and oligomerization states of native pig brain HSP90 studied by mass spectrometry, *Eur J Biochem* 268, 2402–2407.
37. Zhao, Y. G., Gilmore, R., Leone, G., Coffey, M. C., Weber, B., and Lee, P. W. (2001) Hsp90 phosphorylation is linked to its chaperoning function. Assembly of the reovirus cell attachment protein, *J Biol Chem* 276, 32822–32827.
38. Fontana, J., Fulton, D., Chen, Y., Fairchild, T. A., McCabe, T. J., Fujita, N., Tsuruo, T., and Sessa, W. C. (2002) Domain mapping studies reveal that the M domain of hsp90 serves as a molecular scaffold to regulate Akt-dependent phosphorylation of endothelial nitric oxide synthase and NO release, *Circ Res* 90, 866–873.
39. Adinolfi, E., Kim, M., Young, M. T., Di Virgilio, F., and Surprenant, A. (2003) Tyrosine phosphorylation of HSP90 within the P2X7 receptor complex negatively regulates P2X7 receptors, *J Biol Chem* 278, 37344–37351.
40. Wang, C., and Chen, J. (2003) Phosphorylation and hsp90 binding mediate heat shock stabilization of p53, *J Biol Chem* 278, 2066–2071.
41. Duval, M., Le Boeuf, F., Huot, J., and Gratton, J. P. (2007) Src-mediated phosphorylation of Hsp90 in response to vascular endothelial growth factor (VEGF) is required for VEGF receptor-2 signaling to endothelial NO synthase, *Mol Biol Cell* 18, 4659–4668.
42. Miyata, Y., and Yahara, I. (1992) The 90-kDa heat shock protein, HSP90, binds and protects casein kinase II from self-aggregation and enhances its kinase activity, *J Biol Chem* 267, 7042–7047.
43. Ogiso, H., Kagi, N., Matsumoto, E., Nishimoto, M., Arai, R., Shirouzu, M., Mimura, J., Fujii-Kuriyama, Y., and Yokoyama, S. (2004) Phosphorylation analysis of 90 kDa heat shock protein within the cytosolic arylhydrocarbonreceptor complex, *Biochemistry* 43, 15510–15519.

Chapter 6

Role of Molecular Chaperones in Biogenesis of the Protein Kinome

Atin K. Mandal, Maria A. Theodoraki, Nadinath B. Nillegoda, and Avrom J. Caplan

Abstract

Molecular chaperones promote polypeptide folding in cells by protecting newly made and otherwise misfolded proteins against aggregation or degradation by the ubiquitin proteasome pathway. The roles of *Saccharomyces cerevisiae* Cdc37 and Ydj1 molecular chaperones are described in this chapter. We focus on biogenesis of protein kinases that require several different molecular chaperones for their proper folding. Specific among these is Cdc37, which binds directly to its kinase clients either during or shortly after translation and protects them against rapid proteasomal degradation. Ydj1 has a similar role, but is less specific for protein kinases in its role as a molecular chaperone. The method that we describe uses pulse chase and immunoprecipitation to analyze the fate of newly made proteins. Two kinetically distinct pathways of degradation can be discerned using this methodology that is dependent on the presence of an Hsp90 inhibitor or occurs in mutants of the molecular chaperones under study. The first is "zero-point" degradation that occurs either during or immediately after translation. The second is a slower pathway, where the half-life of kinase is approximately 20 min after translation.

Key words: Molecular chaperone, Protein folding, Protein kinase, Ubiquitin proteasome system, Quality control

1. Introduction

Cdc37 was originally identified as being important for the yeast cell cycle. Subsequent studies led to its characterization as a molecular chaperone that functioned with Hsp90 in the biogenesis of protein kinases. This activity was thought to be specific, but its role in chaperoning other proteins including transcription factors was later recognized. Cdc37 is essential and highly conserved. Furthermore, its mammalian homolog acts as an oncogene (see ref. 1 for review).

Stuart K. Calderwood and Thomas L. Prince (eds.), *Molecular Chaperones: Methods and Protocols*,
Methods in Molecular Biology, vol. 787, DOI 10.1007/978-1-61779-295-3_6,

Initial studies focused on the role that Cdc37 played as a recruiting factor for Hsp90. This was based on the finding that binding of protein kinases to Hsp90 required Cdc37. Dissection of the functional domains of Cdc37 revealed that it contained distinct protein kinase-binding and Hsp90 binding domains at the N-terminus and in the middle of the protein, respectively (2, 3). A function for Cdc37 in recruitment of client protein kinases to Hsp90 placed the chaperone in a similar role of Hop/Sti1, which was known for its ability to help recruit nuclear receptors bound to Hsp70 to the Hsp90 chaperone complex (4). Such a view was contrasted by other studies showing that Cdc37 could function independently of Hsp90 and that the kinase-binding domain alone constituted the essential part of the chaperone (5). To determine what this function might be, our group analyzed very early events in the biogenesis of protein kinases. Initial studies focused on a mutant form of *CDC37* that failed to properly interact with its client kinases (6), although the Hsp90-binding domain of the chaperone was intact. A global analysis of protein kinase levels by Western blot revealed that Cdc37 was required to maintain normal levels of over 75% of the protein kinome based on a sampling of over 50% of yeast protein kinases (7). Subsequent studies investigated the mechanism underlying this effect and our findings revealed that Cdc37 played an important role in protecting newly synthesized protein kinases against rapid degradation by the ubiquitin/proteasome system (UPS). This role was similar to that played by Hsp90, whose inhibition with geldanamycin also leads to rapid degradation of clients via the UPS (8). Importantly, we observed two kinetically distinct pathways for the degradation of newly made protein kinases. In the *cdc37* mutant, we observed "zero-point" degradation, where the disposal was so rapid after translation that we were unable to trap the undegraded polypeptide in the absence of proteasome inhibitor. Such an effect was observed previously by Varshavsky and colleagues while studying the N-end rule degradation via Ubr1 (9–11). There was one exception to this rule; we observed that Cdc28, the main cyclin-dependent kinase regulating the cell cycle, was always degraded via the slow pathway even in the *cdc37* mutant strain (7).

In subsequent studies, we also discovered that Ydj1, an Hsp40 protein, played a similar protective role as Cdc37 (12). In *ydj1Δ* cells, we also observed rapid zero-point degradation of protein kinases, except for Cdc28. Together, our studies with Cdc37 and Ydj1 suggest that the quality control machinery uses molecular chaperones to protect newly made proteins from rapid degradation. In their absence, there is a rapid degradation that appears to be closely coupled to translation.

In contrast to the studies described above, protein kinases are degraded via a slower pathway in cells treated with the Hsp90 inhibitor, geldanamycin. In this case, zero-point degradation was not observed, and the half-life of protein kinases was approximately 20 min. The molecular basis for the two kinetically different pathways is not understood, although recent studies have shown that Ubrl and Ubr2 ubiquitin ligases function in both zero-point and slow pathway degradation. Furthermore, the nuclear-localized ubiquitin ligase, San1, also plays a role in the degradation of protein kinases in geldanamycin-treated cells (13).

The following describes our methods for analyzing protein kinase stability using pulse chase methodology and immunoprecipitation in the yeast, *Saccharomyces cerevisiae*.

2. Materials

2.1. Antisera

1. Rabbit polyclonal TAP (tandem affinity purification) antiserum for TAP-tagged kinase immunoprecipitation.
2. Rabbit polyclonal anti-PSTAIRE (Santa Cruz SC-53).

2.2. Reagents

1. 50% protein-A/G agarose resin (Pierce/Thermoscientific 20422).
2. Cycloheximide (Sigma Chemical Company) in water at –20°C at 10 mg/ml.
3. MG132 (CalBiochem) stored at –20°C (50 mM in DMSO).
4. Geldanamycin (Invivogen) stored at –20°C at 10 mM in DMSO.
5. Sodium salicylate (Mallinckrodt Chemicals) stored at room temperature in a dark bottle as a 1 M solution.
6. Phenylmethanesulfonyl fluoride (PMSF) was stored at –20°C in isopropanol at 100 mM.

2.3. Buffers and Solutions

1. Sterile water.
2. Synthetic dextrose (SD, 0.67% yeast nitrogen base, 2% glucose) dropout medium.
3. Amino acid mixtures from Bio101.
4. Protein-labeling solution: 100 μCi/ml ^{35}S-methionine.
5. Quenching solution: 200 μg/ml cycloheximide and cold methionine (1 mM).
6. Cell-fixation solution: Ice-cold 20% (v/v) trichloroacetic acid.
7. Cell lysis buffer: 50 mM Tris–HCl, pH 7.5, 1% sodium dodecyl sulfate (SDS), 1 mM EDTA, 1 mM PMSF, and 1× complete protease inhibitor cocktail – Roche.

8. IP dilution buffer (60 mM Tris–HCl, pH 7.5, 190 mM NaCl, 1.25% Triton X-100, and 6 mM EDTA).
9. Sample buffer: 2× SDS (62.5 mM Tris–HCl, pH 6.8, 10% (v/v) glycerol, 2% SDS, 0.01 mg/ml bromophenol blue, 5% β-mercaptoethanol).
10. Gel fixation solution: 10% acetic acid and 30% methanol.
11. 1 M sodium salicylate solution.

2.4. Hardware

1. 50-ml Falcon centrifuge tubes.
2. Glass beads (Biospec products).
3. 0.5-ml microcentrifuge tubes.
4. Bead beater (Biospec products).
5. 25-gauge syringe needles.
6. 1.5-ml microcentrifuge tubes.
7. X-ray film.

2.5. Yeast Strains

Strains used in experiments with geldanamycin were deleted for *ERG6* to facilitate drug incorporation into cells. The *erg6*Δ::Kan-MX6 module was amplified from genomic DNA from BY4741 MAT a strain (Open Biosystems) using primers beginning 200 bp upstream and 200 bp downstream.

Forward primer: 5′-TGC GAA GAT TGG TGA GAA ACC-3′ and reverse primer: 5′-TGG TCG TTT GCC ACG ACA TGA TGA AA-3′.

3. Methods

Yeast strains carrying TAP-tagged protein kinases were converted to *erg6*Δ::KanMX6 by one-step yeast transformation. Yeasts were grown in synthetic dextrose (0.67% yeast nitrogen base, 2% glucose) dropout media.

Pulse chase:

1. Yeast cells were grown in SD media to mid-log phase (OD_{600} = 0.4–0.6), washed twice in sterile water, and resuspended in SD media lacking methionine at a concentration of OD_{600} = 6/ml of culture.
2. The cells were starved for 45 min with shaking at 30°C in a water bath in a 50-ml Falcon centrifuge tube (see Note 1). Pulse labeling of newly synthesized proteins was performed at the end of the starvation period with ^{35}S-methionine. ^{35}S-methionine-treated cells were incubated for 10 min at 30°C in a shaker before stopping the labeling reaction with quenching solution. After the quenching and at selected time points

thereafter, 400 μl of the culture was taken and mixed with an equal volume of ice-cold cell fixation solution. Trichloroacetic acid-fixed cells were pelleted at 10,000 rpm (9,000 × *g*) and washed twice in ice-cold acetone before vacuum drying.

3. For cell lysate preparation, the dried cell pellets were resuspended in 200 μl ice-cold cell lysis buffer and an equal volume of 0.4-mm glass beads was added (Biospec products) in 0.5-ml microcentrifuge tubes. These tubes were sealed with parafilm and placed inside an empty 1.5-ml screw cap tube (minus the lid). The 0.5-ml tube was secured inside the larger tube with parafilm. Cell lysis was carried out using a bead beater (Biospec products) with 2× 30 s pulses with a 30-s interval at 4°C. The 0.5-ml tubes were punctured with a 25-gauge needle at both top and bottom, and placed back inside the 1.5-ml microcentrifuge tube. The lysates were then spun at 14,000 rpm (18,000 × *g*) for 15 min at 4°C to extract them from the 0.5-ml tube and collect them in the 1.5-ml tube. The supernatants were removed and a 5-μl aliquot quantified for ^{35}S-methionine incorporation using a scintillation counter. Cell extracts carrying equal amounts of counts (see Note 2) were used for subsequent immunoprecipitations.

Immunoprecipitation:

Cell extracts containing equal amounts of counts were diluted tenfold in IP dilution buffer to adjust the final concentration of SDS to 0.1% or less. To this, 0.5–1 μg of antisera (depending on the antisera used) was added, and the lysates were incubated overnight on a rotator or rocker at 4°C. The next day, 30 μl of 50% protein A/G agarose beads was added to each immunoprecipitation reaction and the samples were incubated at 4°C for 1 h with rocking. The resin was washed 5× in 1 ml of ice-cold IP dilution buffer. 20 μl of 2× SDS sample buffer was added to the washed resin and the samples were boiled for 3 min. Proteins in the supernatants were resolved by denaturing gel electrophoresis. The gels were fixed in gel fixation buffer for 30 min with shaking, washed twice in water for 15 min, and incubated in 1 M sodium salicylate solution for 30 min prior to drying and exposing to X-ray film (see Note 3).

4. Notes

1. Geldanamycin (50 μM) was added to cells at the beginning of the starvation period. MG132 (100 μM) was added at 30 min prior to pulse labeling.
2. Equal amounts of cpm in all samples are essential for the immunoprecipitation step. The final amount of cpm used in any

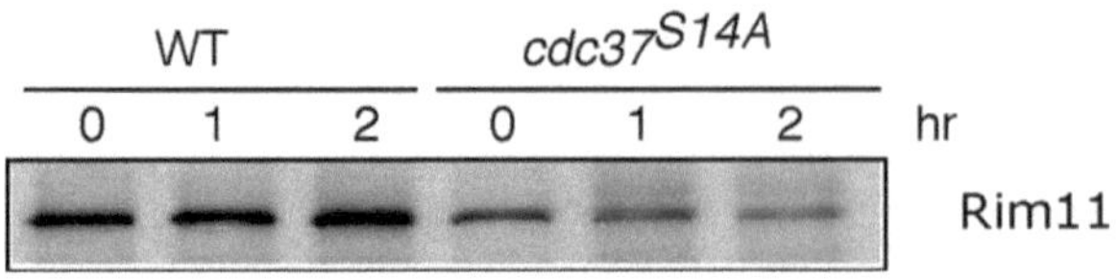

Fig. 1. Pulse chase and immunoprecipitation of C-terminally TAP-tagged Rim11 protein kinase. Chase was over a 2-h period as indicated. Note that there was much less Rim11 immunoprecipitated at "0" time in the *cdc37S14A* mutant compared with the wild type. Previous studies showed that this was not due to impaired transcription, and the levels of protein kinase at "0" time could be recovered by treatment with MG132, the proteasome inhibitor.

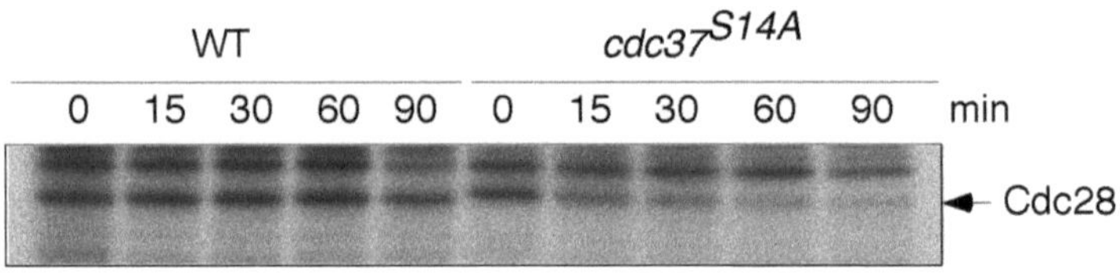

Fig. 2. Pulse chase and immunoprecipitation of Cdc28 protein kinase. Chase was over a 1.5-h period as indicated. Note that there were similar amounts of Cdc28 (labeled with an *arrow*; the upper band is nonspecific for the PSTAIRE antisera used) at "0" time, but that it decayed over the 90-min period. Note also the slight change in Cdc28 when derived from the different strains. Its migration was slightly slower in the *cdc37S14A* mutant.

experiment is normalized to the sample having the lowest counts. We aimed to have 0.5×10^6–1×10^7 cpm per sample at a minimum for the immunoprecipitation for an efficient signal. You can check that equal amounts were in each immunoprecipitation by running some of the samples directly by SDS-PAGE and following directions for drying the gel and exposing it for autoradiography.

3. Typical results from these experiments showed zero-point degradation (Fig. 1) and slow degradation of protein kinases in the *cdc37S14A* mutant (Fig. 2). Control experiments can involve immunoprecipitation of proteins whose levels do not change when cells are treated with geldanamycin or in *cdc37S14A* mutants. We use TAP-tagged Cmk2 for these controls as an example of a protein kinase whose levels do not change when chaperone function is compromised.

Acknowledgments

This work was supported by NIH grants RO1GM70596, U54CA132378 (to A.J.C.), and NCRR 5 G12-RR03060 (CCNY).

References

1. L.H. Pearl, Hsp90 and Cdc37 – a chaperone cancer conspiracy. Curr Opin Genet Dev 15 (2005) 55–61.
2. N. Grammatikakis, J.H. Lin, A. Grammatikakis, P.N. Tsichlis, and B.H. Cochran, p50(cdc37) acting in concert with Hsp90 is required for Raf-1 function. Mol Cell Biol 19 (1999) 1661–72.
3. J. Shao, A. Irwin, S.D. Hartson, and R.L. Matts, Functional dissection of cdc37: characterization of domain structure and amino acid residues critical for protein kinase binding. Biochemistry 42 (2003) 12577–88.
4. P. Lee, A. Shabbir, C. Cardozo, and A.J. Caplan, Sti1 and Cdc37 can stabilize Hsp90 in chaperone complexes with a protein kinase. Mol Biol Cell 15 (2004) 1785–92.
5. P. Lee, J. Rao, A. Fliss, E. Yang, S. Garrett, and A.J. Caplan, The Cdc37 protein kinase-binding domain is sufficient for protein kinase activity and cell viability. J Cell Biol 159 (2002) 1051–9.
6. S. Bandhakavi, R.O. McCann, D.E. Hanna, and C.V. Glover, A positive feedback loop between protein kinase CKII and Cdc37 promotes the activity of multiple protein kinases. J Biol Chem 278 (2003) 2829–36.
7. A.K. Mandal, P. Lee, J.A. Chen, N. Nillegoda, A. Heller, S. Distasio, H. Oen, J. Victor, D.M. Nair, J.L. Brodsky, and A.J. Caplan, Cdc37 has distinct roles in protein kinase quality control that protect nascent chains from degradation and promote posttranslational maturation. J Cell Biol 176 (2007) 319–28.
8. A.J. Caplan, A.K. Mandal, and M.A. Theodoraki, Molecular chaperones and protein kinase quality control. Trends Cell Biol 17 (2007) 87–92.
9. R.T. Baker, and A. Varshavsky, Inhibition of the N-end rule pathway in living cells. Proc Natl Acad Sci U S A 88 (1991) 1090–4.
10. T. Suzuki, and A. Varshavsky, Degradation signals in the lysine-asparagine sequence space. Embo J 18 (1999) 6017–26.
11. Y. Xie, and A. Varshavsky, UFD4 lacking the proteasome-binding region catalyses ubiquitination but is impaired in proteolysis. Nat Cell Biol 4 (2002) 1003–7.
12. A.K. Mandal, N. Nillegoda, J.A. Chen, and A.J. Caplan, Ydj1 protects nascent protein kinases from degradation and controls the rate of their maturation. Mol Cell Biol (2008).
13. N.B. Nillegoda, M.A. Theodoraki, A.K. Mandal, K.J. Mayo, H.Y. Ren, R. Sultana, K. Wu, J. Johnson, D.M. Cyr, and A.J. Caplan, Ubr1 and Ubr2 Function in a Quality Control Pathway for Degradation of Unfolded Cytosolic Proteins. Mol Biol Cell.

Chapter 7

Nucleotide Exchange Factors for Hsp70 Chaperones

Heike Rampelt, Matthias P. Mayer, and Bernd Bukau

Abstract

The ATPase cycle of Hsp70 chaperones controls their transient association with substrate and, thus, governs their function in protein folding. Nucleotide exchange factors (NEFs) accelerate ADP release from Hsp70 which results in rebinding of ATP and release of the substrate. This chapter describes several methods suitable to study NEFs of Hsp70 chaperones. On the one hand, steady-state ATPase assays provide information on how the NEF influences progression of the Hsp70 through the entire ATPase cycle. On the other hand, nucleotide release can be measured directly using labeled nucleotides, which enables identification and further characterization of NEFs.

Key words: Nucleotide exchange factor, Hsp70, GrpE, Bag1, HspBP1, Hsp110, Steady-state ATPase, Nucleotide release, Stopped flow

1. Introduction

Hsp70 chaperones accomplish their role as folding helpers in a variety of cellular processes by associating with substrates in a manner controlled by cycles of ATP hydrolysis and ADP-to-ATP exchange (1). Thus, analyzing the ATPase cycle of Hsp70s in molecular detail is crucial for our understanding of Hsp70-assisted protein folding. Hsp70 proteins consist of an N-terminal nucleotide-binding domain (NBD) allosterically coupled to a C-terminal substrate-binding domain (SBD). ATP binding to the NBD induces a conformational change in the SBD leading to a low affinity for substrates which is the result of high association and dissociation rates (Fig. 1). Hydrolysis of ATP to ADP converts the SBD to the high-affinity state, which is characterized by low substrate association and dissociation rates. Cycling of the Hsp70 chaperone through these states is facilitated by the action of co-chaperones

Stuart K. Calderwood and Thomas L. Prince (eds.), *Molecular Chaperones: Methods and Protocols*,
Methods in Molecular Biology, vol. 787, DOI 10.1007/978-1-61779-295-3_7, © Springer Science+Business Media, LLC 2011

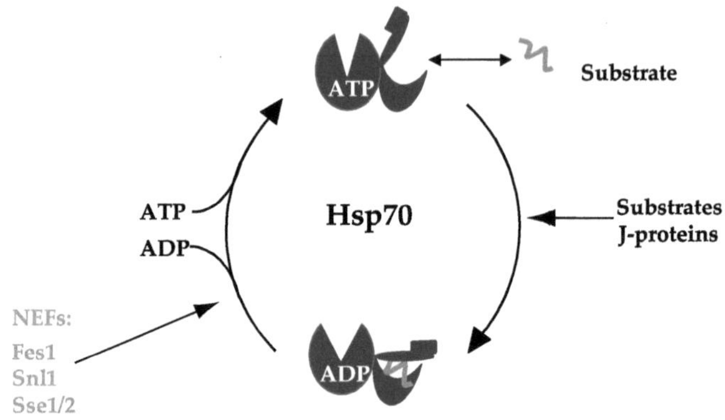

Fig. 1. The ATPase cycle of Hsp70 chaperones. Cytosolic NEFs from *Saccharomyces cerevisiae* are listed.

and by association with chaperone substrates. On the one hand, J-proteins deliver substrates to the chaperone or are located in close proximity of Hsp70 substrates and stimulate ATP hydrolysis to facilitate substrate binding. On the other hand, the release of ADP is accelerated by nucleotide exchange factors (NEFs), whose interaction with the Hsp70 NBD enables ATP rebinding and concomitant substrate release (Fig. 1). The NBD of Hsp70s consists of two lobes which form a deep cleft. The nucleotide is bound at the bottom of this cleft such that it interacts with all four subdomains that form the two lobes.

So far, five structurally unrelated classes of NEFs have been characterized (2–8). Although each of these NEF classes possesses a unique structure and interacts with Hsp70 in a distinct fashion, they can be placed into two groups according to their mode of action. Four NEF classes stimulate ADP release by tilting out subdomain IIB of the Hsp70 NBD which results in an opening of the nucleotide-binding cleft: the prokaryotic/organellar GrpE; the eukaryotic Bag-1 type as well as the recently characterized Bag-2; and finally the eukaryotic Hsp110 and Grp170 proteins, which are divergent members of the Hsp70 family themselves. In contrast to these NEF classes, eukaryotic HspBP1-like NEFs displace ADP by wrapping around subdomain IIB of the Hsp70 NBD resulting in a massive destabilization of subdomain IA and IB.

Two complementary approaches lend themselves to study how co-chaperones affect the Hsp70 ATPase cycle: one can assess how co-chaperones affect the rate-limiting step of the Hsp70 ATPase cycle by studying steady-state kinetics or alternatively, an individual step of the cycle, for instance ADP release, can be monitored in isolation.

2. Materials

2.1. Steady-State ATPase Assay

1. HKM buffer: 25 mM HEPES/KOH, pH 7.6, 50 mM KCl, 5 mM $MgCl_2$ (see also note 1).
2. NADH, 30 mM stock solution.
3. ATP, 20 mM stock solution: Adjust the pH to 7.
4. Phosphoenolpyruvate (PEP), 50 mM stock solution.
5. Pyruvate kinase/lactate dehydrogenase (LDH) mix: Commercially available (Sigma Inc.).
6. Proteins: Hsp70, J-protein, NEF (see also note 3).

2.2. Release of Radiolabeled Nucleotide

1. HKM buffer: 25 mM HEPES/KOH, pH 7.6, 50 mM KCl, 5 mM $MgCl_2$.
2. ATP, 100 mM and 0.1 mM stock solutions: Adjust the pH to 7.
3. ADP, 100 mM stock solution: Adjust the pH to 7.
4. [α-^{32}P]-ATP 3,000 Ci/mmol.
5. 1 mg/ml BSA.
6. 10% (v/v) acetic acid/400 mM LiCl.
7. Proteins: Hsp70, NEF.

2.3. Nucleotide Release by Stopped-Flow Instrumentation

1. HKM buffer: 25 mM HEPES/KOH, pH 7.6, 50 mM KCl, 5 mM $MgCl_2$.
2. MABA-ADP, 100 mM stock solution.
3. ATP, 100 mM stock solution: Adjust the pH to 7.
4. Proteins: Hsp70, NEF.

3. Methods

3.1. NEF Effects on Hsp70 Steady-State ATPase Activity

In the ATPase cycle of an Hsp70 chaperone whose ATP hydrolysis is maximally stimulated by the concerted action of a J-protein and a substrate, ADP release becomes the rate-limiting step. Therefore, in the presence of a stimulatory J-protein, the addition of an NEF increases the Hsp70 ATPase rates in a steady-state assay. However, differences among J-proteins must be taken into account. The bacterial DnaJ, when present at equimolar concentration relative to DnaK, is sufficient for maximal stimulation of ATP hydrolysis (13-fold to a stimulated rate of ca. 0.01/s in the absence of GrpE and 190-fold to a rate of 0.14/s in the presence of GrpE (9)). This is because under the artificial in vitro conditions DnaJ serves as both J-protein and substrate (10). In contrast, many eukaryotic

J-proteins on their own only marginally (two- to fivefold) stimulate their partner Hsp70 to hydrolyze ATP in the absence of a chaperone substrate (11, 12). This presumably reflects that these J-proteins are not serving as substrate.

In contrast, in a single-turnover ATPase assay, many NEFs inhibit ATP hydrolysis because they can displace not only ADP from Hsp70, but also ATP prior to hydrolysis. Two examples are the bacterial NEF GrpE (13) and the yeast Hsp110 protein Sse1 (14). This property can also become apparent under steady-state conditions (both in ATPase and substrate-refolding assays) if the ratio of NEF to Hsp70 is too high (13, 15–17).

The most convenient method to determine the steady-state ATPase activity is a coupled enzymatic test, where pyruvate kinase and LDH convert ADP produced by the Hsp70, along with PEP and NADH into ATP, lactate, and NAD^+ (18). Consumption of NADH is monitored as a decrease in absorption at 340 nm.

1. Prepare a master mix containing PEP and NADH, the PK/LDH mix, and $MgCl_2$. The final concentrations in the assay should be 0.25 mM for NADH; 0.4–1 mM for PEP depending on the ATP hydrolysis rate of the Hsp70; $MgCl_2$ 5 mM; dilute out the commercially available PK/LDH mix 1:100.
2. Dispense the master mix into a microtiter plate and add Hsp70 chaperone and co-chaperones as required. Hsp70 concentrations are typically around 1–3 μM. Monitor background ATP hydrolysis with a reaction lacking Hsp70 and co-chaperones. Also control for ATP hydrolysis by co-chaperones or contaminating ATPases by assaying all co-chaperone concentrations in the absence of Hsp70. It is also advisable to construct an ATPase-deficient mutant of Hsp70, like the K71M mutant of bovine Hsc70 (19), and purify this variant protein using the same protocol to determine the contamination level by copurifying ATPases (see also notes 2–4, 7).
3. Start the reaction by addition of different concentrations of ATP, e.g., 0.25–2 mM ATP.
4. Measure the decrease in A_{340} as a function of time, fit the data to a linear equation using a suitable software package, and calculate the amount of ADP produced by the Hsp70. Determine the k_{cat} [s^{-1}] of the hydrolysis reactions by fitting the data to the Michaelis Menten equation (see ref. 20).

3.2. Determination of Nucleotide Exchange Using Radiolabeled Nucleotide

In a single-turnover assay, nucleotide dissociation is monitored directly which allows to precisely define the action of a potential NEF. The two methods outlined below similarly monitor the dissociation of labeled ADP or ATP from Hsp70 upon addition of excess unlabeled nucleotide and NEF. Either radiolabeled nucleotide (this chapter) or a fluorescent nucleotide analog (Subheading 3.3) may be used.

To assess the release of radiolabeled nucleotide, complexes of Hsp70 with [α-^{32}P]-ATP are incubated with excess unlabeled nucleotide in the absence or presence of an NEF. A rapid gel filtration step then separates dissociated from Hsp70-bound nucleotide, and the radioactivity in the protein peak is measured by scintillation counting. Additionally, nondissociated nucleotide may be analyzed by thin-layer chromatography (TLC) (17, 21).

1. To form Hsp70–nucleotide complexes, prepare per reaction and per time point 5 μl samples each containing 5 μM of the Hsp70 chaperone with 5 μM ATP, including 400 Bq of [α-^{32}P]-ATP, and incubate for 2 min on ice. In order to obtain Hsp70–[α-^{32}P]-ADP complexes by ATP hydrolysis, the incubation is performed for 30 min at 30°C.
2. Prewarm the reaction at 30°C and add 5 μl of prewarmed buffer containing 20–100 mM cold ATP with or without an NEF at a two- to fourfold stoichiometric excess to the Hsp70.
3. At different time points, rapidly add 40 μl of ice-cold buffer to the sample and immediately load it onto a G-50 Nick gel filtration column (GE Healthcare) to separate Hsp70–nucleotide complexes from unbound labeled nucleotide. Before use, the column should be prepared in the following manner: equilibrate the column with HKM buffer, then load 1 ml of 1 mg/ml BSA onto the column, and elute with 3-column volumes of HKM buffer. This measure decreases nonspecific interactions of the Hsp70 with the column material. Collect fractions of two drops in scintillation vials and determine the radioactivity of the fractions.
4. Calculate the amount of protein-associated radioactivity per total radioactivity for each time point and plot the decrease of bound nucleotide over time. In order to obtain basal and stimulated apparent release rates, fit a single exponential decay function to the data.
5. In order to analyze the release of ATP and ADP separately, pool protein fractions from the gel filtration column and spot the eluate onto a TLC plate that has been prepared in the following manner: spot 2 μl of a 5 mM ADP + 5 mM ATP solution on each start position of the TLC plate. The unlabeled nucleotides serve as a carrier in the chromatographic separation of radiolabeled ATP and ADP.
6. Develop the TLC plate in a running buffer of 10% (v/v) acetic acid/400 mM LiCl, dry the plate, and expose it to a phosphoimager screen.
7. Quantitate amounts of radioactive ADP and ATP using software suited for densitometric analysis and fit exponential equations to the data.

3.3. Determination of Nucleotide Release Rates by Stopped-Flow Instrumentation

NEF-mediated stimulation of ADP release from an Hsp70 chaperone can be determined directly and elegantly using stopped-flow instrumentation. This instrumentation is ideally suited for monitoring fast dissociation kinetics with halftimes of seconds to milliseconds since the instrument's dead time is approximately 1 ms. In principle, rapid mixing of two solutions starts the reaction whose progress is monitored as changes in fluorescence emission. The fluorescent ADP-analog MABA-ADP (8-(*N*′-methylanthraniloylaminobutyl)-aminoadenosine 5′-diphosphate) (22) is ideally suited to analyze nucleotide release from Hsp70 chaperones. Earlier studies have established that the fluorescent group in both MABA-ADP and MABA-ATP does not affect the affinity of the nucleotide analogs to DnaK (22) and, by extension, to other Hsp70 chaperones. However, when analyzing a different combination of nucleotide analog and ATPases, the equivalence of the fluorophore-labeled and unlabeled nucleotide needs to be established. To this end, the K_d of the nucleotide analog should be determined, and additionally a competition experiment can be performed, where the nucleotide analog is displaced by titrating in unlabeled nucleotide (see ref. 22). Below, the experimental setup for measuring MABA-ADP dissociation is described.

After preincubating the Hsp70 chaperone with MABA-ADP, the sample is mixed with either excess (≥100-fold) unlabeled nucleotide alone or with excess nucleotide and different concentrations of the potential NEF to determine basal and stimulated ADP release rates, respectively. Since protein-bound MABA-ADP exhibits a higher fluorescence than the free molecule (e.g., twofold increase in fluorescence for DnaK-bound MABA-ADP (23)), its dissociation from the Hsp70 chaperone translates into a decreased fluorescence.

1. Equilibrate the stopped-flow instrument in assay buffer at the assay temperature. We employ a fluorescence excitation wavelength of 360 nm and a 420 nm cutoff filter for detection of emission (22). Use filtrated buffer and water throughout.
2. Preincubate the Hsp70 chaperone with MABA-ADP at assay temperature. We commonly use 0.5 μM of Hsp70 and nucleotide each; however, the concentrations may require adjusting in case the affinity of the Hsp70 for ADP is too low to allow for accurate measurement or the fluorescence increase upon binding is lower than for DnaK (see also note 5).
3. Preincubate solutions with excess unlabeled nucleotide (e.g., 250 μM ATP) +/− different concentrations of NEF. If the NEF of interest possesses an NBD itself, a potential role of nucleotide binding for its activity can be assayed by varying the nucleotide or changing its concentration in this step (see also note 6).
4. Load the Hsp70–MABA-ADP sample and the chase solution (excess unlabeled nucleotide) into the two reservoir syringes of the stopped-flow instrument.

5. Make a test measurement to collect the necessary information for setting the measurement parameters, like the reaction time (record the reaction for 10× the reaction halftime).
6. Perform multiple measurements for every reaction. Average and fit the data to a single exponential equation to obtain apparent off-rates for basal and stimulated nucleotide release.

4. Notes

1. It is vital that the buffers used in Hsp70 experiments contain KCl, but not NaCl. The Hsp70 NBD requires the presence of two potassium ions, which coordinate the bound nucleotide; ATP hydrolysis is minimal in the absence of potassium ions. Sodium ions cannot replace potassium due to their significantly smaller ionic radius, resulting in a fivefold reduction of the ATPase rate and altered nucleotide association and dissociation kinetics (24, 25).
2. Since the steady-state ATPase assay also measures ATP hydrolysis by contaminating ATPases, the protein preparations should be highly pure. This applies especially to the Hsp70 protein.
3. Hsp70s are generally purified using several chromatography steps (26, 27) because N- and C-terminal tags have been shown to influence the steady-state ATPase activity. However, an N-terminal His-tag has been used for purification of the endoplasmic reticulum Hsp70 BiP/Grp78 and subsequently cleaved with thrombin to give the authentic N-terminus (28). Similarly, an N-terminal fusion to SUMO was employed to purify the yeast Hsp70 chaperone Ssa1 (29).

   For purification procedures for the *Escherichia coli* NEF GrpE and the human Bag1, refer to refs. 13 and 30, respectively. HspBP1 and FES1 were purified using an N-terminal His-tag (17). The Hsp110 proteins were purified with a TEV-cleavable N-terminal Histag or as an N-terminal fusion to SUMO (29, 31).
4. Performing the steady-state ATPase assay in the presence of inorganic phosphate (e.g., 10 mM) enhances the stimulatory effect of NEFs because dissociation of ADP*P_i proceeds more slowly under these circumstances. This is due to the ordered product release mechanism of Hsp70s, where inorganic phosphate is released before the ADP (20).
5. Dissociation of MABA-ATP from an Hsp70 chaperone can be monitored in a manner similar to that of MABA-ADP; however, due to hydrolysis of the ATP, a double-mixing experiment has to be performed. In this case, Hsp70 protein is mixed

in the stopped-flow apparatus with a MABA-ATP solution, and after an incubation time of, e.g., 30 s, the resulting solution is mixed with chase nucleotide and NEF and nucleotide release is measured.

6. As mentioned earlier, the NEFs from the Hsp110 and Grp170 protein families belong to the Hsp70 superfamily and thus bind and hydrolyze ATP themselves. The nucleotide requirements of these proteins for their NEF activity can be assessed in the stopped-flow nucleotide release assay by varying the chase nucleotide: you may preincubate the NEF with either ATP or ADP at differing concentrations. While the nature of the chase nucleotide does not have an influence on the nucleotide dissociation step, it may affect the activity of the NEF.
7. Similarly, the rate of ATP hydrolysis by these Hsp70-like NEFs may be stimulated upon interaction with their partner Hsp70 chaperone, as has been demonstrated for the yeast Grp170 protein Lhs1 (32). To account for such a contribution to the total ATP hydrolysis measured in a steady-state assay, control measurements with the NEF and an Hsp70 mutant deficient in either nucleotide binding or ATP hydrolysis should be performed.

References

1. Mayer, M. P., and Bukau, B. (2005) Hsp70 chaperones: cellular functions and molecular mechanism, *Cell Mol Life Sci* 62, 670–684.
2. Harrison, C. J., Hayer-Hartl, M., Di Liberto, M., Hartl, F., and Kuriyan, J. (1997) Crystal structure of the nucleotide exchange factor GrpE bound to the ATPase domain of the molecular chaperone DnaK, *Science* 276, 431–435.
3. Sondermann, H., Scheufler, C., Schneider, C., Hohfeld, J., Hartl, F. U., and Moarefi, I. (2001) Structure of a Bag/Hsc70 complex: convergent functional evolution of Hsp70 nucleotide exchange factors, *Science* 291, 1553–1557.
4. Shomura, Y., Dragovic, Z., Chang, H. C., Tzvetkov, N., Young, J. C., Brodsky, J. L., Guerriero, V., Hartl, F. U., and Bracher, A. (2005) Regulation of Hsp70 function by HspBP1: structural analysis reveals an alternate mechanism for Hsp70 nucleotide exchange, *Mol Cell* 17, 367–379.
5. Polier, S., Dragovic, Z., Hartl, F. U., and Bracher, A. (2008) Structural basis for the cooperation of Hsp70 and Hsp110 chaperones in protein folding, *Cell* 133, 1068–1079.
6. Schuermann, J. P., Jiang, J., Cuellar, J., Llorca, O., Wang, L., Gimenez, L. E., Jin, S., Taylor, A. B., Demeler, B., Morano, K. A., Hart, P. J., Valpuesta, J. M., Lafer, E. M., and Sousa, R. (2008) Structure of the Hsp110:Hsc70 nucleotide exchange machine, *Mol Cell* 31, 232–243.
7. Andreasson, C., Fiaux, J., Rampelt, H., Druffel-Augustin, S., and Bukau, B. (2008) Insights into the structural dynamics of the Hsp110-Hsp70 interaction reveal the mechanism for nucleotide exchange activity, *Proc Natl Acad Sci U S A* 105, 16519–16524.
8. Xu, Z., Page, R. C., Gomes, M. M., Kohli, E., Nix, J. C., Herr, A. B., Patterson, C., and Misra, S. (2008) Structural basis of nucleotide exchange and client binding by the Hsp70 cochaperone Bag2, *Nat Struct Mol Biol* 15, 1309–1317.
9. McCarty, J. S., Buchberger, A., Reinstein, J., and Bukau, B. (1995) The role of ATP in the functional cycle of the DnaK chaperone system, *J Mol Biol* 249, 126–137.
10. Laufen, T., Mayer, M. P., Beisel, C., Klostermeier, D., Mogk, A., Reinstein, J., and Bukau, B. (1999) Mechanism of regulation of

hsp70 chaperones by DnaJ cochaperones, *Proc Natl Acad Sci U S A* 96, 5452–5457.
11. Jiang, R. F., Greener, T., Barouch, W., Greene, L., and Eisenberg, E. (1997) Interaction of auxilin with the molecular chaperone, Hsc70, *J Biol Chem* 272, 6141–6145.
12. Bimston, D., Song, J., Winchester, D., Takayama, S., Reed, J. C., and Morimoto, R. I. (1998) BAG-1, a negative regulator of Hsp70 chaperone activity, uncouples nucleotide hydrolysis from substrate release, *Embo J* 17, 6871–6878.
13. Packschies, L., Theyssen, H., Buchberger, A., Bukau, B., Goody, R. S., and Reinstein, J. (1997) GrpE accelerates nucleotide exchange of the molecular chaperone DnaK with an associative displacement mechanism, *Biochemistry* 36, 3417–3422.
14. Raviol, H., Sadlish, H., Rodriguez, F., Mayer, M. P., and Bukau, B. (2006) Chaperone network in the yeast cytosol: Hsp110 is revealed as an Hsp70 nucleotide exchange factor, *Embo J* 25, 2510–2518.
15. Raynes, D. A., and Guerriero, V., Jr. (1998) Inhibition of Hsp70 ATPase activity and protein renaturation by a novel Hsp70-binding protein, *J Biol Chem* 273, 32883–32888.
16. Gassler, C. S., Wiederkehr, T., Brehmer, D., Bukau, B., and Mayer, M. P. (2001) Bag-1 M accelerates nucleotide release for human Hsc70 and Hsp70 and can act concentration-dependent as positive and negative cofactor, *J Biol Chem* 276, 32538–32544.
17. Kabani, M., Beckerich, J. M., and Brodsky, J. L. (2002) Nucleotide exchange factor for the yeast Hsp70 molecular chaperone Ssa1p, *Mol Cell Biol* 22, 4677–4689.
18. Ali, J. A., Jackson, A. P., Howells, A. J., and Maxwell, A. (1993) The 43-kilodalton N-terminal fragment of the DNA gyrase B protein hydrolyzes ATP and binds coumarin drugs, *Biochemistry* 32, 2717–2724.
19. O'Brien, M. C., Flaherty, K. M., and McKay, D. B. (1996) Lysine 71 of the chaperone protein Hsc70 Is essential for ATP hydrolysis, *J Biol Chem* 271, 15874–15878.
20. Ha, J. H., and McKay, D. B. (1994) ATPase kinetics of recombinant bovine 70 kDa heat shock cognate protein and its amino-terminal ATPase domain, *Biochemistry* 33, 14625–14635.
21. Liberek, K., Marszalek, J., Ang, D., Georgopoulos, C., and Zylicz, M. (1991) Escherichia coli DnaJ and GrpE heat shock proteins jointly stimulate ATPase activity of DnaK, *Proc Natl Acad Sci U S A* 88, 2874–2878.
22. Theyssen, H., Schuster, H. P., Packschies, L., Bukau, B., and Reinstein, J. (1996) The second step of ATP binding to DnaK induces peptide release, *J Mol Biol* 263, 657–670.
23. Leskovar, A., and Reinstein, J. (2008) Photophysical properties of popular fluorescent adenosine nucleotide analogs used in enzyme mechanism probing, *Arch Biochem Biophys* 473, 16–24.
24. O'Brien, M. C., and McKay, D. B. (1995) How potassium affects the activity of the molecular chaperone Hsc70. I. Potassium is required for optimal ATPase activity, *J Biol Chem* 270, 2247–2250.
25. Wilbanks, S. M., and McKay, D. B. (1995) How potassium affects the activity of the molecular chaperone Hsc70. II. Potassium binds specifically in the ATPase active site, *J Biol Chem* 270, 2251–2257.
26. O'Brien, M. C., and McKay, D. B. (1993) Threonine 204 of the chaperone protein Hsc70 influences the structure of the active site, but is not essential for ATP hydrolysis, *J Biol Chem* 268, 24323–24329.
27. Buchberger, A., Valencia, A., McMacken, R., Sander, C., and Bukau, B. (1994) The chaperone function of DnaK requires the coupling of ATPase activity with substrate binding through residue E171, *Embo J* 13, 1687–1695.
28. Wei, J., and Hendershot, L. M. (1995) Characterization of the nucleotide binding properties and ATPase activity of recombinant hamster BiP purified from bacteria, *J Biol Chem* 270, 26670–26676.
29. Andreasson, C., Fiaux, J., Rampelt, H., Mayer, M. P., and Bukau, B. (2008) Hsp110 is a nucleotide-activated exchange factor for Hsp70, *J Biol Chem* 283, 8877–8884.
30. Hohfeld, J., and Jentsch, S. (1997) GrpE-like regulation of the hsc70 chaperone by the anti-apoptotic protein BAG-1, *Embo J* 16, 6209–6216.
31. Dragovic, Z., Broadley, S. A., Shomura, Y., Bracher, A., and Hartl, F. U. (2006) Molecular chaperones of the Hsp110 family act as nucleotide exchange factors of Hsp70s, *Embo J* 25, 2519–2528.
32. Steel, G. J., Fullerton, D. M., Tyson, J. R., and Stirling, C. J. (2004) Coordinated activation of Hsp70 chaperones, *Science* 303, 98–101.

Chapter 8

Reconstitution of CHIP E3 Ubiquitin Ligase Activity

Hong Yu Ren, Cam Patterson, Douglas M. Cyr, and Meredith F.N. Rosser

Abstract

CHIP, the carboxyl-terminus of Hsp70 interacting protein, is both an E3 ubiquitin ligase and an Hsp70 co-chaperone and is implicated in the degradation of cytosolic quality control and numerous disease substrates. CHIP has been shown to monitor the folding status of the CFTR protein, and we have successfully reconstituted this activity using a recombinant CFTR fragment consisting of the cytosolic NBD1 and R domains. We have found that efficient ubiquitination of substrates requires chaperone activity to either deliver the substrate to CHIP or to maintain the substrate in a ubiquitination-competent conformation. This chaperone activity can be provided by the Hsp70/Hsp40 molecular chaperone system as seen in the NBD1–R ubiquitination assay. Alternatively, heat treatment of CHIP can activate its own innate substrate-binding activity and allow for efficient ubiquitination of model substrates, such as denatured luciferase. Here, we describe methods for purifying the recombinant proteins necessary for in vitro reconstitution of CHIP ubiquitin ligase activity, as well as two methods used to monitor CHIP ligase activity. One method allows for the measurement of the Hsp70- and Hsp40-dependent CHIP activity while the other measures the Hsp40- and Hsp70-independent activity of heat-activated CHIP.

Key words: Carboxy-terminus of Hsp70 interacting protein, Ubiquitination, Assay, In vitro reconstitution, Hsp40, E3 ubiquitin ligase

1. Introduction

CHIP, the carboxyl-terminus of Hsp70 interacting protein, is both an E3 ubiquitin ligase and an Hsp70 co-chaperone which has been implicated in the degradation of cytosolic quality control and numerous disease substrates (1–4). CHIP contains a TPR domain, which allows for interaction with the C-terminus of Hsp70, and a U-box domain, which is responsible for the E3 ubiquitination activity (5, 6). Chip has both Hsp70-dependent and -independent activities, both of which are important for maintaining cellular

Stuart K. Calderwood and Thomas L. Prince (eds.), *Molecular Chaperones: Methods and Protocols*, Methods in Molecular Biology, vol. 787, DOI 10.1007/978-1-61779-295-3_8, © Springer Science+Business Media, LLC 2011

homeostasis (7). For example, CHIP has been shown to regulate HSF activation in a U-box-independent manner (8) as well as the turnover of Hsp70 after a heat-shock response (4), thereby playing a crucial role in tuning the stress response. The assays described in this chapter allow for the biochemical measurement of both Hsp70-dependent and -independent ubiquitination activity of CHIP. These assays can also be adapted to test the ability of CHIP to ubiquitinate specific substrates of interest or to look for modulators of CHIP activity.

2. Materials

2.1. Protein Purifications

1. LB (Luria Broth): 10% (w/v) tryptone, 5% (w/v) yeast extract, 10% (w/v) NACL (Fisher, Cat. # BP1426-2).
2. IPTG [Isopropyl β-D-thiogalactopyranoside (Fisher, Cat. # BP1755-10)].
3. Lysozyme (Sigma, Cat. # L-6876).
4. Glutathione Sepharose (GE, Cat. # 17-5132-01)).
5. Sarkosyl (Fisher, Cat. # BP234-500).
6. Glutathione (GSH) (Sigma, Cat. # G6529-25G).
7. HA column (Bio-Rad, Cat. # 732-0081).
8. HQ column (Bio-Rad, Cat. # 732-0026).
9. Talon metal affinity resin (Clone Tech, Cat. # 635502).
10. STE buffer (10 mM Tris–HCl, pH 8.0, 1 mM EDTA, 150 mM NaCl).
11. Glutathione elution buffer (50 mM Tris–HCl, pH 8.0, 20 mM GSH).
12. Buffer A (20 mM Hepes, pH 7.4, 20 mM NaCl).
13. Buffer B (20 mM Hepes, pH 7.4, 500 mM NaCl).
14. Buffer C (10 mM potassium phosphate buffer, pH 7.4).
15. Buffer D (500 mM potassium phosphate buffer, pH 7.4).
16. Buffer E (20 mM Hepes, 150 mM NaCl, pH 7.4).
17. Buffer F (20 mM Tris–HCl, pH 8.0, 150 mM NaCl).
18. Buffer G (20 mM Tris–HCl, pH 8.0, 150 mM NaCl, 150 mM imidazole).
19. Dialysis tubing (12,000–14,000 molecular weight cutoff).
20. Complete Protease Inhibitor (PI) Tablets (Roche).
21. PMSF (stock solution made at 100 mM in ethanol).
22. 2× SDS-PAGE sample buffer [(100 mM Tris base, pH 6.8, 4% SDS, 20% glycerol, and Coomassie blue (2 mg/50 ml or to desired darkness)].

2.2. Ubiquitination Reaction

1. Reaction buffer (20 mM Hepes, pH 7.4, 50 mM NaCl, 5 mM $MgCl_2$, 2.5 mM ATP, 2 mM DTT).
2. Bovine ubiquitin.
3. Rabbit E1.
4. Purified recombinant human UbcH5a protein.
5. Purified recombinant human CHIP protein.
6. Purified recombinant human Hsp70 protein.
7. Purified recombinant human Hsp40 (Hdj-2) protein.
8. Luciferase (Promega, 14.4 mg/ml).
9. Gdm–HCl (6 M).
10. 2× SDS-PAGE sample buffer.
11. SDS-PAGE gel.
12. Western blot transfer buffer (20 mM Tris–HCl, 150 mM glycine, 20% methanol, 0.038% SDS).
13. Nitrocellulose.
14. PBS-T (1% Triton) (for 1 l PBS: 8 g NaCl, 0.2 g KCl, 1.44 g Na_2HPO_4, 0.24 g of KH_2PO_4).
15. Blocking solution (10% nonfat milk in PBS-T).
16. Primary antibody for immunoblotting (α-CFTR R-domain (mouse, R&D Systems) or α-luciferase (goat, Chemicon International) (diluted 1:1,000 in 1% nonfat milk made in PBS-T).
17. Secondary horseradish peroxidise-conjugated goat anti-mouse (Bio-Rad) or rabbit anti-goat antibody (Chemicon International) (diluted 1:3,000 in 1% nonfat milk made in PBS-T).
18. ECL (GE Healthcare).
19. Ubiquitination buffer (20 mM Hepes, pH 7.4, 50 mM NaCl, 5 mM $MgCl_2$, 2.5 mM ATP, 2 mM DTT, 10 μM bovine ubiquitin) (see Note 13).
20. Recombinant firefly luciferase (Promega).

3. Methods

The methods described herein include the purifications of recombinant proteins necessary for in vitro reconstitution of CHIP ubiquitin ligase activity, as well as two methods used to monitor CHIP ligase activity. One method allows for the measurement of the Hsp70- and Hsp40-dependent CHIP activity (5, 6) while the other measures the Hsp40- and Hsp70-independent activity of heat-activated CHIP (5). CHIP has been shown to monitor the folding status of the CFTR protein, and we have successfully reconstituted this activity using a recombinant CFTR fragment consisting of the cytosolic NBD1 and R domain (5). We have found that efficient ubiquitination of substrates requires chaperone activity to either deliver the substrate to CHIP or to maintain the substrate in a ubiquitination-competent conformation. This chaperone activity can be provided by the Hsp70/Hsp40 molecular chaperone system as seen in the NBD1-R ubiquitination assay. Alternatively, heat treatment of CHIP can activate its own innate substrate-binding activity and allow for efficient ubiquitination of model substrates, such as denatured luciferase (9).

3.1. Protein Purification

3.1.1. Purification of GST–NBD1–R Fusion Protein from E. coli Strain BL21(DE3)

1. Day 1: Prepare a 50-ml overnight culture by adding 100 μg/ml ampicillin and inoculating with BL21(DE3) *E. coli* harbouring the GST–NBD1–R expression vector. Incubate overnight at 37°C, shaking.
2. Day 2: Seed 20 ml of overnight culture into 600 ml LB with 100 μg/ml ampicillin. Grow at 37°C until the culture reaches an OD_{600} of 0.6 (see Note 1).
3. Induce with 0.2 mM of IPTG. Grow for 16 h at 30°C (see Note 1).
4. Harvest cells by centrifugation at 3,000×*g* for 10 min at 4°C. Decant media (see Note 2).
5. Resuspend pellet in 10 ml of ice-cold STE buffer (10 mM Tris–HCl, pH 8.0, 1 mM EDTA, 150 mM NaCl) with fresh PI (1×) and PMSF (1 mM).
6. Add 100 μl of freshly prepared lysozyme solution (10 mg/ml in water) and incubate on ice for 30 min.
7. Just before sonication, add 100 μl of 1 M DTT and 1.4 ml of 10% sarkosyl. Mix and sonicate for four times, each for a period of 15 s (see Note 3).
8. Centrifuge at 16,000 rpm for 20 min to pellet debris (see Note 4).
9. Transfer supernatant to a 50-ml conical tube and discard the pellet. Add 4 ml of 10% Triton X-100 and bring up to a final volume of 20 ml with STE buffer. The final concentrations of sarkosyl and Triton X-100 are 0.7 and 2%, respectively. Incubate at room temperature for 30 min.
10. Pour the lysate into a tube with 1 ml of prepared glutathione sepharose (50% slurry) in PBS. Incubate at 4°C for 1 h with rotation.
11. Wash the beads three times with 30 ml of cold PBST.
12. Remove last portion of PBS-T and add 1 ml of elution buffer (50 mM Tris–HCl, pH 8.0, 20 mM GSH).

13. Incubate at 4°C for 10 min with rotation.
14. Centrifuge at 6,000 rpm for 1 min, remove supernatant to 1.6-ml tube (see Note 4).
15. Determine the concentration of protein with protein assay, such as Bio-Rad DC Assay (see Note 5).
16. Add DTT to final concentration of 10 mM. Aliquot protein and store at −80°C (see Note 6).

3.1.2. Purification of Recombinant Human Hsp70, Hdj-2, or CHIP Protein from E. coli Strain BL21(DE3)

1. Day 1: Prepare a 50-ml overnight culture by adding 100 μg/ml ampicillin and inoculating with BL21(DE3) *E. coli* harbouring the appropriate expression vector. Incubate overnight at 37°C, shaking.
2. Day 2: Seed 20 ml of overnight culture into 600 ml LB with 100 μg/ml ampicillin. Grow at 37°C until the culture reaches an OD_{600} of 0.6 (see Note 1).
3. Induce with 0.2 mM of IPTG. Grow for 3 h at 30°C.
4. Centrifuge at 3,000×*g* at 4°C for 10 min. Decant media (see Note 2).
5. Resuspend pellet in 30 ml of cold buffer A (20 mM Hepes, pH 7.4, 20 mM NaCl), put on ice, and sonicate ten times for a period of 20 s each time (see Note 3).
6. Centrifuge at 12,000 rpm for 10 min to pellet debris (see Note 4).
7. Load supernatant on HQ column (Bio-Rad), and then wash column with buffer A for 4-column volumes.
8. Perform gradient elution with buffer A and buffer B (0→100% buffer B) (20 mM Hepes, pH 7.4, 500 mM NaCl). Collect 1.3-ml fractions and run SDS-PAGE on each fraction (see Note 7).
9. Combine fractions that contain your protein of interest and dialyze in buffer C (10 mM KH_2PO_4, pH 7.4) (see Note 7). Alternatively, you can dilute the combined fractions tenfold in buffer C.
10. Load dialyzed protein on HA column (Bio-Rad), wash column with 4-column volumes of buffer C.
11. Perform gradient elution with buffer C and buffer D (0→100% buffer D) (500 mM KH_2PO_4), and collect 1.3-ml fractions.
12. Run SDS-PAGE to check protein purity, and combine appropriate fractions (see Note 7). Dialyze combined fractions in buffer E (20 mM Hepes, 150 mM NaCl, pH 7.4) (see Note 8).
13. Determine the concentration of protein with protein assay, such as Bio-Rad DC Assay (see Note 4).

14. Aliquot protein and store at –80°C (see Note 6).

3.1.3. Purification of Recombinant His-UbcH5a Protein from E. coli Strain BL21(DE3)

1. Collect bacteria pellets as above in Subheading 3.1.2 (steps 1–4).
2. Resuspend pellet in 30 ml of ice-cold buffer F (20 mM Tris–HCl, pH 8.0, 150 mM NaCl). Sonicate on ice ten times for a period of 20 s each time (see Note 3).
3. Centrifuge at 12,000 rpm for 10 min to pellet debris and then collect supernatant (see Note 4).
4. Load supernatant on column containing Talon affinity residue (Covance). Wash column with buffer F (20 mM Tris–HCl, pH 8.0, 150 mM NaCl) for 4-column volumes.
5. Perform gradient elution with buffer F and buffer G ($0 \rightarrow 100\%$ buffer G) (20 mM Tris–HCl, pH 8, 150 mM NaCl, 150 mM imidazole), and collect 1.3-ml fractions.
6. Run SDS-PAGE to check protein purity, and combine appropriate fractions. Dialyze in buffer E (20 mM Hepes, 150 mM NaCl, pH 7.4) (see Note 8).
7. Determine the concentration of protein with protein assay, such as Bio-Rad DC Assay (see Note 5).
8. Aliquot protein and store at –80°C (see Note 6).

3.1.4. In Vitro Reconstitution of the Hsp70/Hsp40-Dependent Ubiquitination Activity of the E3 Ubiquitin Ligase, CHIP

This assay reconstitutes the ubiquitination of a CFTR fragment (NBD1-R) and requires the action of the Hsp70 and Hsp40 molecular chaperones. The Hsp40/Hsp70 chaperone pair is thought to deliver substrate to CHIP. CHIP binds the Hsp70 chaperone through its TPR domains, and then promotes polyubiquitination of the substrate through action of the U-box domain (Fig. 1).

1. Combine appropriate amounts of a 10× reaction buffer (200 mM Hepes, pH 7.4, 500 mM NaCl, 50 mM $MgCl_2$, 25 mM ATP, and 20 mM DTT) (see Notes 9 and 13) with 10 μM bovine ubiquitin, 0.1 μM rabbit E1, and 1 μM GST–NBD1–R.
2. Add purified E2 (4 μM UbcH5a), E3 (3 μM CHIP), chaperones (4 μM of Hdj2 and 2 μM of Hsp70), or buffer controls to appropriate samples (see Note 11). Make sure that proteins are concentrated enough so that total reaction volume can be kept at 25 μl (see Note 5).
3. Incubate reaction at 37°C for 2 h.
4. Terminate reaction by adding 25 μl of 2× SDS sample buffer with BME (see Note 12) to each 25 μl reaction.
5. Load 25 μl of reaction on 7% SDS-PAGE gels and after electrophoresis, transfer to nitrocellulose membranes.

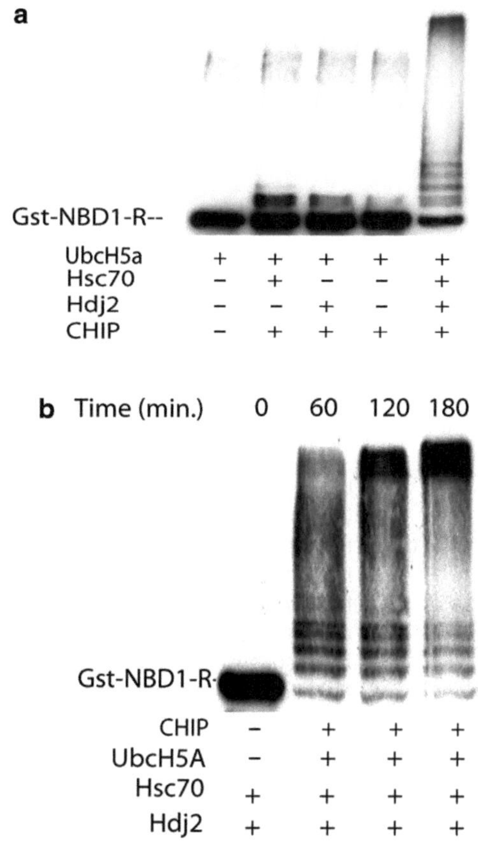

Fig. 1. Hsp40/Hsp70-dependent ubiquitination of GST–NBD1–R. (**a**) Efficient ubiquitination of GST–NBD1–R is only achieved in the presence of the Hsp40 and Hsp70 chaperone pair. All reactions contain E1 enzyme and E2 (UbcH5a) while the reaction in *lane 1* is lacking the E3 CHIP as a negative control. (**b**) Time course of ubiquitination. A time-dependent increase in ubiquitination can be observed over the course of 3 h. *Lane 1* represents the negative control in which CHIP and UbcH5a are not added to the reaction mixture.

6. Block membranes with blocking solution for 1 h at room temperature with gentle agitation.
7. Rinse membranes with PBS-T to remove the blocking solution.
8. Incubate membranes with αR domain antibody (1:1,000 dilution in 1% nonfat milk) for 1 h at room temperature with gentle agitation.
9. Wash membranes three times for 5 min each with PBS-T.
10. Add HRP-conjugated secondary antibody (1:3,000 dilution in 1% nonfat milk) and incubate for 1 h at room temperature with gentle agitation.
11. Wash membranes three times for 5 min each with PBS-T.
12. Add ECL and expose to film.

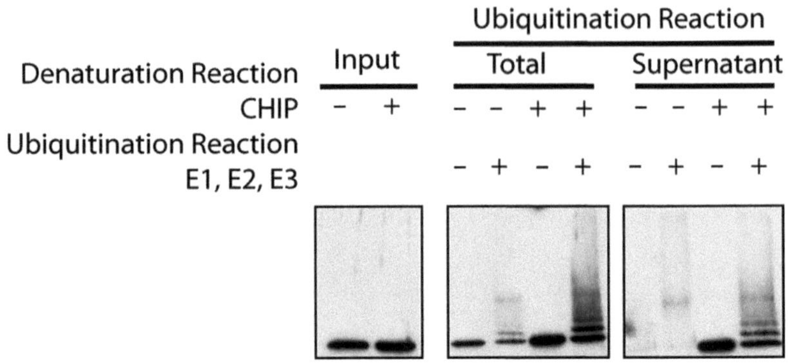

Fig. 2. Hsp40/Hsp70-independent ubiquitination of denatured luciferase. When CHIP is present during the reaction in which luciferase is heat denaturated by incubation at 42°C for 15 min, the luciferase is maintained in a soluble, ubiquitination-competent state. The total ubiquitination reaction (*lanes 3–6*) represents the amount of ubiquitination that occurs when CHIP is included or excluded from the denaturation reaction. In *lanes 7–10*, the denaturation reaction is spun at 20,000 rpm for 10 min at 4°C and the supernatant is removed to a different tube. Ubiquitination enzymes are then added to this supernatant fraction. As apparent from *lanes 7* and *8*, when luciferase is denatured in the absence of CHIP, the majority aggregates, pellets after centrifugation, and is therefore not detectable in the ubiquitination reactions performed on the supernatant fraction alone. Alternatively, when CHIP is present during the initial denaturation reaction (*lanes 9* and *10*), the CHIP is maintained soluble and efficiently ubiquitinated in the presence of the ubiquitination enzymes.

3.1.5. *In Vitro* Reconstitution of the Hsp70- and Hsp40-Independent CHIP Ubiquitin Ligase Activity

Recently, we have demonstrated that CHIP can act as a chaperone by binding to denatured substrates and preventing their aggregation (9). This activity of CHIP is enhanced after heat shock, as is seen for many other chaperone proteins. The following assay allows one to test for the ability of CHIP to ubiquitinate substrates independent of the Hsp40/Hsp70 chaperone pair, can be used to test the ability of proteins to modulate CHIP's activity, or can be adapted to directly test the ability of CHIP to ubiquitinate a specific substrate. In this assay, when luciferase is heat denatured in the presence of CHIP, the CHIP protein is able to prevent the aggregation of the luciferase which maintains the luciferase in a ubiquitination-competent state. When the luciferase is denatured in the absence of the CHIP protein, the majority aggregates, and is not ubiquitinated (Fig. 2).

1. Make 50 μl of a luciferase cocktail containing the following: 10 mM ATP, 20 mM Hepes, pH 7.4, 50 mM NaCl, 5 mM $MgCl_2$, 10 mM DTT, and 28 μg/ml luciferase.
2. Make two different denaturation buffers: A: 40 μl of dialysis buffer from CHIP purification (see Note 8) and B: 40 μl of CHIP diluted in dialysis buffer to a concentration of 0.25 mg/ml.
3. Add 10 μl of the luciferase cocktail to denaturation reaction A or denaturation reaction B to make your final volume 50 μl. Mix gently by pipetting.

4. Take three 10-μl aliquots of each denaturation reaction. You have six reactions (A1, A2, A3, B1, B2, B3).
   (a) Heat at 42°C for 15 min – this sample is your denatured input.
   (b) Heat at 42°C for 15 min – this sample receives ubiquitination enzymes.
   (c) Store on ice – this represents your nondenatured input.
5. After the 42°C heat denaturation, put tubes on ice.
6. Add 10 μl of 2× ubiquitination buffer 1 (40 mM Hepes, pH 7.4, 100 mM NaCl, 10 mM $MgCl_2$, 5 mM ATP, 4 mM DTT, and 20 μM bovine ubiquitin) to samples A1 and B1, and 10 μl of 2× ubiquitination buffer 2 (40 mM Hepes, pH 7.4, 100 mM NaCl, 10 mM $MgCl_2$, 5 mM ATP, 4 mM DTT, 20 μM bovine ubiquitin, 0.2 μM rabbit E1, 8 μM UbcH5a, and 6 μM CHIP) to samples A2 and B2 (see Note 13).
7. Incubate at 37°C for 1 h.
8. Terminate reactions by adding 2× sample buffer with BME to samples, heat at 37°C for 5 min, and run on 10% SDS-PAGE gel.
9. Transfer to nitrocellulose and follow the Western blotting procedure above (steps 6–12 in Subheading 3.1.4).

4. Notes

1. In order to verify that the protein of interest has been induced sufficiently, you can collect a sample of bacteria before and after induction, lyse in sample buffer by sonication, and run on an SDS-PAGE gel. The induced protein should be visible by Coomassie stain.
2. The bacterial pellets can be frozen and stored at –20°C before proceeding with the purification.
3. When sonicating, ensure that the samples remain cold. In order to do this, pack the samples on ice during sonication and wait for at least 30 s in between each sonication.
4. Samples can be saved at each step of the purification to monitor the purity and determine if the protein of interest is being lost at any particular step.
5. When determining protein concentration, always make BSA standards for standard line in the same buffer as protein of interest. If the protein of interest is not as concentrated as desired, then it can be concentrated using concentrators, such as the Amicon Ultra-15 centrifugal filter device with a 10,000 molecular weight cutoff. Typically, a concentration of 1–4 mg/ml is ideal for the subsequent reactions.

6. The protein can be stored at −80°C for an extended period of time, but should be subjected to minimal freeze thaw cycles.
7. If the purification is being done with the use of a detector, then the elution can be monitored at an absorbance of 280 nm. Then, fractions which correspond to the protein peaks can be assayed by SDS-PAGE. Otherwise, every other fraction of the entire elution can be tested in order to determine which fractions are the most pure for the protein of interest.
8. A portion of dialysis buffer can be saved after dialysis, aliquoted, and frozen in order to have a buffer-matched control for future assays.
9. The pH of ATP stocks should be adjusted to 7.5 with NaOH.
10. Making a cocktail containing all the common reagents for a series of samples within an experiment reduces the pipetting error that would come from adding each reagent individually.
11. As a negative control, the E2 or E3 proteins can be left out of the assay.
12. Add 80 μl of BME and 10 μl PI per ml of 2× SB fresh before use.
13. We typically make a 10× stock of the ubiquitination buffer without ATP, DTT, and ubiquitin, then dilute it to the desired working concentration, and add ATP, DTT, and ubiquitin fresh.

Acknowledgments

DMC is supported by NIH GM056981. CP is supported by NIH AG024282 and GM61728.

References

1. Meacham GC, Patterson C, Zhang W, Younger JM, Cyr DM. (2001) The Hsc70 co-chaperone CHIP targets immature CFTR for proteasomal degradation. Nat Cell Biol 3:100–5.
2. Cyr DM, Hohfeld J, Patterson C. (2002) Protein quality control: U-box-containing E3 ubiquitin ligases join the fold. Trends Biochem Sci 27:368–75.
3. Younger JM, Chen L, Ren HY, Rosser MNF, Fan C-Y, Patterson C and Cyr, DM (2006) Sequential quality-control checkpoints triage misfolded cystic fibrosis transmembrane conductance regulator. Cell 126:571–82.
4. Qian SB, McDonough H, Boellmann F, Cyr DM, Patterson C. (2006) CHIP-mediated stress recovery by sequential ubiquiti nation of substrates and Hsp70. Nature 440:551–5.
5. Younger JM, Ren HY, Chen L, Fields A, Patterson, C, and Cyr, DM. (2004) A foldable CFTR{Delta}F508 biogenic intermediate accumulates upon inhibition of the Hsc70-CHIP E3 ubiquitin ligase. J Cell Biol 167:1075–85.
6. Jiang J, Ballinger CA, Wu Y, Cyr DM, and Patterson, C (2001) CHIP is a U-box-dependent E3 ubiquitin ligase: identification of Hsc70 as a target for ubiquitylation. J Biol Chem 276:42938–44.
7. Min JN, Whaley RA, Sharpless NE, Lockyer P, Portbury AL, and Patterson C. (2008) CHIP

deficiency decreases longevity, with accelerated aging phenotypes accompanied by altered protein quality control. Mol Cell Biol 28:4018–25.

8. Dai Q, Zhang C, Wu Y,Cyr DM, and Patterson C (2003) CHIP activates HSF1 and confers protection against apoptosis and cellular stress. EMBO J 22:5446–58.

9. Rosser MF, Washburn E, Muchowski PJ, Patterson C, and Cyr DM. (2007) Chaperone functions of the E3 ubiquitin ligase CHIP. J Biol Chem 282:22267–77.

Chapter 9

Structure–Functions of HspB1 (Hsp27)

André-Patrick Arrigo

Abstract

Human HspB1 (also denoted Hsp27) is a well-known member, together with alphaB-crystallin, of the small heat-shock (or stress) proteins (sHsps) (20–40 kDa). In this chapter, I describe procedures for testing the oligomeric and phosphorylation patterns of HspB1 as well as its interaction with specific partner/client polypeptides using tissue culture cells genetically modified to express different levels of this protein. The procedures have been developed in my laboratory and could be used in any well-established cellular laboratory. In addition, the different procedures presented here could be extended to test the nine other human sHsp members as well as sHsps from other species.

Key words: Small stress proteins, Heat-shock proteins, Hsp27/B1, Oligomerization, Phosphorylation

1. Introduction

Human HspB1 (also denoted Hsp27) is a well-known member, together with alphaB-crystallin, of the small heat-shock (or stress) proteins (sHsps) (20–40 kDa). HspB1 synthesis is stimulated when cells are exposed to sublethal stress conditions that alter the folding of proteins (1). This leads to enhanced resistance of cells to heat-shock and oxidative injuries. HspB1 is also constitutively expressed in numerous human cells, particularly in pathological conditions. In human, HspB1 is part of a family of ten different proteins (denoted HspB1 to HspB10) (2), plus an additional less-conserved polypeptide (Hsp16.2). The different members of this family have in common a C-terminal alpha-crystallin domain (about 40% of the protein) found in alphaA,B-crystallin polypeptides from the vertebrate eye (1, 3, 4) (Fig. 1A). sHsps are also characterized by a flexible C-terminal tail and a less-conserved N-terminal domain containing a hydrophobic WDPF motif. As a consequence of their

Stuart K. Calderwood and Thomas L. Prince (eds.), *Molecular Chaperones: Methods and Protocols*,
Methods in Molecular Biology, vol. 787, DOI 10.1007/978-1-61779-295-3_9, © Springer Science+Business Media, LLC 2011

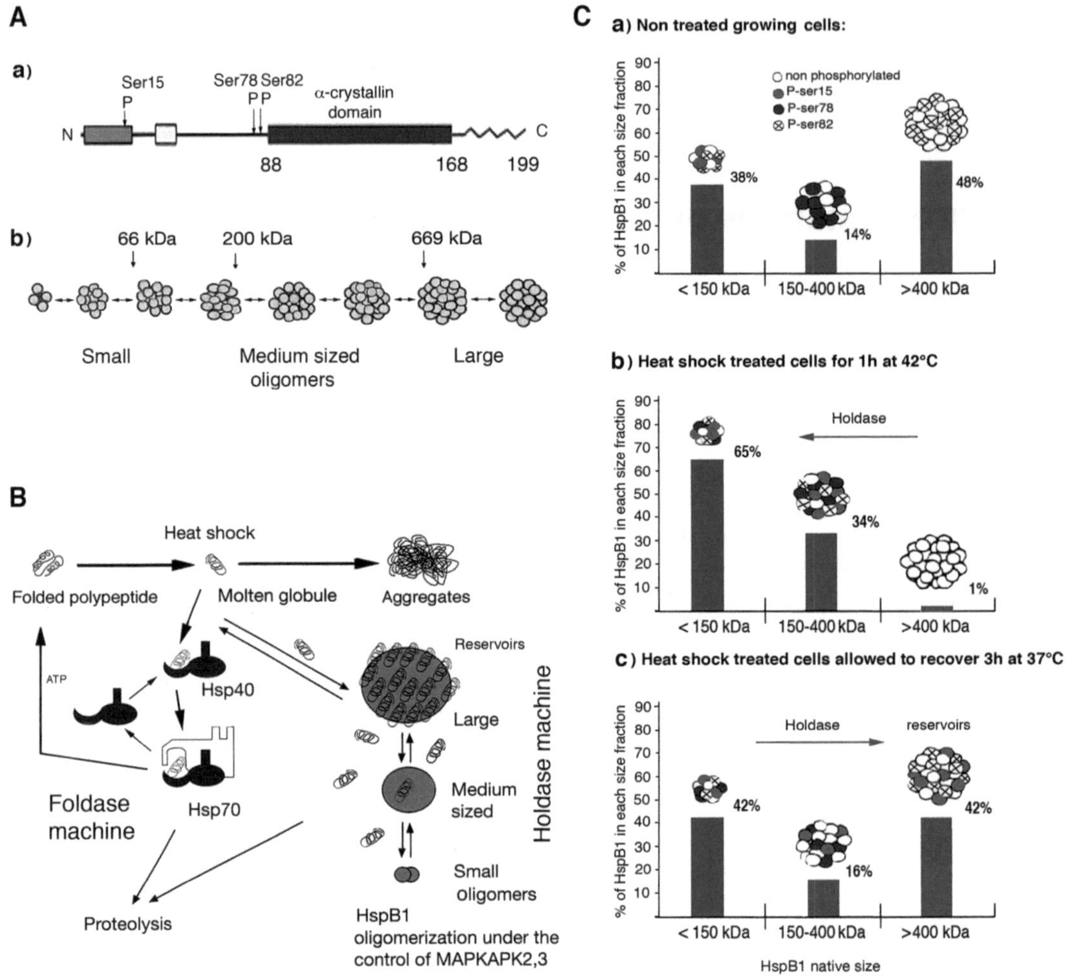

Fig. 1. (**A**) HspB1. (**a**) Organization of human HspB1 protein sequences. *Gray box*: N-terminal WD/EPF domain; *light box*: conserved region; *dark gray box*: alpha crystallin domain (aa 88–168); *wave-like line*: flexible C-terminal domain; P: phosphorylated serine residues. Amino acid number is indicated. (**b**) Oligomeric structure of HspB1 (from dimers to molecular masses larger than 700 kDa). Mosaic heterooligomers formed by different sHsps and/or their phosphorylated forms have been detected. (**B**) Exposure to stress alters the folding of polypeptides, which then accumulate as molten globules having the tendency to form deleterious aggregates. Molten globules are recognized by Hsp40 and are immediately refolded by ATP-dependent foldase chaperones (Hsp70), stored by HspB1 holdase chaperone to avoid their aggregation, or eliminated through proteolytic degradation. Rapid changes in HspB1 oligomerization and phosphorylation are observed during and after heat shock that probably reflect the capture and storage of the misfolded polypeptides. (**C**) Analysis of HspB1 oligomerization and phosphorylation in response to a heat stress of 1 h at 42°C. The % of HspB1 displaying native molecular masses smaller than 150 kDa in the 150–400 kDa range and larger than 400 kDa is presented. Note that heat shock rapidly shifts HspB1 toward small oligomers. The phenomenon is transient, and 3 h after the heat shock HspB1 reforms large oligomers. However, the pattern of phosphorylation is drastically different from that observed in nontreated cells. This reflects changes in HspB1 holdase activity to better cope with misfolded polypeptides.

particular amino acid sequence, an intriguing property of sHsps is their ability to oligomerize in a complex and heterogeneous way (1) (Fig. 1A). The phenomenon is controlled, at least in part, by stress-mediated phosphorylation-sensitive interactions in the N-terminal part of sHsps (5). Tetramers assembled from dimers appear as the building blocks of sHsp polydispersed oligomeric complexes (6). Oligomerization of sHsps is dynamic and, as shown in the case of HspB1, the phenomenon is linked to the physiology of the cells (1, 7–9). The rapid changes in the oligomerization status of sHsps can, therefore, be considered as a sensor of the physiological status of cells (see Fig. 1B). Several sHsps (HspB1, Hsp4, Hsp5, and HspB8 and Hsp16.2) are molecular chaperones that share an ATP-independent "holdase" activity. In stress conditions that alter protein folding (i.e., heat shock), the holdase activity favors interaction with misfolded polypeptides and subsequently their storage in a refolding-competent state (10–15). The phenomenon attenuates and/or suppresses irreversible protein aggregation that could be deleterious to the cell (see Fig. 1B). HspB1 holdase activity is modulated by the dynamic ability of this protein to change its oligomerization profile to trap denatured polypeptides (Fig. 1B). This activity differs from the ATP-dependent "foldase" chaperones (Hsp70, Hsp90, and Hsp60) that refold misfolded polypeptides (16–18), such as those trapped within large HspB1 large oligomeric structures (19–21). Trapped polypeptides can also end up being degraded, as for example in the case of irreversibly oxidized proteins.

Recently, intriguing observations, made both in stressed and nonstressed cells, point to the fact that sHsps are involved in many different cellular mechanisms suggesting that they have a large number of functions in the cell. The apparent pleiotropic activity of sHsps probably results from their holdase activity which allows them to bind and stabilize the folding of a large number of protein targets. Consequently, they stimulate their activity and/or stimulate their degradation. For example, caspase-3, which interacts with HspB1, shows a proteasome-dependent proteolytic degradation in HspB1 immunodepleted cells (22). Alteration in sHsp holdase activity can, therefore, deregulate cellular homeostasis by misfolding targeted proteins prior to their degradation. The phenomenon is similar to the already described "chaperone/client protein concept" mediated by Hsp90 (23, 24). Future studies will let us know whether this is a fundamental property of chaperones or if it is restricted to some heat-shock proteins. This property indirectly links sHsps to numerous unrelated cellular functions and may explain the large number of effects associated to the expression of these proteins described in the current literature. For example, sHspB1 indirectly modulates mRNA translation through interaction with the initiation translation factor eIF4G (25). It also stabilizes the cytoskeleton (26). In cells exposed to oxidative (27) or apoptotic (28) conditions as well as in differentiating cells (29), the protective effects of HspB1 can also originate from its interaction with several crucial "client" regulators, as for example caspase-3, cytochrome c, Daxx, GATA-1, ubiquitin, UBC-9, eIF4E, and Stat polypeptides. The holdase activity may also be responsible for

HspB1 and/or HspB5 direct or indirect effects toward Akt, Bax, Bcl-xs polypeptides, PKCa, Raf/MEK/ERK, and Akt-signaling pathways (30–32). Other examples are HspB8 which, by interacting with Bag-3, indirectly modulates autophagy (33) and the well-characterized reciprocal holding of αB-crystallin (HspB5) and αA-crystallin (HspB4) in the lens eye. In differentiating muscle cells, HspB1 could also modulate the switch from the keratin MK5/MK14 network to the keratin MK1/MK10 network (34).

It is not yet known whether the holdase activity of HspB1 toward different substrates depends or not on the classical formation of large oligomeric structures, such as those trapping the heat shock-induced misfolded substrates (Fig. 1C). The problem is rather complex since the relationship among oligomerization, phosphorylation, and holdase activity appears to be sHsp specific (35–37). Indeed, in response to different apoptotic inducers, the dynamic reorganization of the HspB1 oligomeric structure as well as its phosphorylation is inducer-specific phenomena. Moreover, an other important property of sHps is their ability to interact and form complex mosaic oligomeric structures (38) with specific ratios (i.e., HspB4 and HspB5 form a 3 to 1 unique large mosaic oligomer) (39, 40). In cells expressing several sHsps, complex and multiple combinatorial oligomeric structures are formed that may bear specific protein target recognition abilities. sHsp interactions can also exclude some sHsp partners and, therefore, the recognition of distinct molecular targets. Only few of these complex structures have been characterized yet. Hence, sHsp interaction networks, which are under the control of sHsp oligomerization and phosphorylation, are highly modulatable systems that can rapidly adapt to changes in cellular physiology.

In this chapter, we describe procedures for testing the oligomeric and phosphorylation patterns of HspB1 as well as its interaction with specific partner/client polypeptides using tissue culture cells genetically modified to express different levels of this protein. The procedures have been developed in my laboratory and could be used in any well-established cellular laboratory. In addition, the different procedures presented here could be extended to test the nine other human sHsp members as well as sHsps from other species.

2. Materials

All reagents and materials used in the culture of tissue culture cells are sterile.

1. PBS medium: 137 mM NaCl, 2.7 mM KCl, 8 mM Na_2HPO_4, and 1.5 mM KH_2PO_4, pH 7.4.
2. Cell culture medium (HeLa cells): DMEM medium (Life Technologies, #41966-052) containing 4.5 g/l glucose, 1 mM

sodium pyruvate, L-glutamine supplemented with 5,000 U/ml penicillin–streptomycin (Life Technologies, #15070-063), 250 μg/ml Fungizone (Life Technologies, #15290-026). 5–10% fetal calf serum (Life Technologies #10270-106) is added to cell culture medium. Fetal calf serum is stored at −70°C. The cell culture medium is stored at 4°C without serum and used within 5 weeks of supplementation. Medium containing serum is made fresh and used within 1 week of supplementation (storage at 4°C). Geneticin G418 (Gibco BRL, #11811-049) must be present in the medium to select stably transfected clones. The culture medium is filtered in sterile conditions on 0.2-μm filters.

3. Trypsination buffer: PBS containing 0.05% Trypsin (Gibco BRL, #35400-027). The buffer is stored in aliquots at −20°C.
4. Cell culture dishes (TPP, #93060) (Corning, #430167).
5. Inverted photomicroscope equipped with a phase-contrast equipment (i.e., TMS Nikon).
6. Trypan blue (Sigma-Aldrich, #T 0776).
7. Hemocytometer chamber.
8. TEM buffer (Tris–HCl, pH 7.4, 20 mM; NaCl 20 mM; $MgCl_2$ 5 mM; EDTA 0.1 mM).
9. IPP 150 buffer (Tris–HCl, PH 8; NaCl 150 mM; NP-40 0.05%).
10. IPP 300 buffer (Tris–HCl, PH 8; NaCl 300 mM; NP-40 0.05%).
11. Monoclonal anti-human HspB1 (Stressgen, #H00027129-M01) antibody as well as polyclonal antibodies that specifically recognize HspB1 phosphorylated at serine 15 (Stressgen #SPA-525), 78 (Stressgen #SPA-523), or 82 (Stressgen #SPA-524). Normal immunoglobulins (Santa Cruz Biotechnology, #sc-2025). Secondary goat anti-mouse or rabbit antibodies were provided by Santa Cruz Biotechnology (#sc-2031, sc-2030).
12. Recombinant human HspB1/27 (Stressgen, #SPP-715).
13. ECL kit from GE Healthcare (Amersham) (#RPN 2105).
14. Protein G-agarose (Sigma-Aldrich, #P7700).
15. Dithiothreitol (DTT) (Gibco, BRL, #15508-013).
16. Triton X100 (Sigma-Aldrich #T 9284).
17. X-Omat AR films (Eastman Kodak Co, #1651454).
18. Cytochalasin D (Sigma-Aldrich, #C8273).
19. MG-132 (Sigma-Aldrich, #C2211).
20. Pre-swollen Sepharose 6B (Sigma-Aldrich, #6B100).

21. Gel filtration column (1 × 100 cm) (GE Healthcare Pharmacia, #28-4064-15).
22. Fraction collector (Bio-Rad model 2110).
23. Peristaltic pump (Bio-Rad model EP-1 Econo Pump)
24. Kit for Molecular Weights 29,000–700,000 for Gel Filtration Chromatography (Sigma-Aldrich, #MWND500).
25. Ampholines, pH 3–10 and pH 5–7 (Bio-Rad, #163-1113 and 163-1153).
26. Lipofectamine™ reagent (Invitrogen, #18324-020).
27. pCIneohsp27 sense and antisense cDNA vectors (41) were constructed using an *Eco*RI–*Eco*RI DNA fragment of plasmid psvhsp27 (42) that was inserted, in normal or reverse orientation, in the *Eco*RI sites of pCIneo vector (Promega, #E1841). The vectors bear a neomycin gene and the entire coding sequence of the human hspB1 gene placed in normal or reverse orientation downstream of the CMV promoter.
28. Mammalian vector Bluescript pKS vector plain or bearing wild-type human Hsp27/B1 gene (cDNApKS27wt), non-phosphorylatable mutant (three serine sites replaced by alanine, pKS2711-3A), or phosphomimicry mutant (three serine sites replaced by aspartic acid, pKS2711-3D), were already described (8).
29. The pSuper RNAi system (Oligoengine, #VEC-pBS-0004; www.oligoengine.com) was used to direct intracellular synthesis of siRNA transcripts causing efficient and specific HspB1 gene silencing (43). The mismatch, scramble, and HspB1 targeting pSuperneo vectors were prepared as previously described (27).

3. Methods

3.1. Cell Culture

When cells are 80–100% confluent, proceed as follows:

1. Aspirate by pipetting the medium from the cell cultures.
2. Wash with PBS.
3. Add 1 ml of trypsin buffer to the culture dish. Let the dish stand for 2–3 min at 37°C and monitor under the microscope for the detachment of cells. Cell detachment can be accelerated by gently pipetting the buffer up and down.
4. Remove the trypsin buffer containing the detached cells from the dish and place it in a conical centrifuge tube. Rinse the dish with the complete culture medium and add it to the centrifuge tube. Serum is required to inhibit the action of trypsin.

5. Centrifuge the cell suspension at 1,500 × *g* for 3 min and resuspend the cell pellet in complete culture medium. Count cells and record for their viability using Trypan blue staining exclusion assay.

3.2. HspB1 Intracellular Distribution and Native Size

HspB1 is an oligomeric protein characterized by dynamic heterogeneous native sizes ranging between 50 and 700 kDa. In exponentially growing Hela cells, constitutively expressed HspB1 is recovered in two major oligomeric structures with native molecular masses comprised between 50 and 200 kDa (small oligomers, population I) and 200 and 700 kDa (large oligomers, population II) (see Fig. 1B). Heat shock induces a phosphorylation-dependent dynamic redistribution of HspB1 oligomeric structures characterized by the transient and rapid shift toward small oligomers. The phenomenon is transient and followed by the recovery of large HspB1 oligomeric population (population II). In cells recovering from heat shock, the large HspB1 structures have been described to bear the ATP-independent chaperone activity of the protein. The particular behavior observed in response to heat shock contrasts with that observed during oxidative stress, which is characterized by a transient and rapid formation of large oligomers followed by a phosphorylation-dependent disruption of these structures (44). Analysis of the oligomeric structures of HspB1 in cells exposed to apoptotic inducers has revealed complex and inducer-specific changes in HspB1 oligomerization that differed from those observed after heat shock and oxidative stress. These changes may reflect putative interaction of HspB1 with key apoptotic regulators. To detect the dynamic changes in HspB1 distribution upon cell fractionation as well as its native size, cells expressing high loads of HspB1 have to be used, as for example HeLa cells that constitutively express 0.4 ng of HsB1 per μg of total cellular proteins. A fraction of HspB1 associates, in an F-actin integrity-dependent way, with nonionic detergent sensitive structures, such as the plasma membrane cytoskeleton compartment, where actin nucleation occurs (45).

1. Expose cells, kept in 37°C incubator, to various stress inducers. Control nonstressed cells are kept at 37°C. Heat shock should be performed in a temperature-controlled water bath (±0.1°C) rather than in a cell culture incubator.

3.2.1. Cell Fractionation

1. Cells, washed in ice-cold PBS, pH 7.4, are lysed at 4°C in a Dounce homogenizer in TEM lysis buffer in the absence or presence of 0.1% Triton X-100.
2. Centrifuge the lysates at 10,000 × *g* for 10 min. Add SDS sample buffer (1× or 5×) to the resulting pellets and supernatants to obtain samples that have similar volumes. Boil immediately the samples, which can then be stored at –20°C.

3. Process for SDS-PAGE and immunoblot analysis using appropriate antibodies (see below Subheading 4). The presence of HspB1 in the pellet fraction and its absence when cells are lysed in the presence of Triton reflect its association with detergent-sensitive structures.
4. The F-actin integrity-dependent association can be tested by incubating cells for 1 h with 0.5 μM of the F-actin disrupting agent, cytochalasin D, before they are lysed.

3.2.2. Native Size Analysis

To detect changes in the native size of HspB1 in response to various stresses, proceed as follows:

1. Prepare a sepharose 6B gel filtration column (1 × 100 cm) equilibrated in TEM according to the manufactory procedures (void volume: 2,000 kDa). The column should be kept in the cold room. Always keep liquid at the top of the column to avoid it to dry. Changes in temperature ruin the column (bubble formation).
2. Set the peristaltic pump to obtain a column flow of about 2.5 ml/5 min. Wash the column with at least one-volume column of TEM. Set the fraction collector so that each fraction contains about 2.5 ml.
3. Calibrate the column using prestained molecular mass markers (Kit for Molecular Weights 29,000–700,000 for Gel Filtration Chromatography). The kit contains blue dextran (>2,000,000 Da), thyroglobulin (669,000 Da), apoferritin (440,000 Da), β-amylase (200,000 Da), and carbonic anhydrase (29,000 Da). To calibrate the column, remove the TEM solution at the top of the column and load 1 ml of the kit preparation solution. Once the 1-ml solution has penetrated inside the sepharose, the column is developed in TEM. Fractions of 2.5 ml are collected and the optical density is determined using a visible spectrophotometer. Note the fractions containing the different stained markers and their corresponding molecular masses.
4. 10^7 cells (in case of HeLa) are lysed in 1 ml of TEM containing 0.1% Triton X-100. Centrifugate the lysate in the cold at 10,000 × *g* for 10 min and collect the supernatant sample.
5. Remove the TEM solution at the top of the column. Load the 1-ml supernatant sample on the top of the preequilibrated sepharose 6B gel filtration column. Loading a larger volume of sample could alter the separating power of the column.
6. Once the 1 ml of sample has penetrated inside the sepharose, the column is developed in TEM. Fractions of 2.5 ml are collected and kept at 4°C.
7. 50 μl of each column fraction is collected and mixed with 10 μl of 5× SDS sample buffer and boil immediately. No protein

precipitation is required if 10^7 HeLa cells are analyzed per column run (see below Subheading 4). Gel electrophoresis and immunoblots using HspB1 antibody are performed as already described (44) and revealed with the ECL kit from Amersham (see Subheadings 2 and 4). The detection in the different column fractions of phosphorylated HspB1 (see Subheading 3.4, step 2) or other polypeptides of interest can easily be performed by probing the immunoblots with specific antibodies. Interaction of HspB1 with other sHsps and formation of mosaic oligomeric structures have been described (38). It is, therefore, important to test the column fraction with antibodies recognizing other sHsps.

8. Autoradiographs from the immunoblot experiments are recorded onto X-Omat LS films (Eastman Kodak). Films are scanned (4990 Epson film scanner) and analyzed with ImageJ software™ (NIH, Bethesda). The duration of the exposure should be calculated as to be in the linear response of the film.

3.3. Immunoprecipitation, Interaction with Partners, or Client Proteins

The detection, in the above-described column fractions, of polypeptides that share the oligomeric profile of HspB1 can reflect an interaction that can be tested by immunoprecipitation. In this respect, it is well-known that HspB1 can interact with other sHsps and form mosaic oligomeric structures (38). Moreover, partner polypeptides, like DAXX, interact with HspB1 small oligomers. In the case of caspase-3, its interaction with HspB1 enhances its stability probably through the holdase activity of HspB1 which maintains caspase-3 in stable form. In contrast, when the level of HspB1 is artificially decreased (see Subheading 3.5), caspase-3 is degraded through an ubiquitin–proteasome process. It is, therefore, important to test the column fractions described in Subheading 3.4 with antibodies recognizing other sHsps or other putatively interesting polypeptides.

1. Perform immunoblot analysis of the column fractions using antibodies against polypeptides that have the potential to interact with HspB1. Check for common oligomeric structures and define the column fractions that should be analyzed. Several adjacent fractions can be pooled. The fractions should always be kept at 4°C. Freezing may alter protein–protein interactions.
2. Collect a 100-μl aliquot of the pooled fractions and add 25 μl of 5× SDS sample buffer and immediately boil the sample (which can be denoted total nondepleted sample). Collect at least 1 ml of the pooled fractions in two different tubes. One tube receives 5 μl of nonimmune antibody (normal immunoglobulins) and the other 5 μl of immune anti-HspB1 monoclonal antibody (see Subheading 2). Let sit on ice for 3 h. Add 50 μl of a 1:1 slurry (made of water and protein G-agarose

beads) to both tubes. Attach the tubes to a rotating wheel in the cold room. After 3 h, collect the tubes and spin down the protein G-agarose beads ($500 \times g \times 5$ min). In both tubes (nonimmune and immune), collect the supernatant and the pellet. Collect a 100-μl aliquot of the supernatant and add 25 μl of 5× SDS sample buffer and immediately boil the sample (which can be denoted total depleted sample). Resuspend the agarose beads in 1 ml of IPP 150 buffer and repeat the washing of the beads for at least three times. Resuspend the sepharose beads in 50 μl of 1× SDS buffer and boil. Spin the beads and collect the sample buffer on top of the beads. Protein G-agarose beads can be recycled

3. Analyze the samples in SDS-PAGE and transfer gels to nitrocellulose membranes. Probe the immunoblot with anti-HspB1 antiserum and secondary antibody and reveal with ECL (see Subheading 2). Note the decreased signal of HspB1 in the total depleted sample compared to the original total nondepleted sample. If HspB1 is not detectable in the total depleted sample, this means that the immunoprecipitation was quantitative (the antibody was able to immunoprecipitate all HspB1 molecules present in the sample). Verify that HspB1 is indeed immunoprecipitated by anti-HspB1 antibody and not by the nonimmune immunoglobulins. If the detection of nonspecific protein bands is too high, repeat the experiment and include more stringent washes using IPP 300 buffer.

4. Probe a similar immunoblot blot (or the same) with an antibody directed against a polypeptide sharing HspB1 oligomeric property (both antibodies can be used simultaneously). In the total depleted fraction, check for a decrease in the level of the targeted polypeptide. Compare the percentage of decreased signal to that observed in the case of HspB1. A complete immunodepletion of the targeted polypeptide suggests that it forms a 1:1 complex with HspB1. Check for the presence of the polypeptide in the immunoprecipitated fraction.

5. Repeat the experiment, but in this case the antibody directed against the interacting polypeptide is used to perform the immunoprecipitation. The resulting immunoblots are then probed with anti-HspB1 antibody.

6. Protein client: Once the interaction of HspB1 with a polypeptide has been detected, repeat the experiments using cells transfected with a DNA vector encoding an RNAi targeting HspB1 (see below Subheading 3.6, step 2). A direct analysis of cell lysates (without the sizing column step) can be performed. Check for the reduced level of expression of HspB1. No decrease

is observed in control cells transfected with DNA vectors encoding nonfunctional RNAi (scramble or mismatch) (see below Subheading 3.6, step 2). A concomitant decrease in the level of the protein client means that it is protected against degradation by its interaction with HspB1 (holdase activity). The effect can be reversed if cells are pretreated with the proteasome inhibitor, MG-132 (0.1–0.5 μM for 24 h), before being analyzed.

3.4. Dimer Formation

Human HspB1, but not αB-crystallin (HspB5), contains a single cysteine (position 137 in HspB1) that is susceptible to oxidation (or even S-thiolation) and can promote dimer formation.

1. To detect the dimeric form of HspB1 in immunoblots, lysates of cells exposed to oxidative stress must be analyzed, as previously described (27, 46), under nonreducing conditions using SDS sample buffer devoid of DTT.
2. Analysis of cells transfected with HspB1 C137A dominant-negative mutant (see Subheading 3.7, step 2) can be performed to visualize the aberrant oligomeric structure formed by HspB1 when it is lacking its unique cystein residue (46).

3.5. Isoform Composition, Phosphorylation

3.5.1. Isoform Composition

HspB1 phosphorylation is MAPKAP Kinase 2/3-dependent and occurs on three serine sites (Ser 15, 78, and 82). At least three isoforms of HspB1 can be resolved in 2-D immunoblots. The acidic end is to the right. The "a" and more basic isoform represents the unphosphorylated form of the protein. The "b" and "c" isoforms are representative of Hsp27 phosphorylated at one or two sites, respectively. A minor isoform (b′) is often detectable between the "a and b" isoforms. Its nature is still unknown.

1. First dimension of the 2-D gels should be prepared using an ampholine mixture made of 20% ampholines, pH 3–10, and 80% ampholines, pH 5–7. The gel should be run as previously described (47).
2. Following the second SDS-PAGE dimension and transfer on nitrocellulose membrane, HspB1 isoforms are detected in immunoblots using antibodies that recognize HspB1 or its phosphorylated forms (see above Subheading 3.4, step 2).

3.5.2. Direct Analysis of HspB1 Level of Phosphorylation

Immunoblot analysis of HspB1 phosphorylation at the different serine sites is performed by probing immunoblots with antibodies specific for HspB1 phosphorylation at serine 15, serine 78, or serine 82 (see Subheading 2).

3.6. Modulation of HspB1 Level of Expression

Assessing the consequences of changes in HspB1 level of expression is an approach to increase our knowledge of the impact of this protein on cell physiology. This is also an easy way to detect client proteins stabilized by the holdase activity of HspB1 (see above Subheading 3.3, step 6). Indeed, a decreased level of HspB1 enhances the degradation of specific client polypeptides, such as caspase-3 (22).

1. Overexpression: Cells that do not (or weakly) constitutively express HspB1 are better used in this approach. Unfortunately, most human cells (except some lymphocytes, such as HL-60 cells) constitutively express HspB1. Consequently, murine NIH 3T3 and L929 fibroblasts can be used that are devoid of HspB1 expression in the absence of heat shock. Transfection of these cells can be achieved using Lipofectamine™ reagent procedure (Subheading 2) as already described (41).
2. Underexpression: Cells constitutively expressing high loads of HspB1, such as HeLa, T47D, 16-HBE, and COS, can be used to assess the consequences of a decreased level of HspB1. Transfection of cells can be achieved using the lipofectamine procedure (41, 48) using pCIneohsp27 antisense cDNA vector (see Subheading 2). In HeLa cells, this procedure decreases the HspB1 level by at least 40% (41). A limitation of this approach is that HspB1 antisense RNA could potentially inhibit the expression of other sequence-related members of the sHsp family.

   To overcome the problem, an RNAi approach that specifically targets the mRNA encoding HspB1 can be used. In separate culture dishes, cells are transfected with 10 μg of HspB1-, Scramble-, or Mismatch-pSuperneo DNA vectors using the Lipofectamine™ reagents (see Subheading 2) according to manufacturer's instructions. Two to three days after transfection, perform immunoblot analysis to determine the efficiency of the siRNA-induced silencing. In HeLa cells, the level of HspB1 should be reduced by at least 80%. Stable cell lines for siRNA expression can be established using neomycin selection.

3.7. Expression of HspB1 Dominant-Negative Mutants Devoid of Holdase Activity

1. Phosphorylation mutants of HspB1: Effects of phosphorylation on HspB1 function is tested following cell transfection with previously described pKS27wt (wild-type Hsp27), pKS2711-3A (nonphosphorylatable mutant: three serine sites replaced by alanine), or pKS2711-3D (phosphomimicry mutant: three serine sites replaced by aspartic acid) vector (8) (see Subheading 2). Expression of the Ser to Asp mutant shifts HspB1 toward small oligomeric structures and abolishes HspB1 holdase and protective activity (8, 37).
2. The C137A dominant-negative mutant (see Subheading 3.3, step 6) knocks out the holdase activity of HspB1. The complex formed by the mutant protein and endogenous wild-type HspB1 is an aberrant and nonfunctional oligomeric structure (46).

4. Notes

1. To avoid the time-consuming step of protein precipitation of the fractions eluting from the sizing column, the starting material should be at least 10^7 cells expressing about 4 ng of HspB1 per µg of total proteins (as for example HeLa cells).
2. In immunoblots, the level of HspB1 present in the different cell lines can be estimated by comparing the signals to that of serial dilutions of the purified recombinant protein. Autoradiography must be performed within the range of proportionality of the film.

References

1. Arrigo A-P, Landry J. (1994) Expression and Function of the Low-molecular-weight Heat Shock Proteins. In: Morimoto RI, Tissieres A, Georgopoulos C (eds.) *The Biology of Heat Shock Proteins and Molecular Chaperones*. Cold Spring Harbor Laboratory Press, Cold Spring Harbor, NY, pp. 335–373.
2. Kappe G, Franck E, Verschuure P, Boelens WC, Leunissen JA, de Jong WW. (2003) The human genome encodes 10 alpha-crystallin-related small heat shock proteins: HspB1-10. *Cell Stress Chaperones*. **8**, 53–61.
3. Ingolia TD, Craig EA. (1982) Four small heat shock proteins are related to each other and to mammalian a-crystallin. *Proc. Natl. Acad. Sci. USA*. **79**, 2360–2364.
4. de Jong WW, Caspers GJ, Leunissen JA. (1998) Genealogy of the alpha-crystallin – small heat-shock protein superfamily. *Int. J. Biol. Macromol.* **22**, 151–162.
5. Lambert H, Charette SJ, Bernier AF, Guimond A, Landry J. (1999) HSP27 multimerization mediated by phosphorylation-sensitive intermolecular interactions at the amino terminus. *J. Biol. Chem.* **274**, 9378–9385.
6. Ehrnsperger M, Lilie H, Gaestel M, Buchner J. (1999) The dynamics of hsp25 quaternary structure. Structure and function of different oligomeric species. *J. Biol. Chem.* **274**, 14867–14874.
7. Arrigo A-P, Suhan JP, Welch WJ. (1988) Dynamic changes in the structure and intracellular locale of the mammalian low-molecular-weight heat shock protein. *Mol. Cell. Biol.* **8:** 5059–5071.
8. Mehlen P, Hickey E, Weber L, Arrigo A-P. (1997) Large unphosphorylated aggregates as the active form of hsp27 which controls intracellular reactive oxygen species and glutathione levels and generates a protection against TNFα in NIH-3T3-ras cells. *Biochem. Biophys. Res. Comm.* **241**, 187–192.
9. Garrido C. (2002) Size matters: of the small HSP27 and its large oligomers. *Cell Death Differ.* **9**, 483–485.
10. Horwitz J, Huang Q-L, Ding L-L. (1992) Alpha-crystallin can function as a molecular chaperone. *Proc. Natl. Acad. Sci. USA*. **89**, 10449–10453.
11. Jakob U, Gaestel M, Engels K, Buchner J. (1993) Small heat shock proteins are molecular chaperones. *J. Biol. Chem.* **268**, 1517–1520.
12. Ganea E. (2001) Chaperone-like activity of alpha-crystallin and other small heat shock proteins. *Curr. Protein Pept. Sci.* **2**, 205–225.
13. Carra S, Sivilotti M, Chavez Zobel AT, Lambert H, Landry J. (2005) HspB8, a small heat shock protein mutated in human neuromuscular disorders, has in vivo chaperone activity in cultured cells. *Hum. Mol. Genet.* **14**, 1659–1669.
14. Bellyei S, Szigeti A, Pozsgai E, Boronkai A, Gomori E, Hocsak E, Farkas R, Sumegi B, Gallyas F. (2007) Preventing apoptotic cell death by a novel small heat shock protein. *Eur. J. Cell. Biol.* **86**, 161–171.
15. Markossian KA, Yudin IK, Kurganov BI. (2009) Mechanism of Suppression of Protein Aggregation by alpha-Crystallin. *Int. J. Mol. Sci.* **10**, 1314–1345.
16. Freeman BC, Morimoto RI. (1996) The human cytosolic molecular chaperones hsp90, hsp70 (hsc70) and hdj-1 have distinct roles in recognition

of a non-native protein and protein refolding. *Embo. J.* **15**, 2969–2979.
17. Bukau B, Horwich AL. (1998) The Hsp70 and Hsp60 chaperone machines. *Cell.* **92**, 351–366.
18. Buchner J. (1999) Hsp90 & Co. - a holding for folding. *Trends Biochem. Sci.* **24,** 136–141.
19. Lee GJ, Roseman AM, Saibil HR, Vierling E. (1997) A small heat shock protein stably binds heat-denatured model substrates and can maintain a substrate in a folding-competent state. *EMBO. J.* **16**, 659–671.
20. Ehrnsperger M, Graber S, Gaestel M, Buchner J. (1997) Binding of non-native protein to Hsp25 during heat shock creates a reservoir of folding intermediates for reactivation. *EMBO. J.* **16**, 221–229.
21. Lee GJ, Vierling E. (2000) A small heat shock protein cooperates with heat shock protein 70 systems to reactivate a heat-denatured protein. *Plant Physiol.* **122**, 189–198.
22. Pandey P, Farber R, Nakazawa A, Nakazawa A, Kumar S, Bharti A, Nalin C, Weichselbaum R, Kufe D, Kharbanda S. (2000) Hsp27 functions as a negative regulator of cytochrome c-dependent activation of procaspase-3. *Oncogene.* **19**, 1975–1981.
23. Neckers L, Mimnaugh E, Schulte TW. (1999) Hsp90 as an anti-cancer target. *Drug Resist. Updat.* **2**, 165–172.
24. Georgakis GV, Younes A. (2005) Heat-shock protein 90 inhibitors in cancer therapy: 17AAG and beyond. *Future Oncol.* **1**, 273–281.
25. Cuesta R, Laroia G, Schneider RJ. (2000) Chaperone Hsp27 inhibits translation during heat shock by binding eIF4G and facilitating dissociation of cap-initiation complexes. *Genes Dev.* **14**, 1460–1470.
26. Nicholl ID, Quinlan RA. (1994) Chaperone Activity of alpha-Crystallins Modulates Intermediate Filament Assembly. *EMBO. J.* **13**, 945–953.
27. Arrigo AP, Firdaus WJ, Mellier G, Moulin, M., Paul, C. Diaz-Latoud, C., and C. Kretz-Remy. (2005) Cytotoxic effects induced by oxidative stress in cultured mammalian cells and protection provided by Hsp27 expression. *Methods.* **35**, 126–138.
28. Bruey JM, Ducasse C, Bonniaud P, Ravagnan, L., Susin, S.A., Diaz-Latoud, C., Arrigo, A.-P., Kroemer, G., Solary, E., Garrido, C. (2000) Hsp27 negatively regulates cell death by interacting with cytochrome c. *Nat. Cell. Biol.* **2**, 645–652.
29. Arrigo AP. (2005) In search of the molecular mechanism by which small stress proteins counteract apoptosis during cellular differentiation. *J. Cell. Biochem.* **94**, 241–246.
30. Mao YW, Liu JP, Xiang H, Li DW. (2004) Human alphaA- and alphaB-crystallins bind to Bax and Bcl-X(S) to sequester their translocation during staurosporine-induced apoptosis. *Cell Death Differ.* **11**, 512–526.
31. Liu JP, Schlosser R, Ma WY, Dong Z, Feng H, Liu L, Huang XQ, Liu Y, Li DW. (2004) Human alphaA- and alphaB-crystallins prevent UVA-induced apoptosis through regulation of PKCalpha, RAF/MEK/ERK and AKT signaling pathways. *Exp. Eye Res.* **79**, 393–403.
32. Rane MJ, Pan Y, Singh S, Poell D, Wu R, Cummins T, Chen Q, McLeish KR, Klein JB. (2003) Heat shock protein 27 controls apoptosis by regulating akt activation. *J. Biol. Chem.* **278**, 27828–27835.
33. Carra S. (2009) The stress-inducible HspB8-Bag3 complex induces the eIF2alpha kinase pathway: implications for protein quality control and viral factory degradation? *Autophagy.* **5**, 428–429.
34. Duverger O, Paslaru L, Morange M. (2004) HSP25 is involved in two steps of the differentiation of PAM212 keratinocytes. *J. Biol. Chem.* **279**, 10252–10260.
35. Koteiche HA, McHaourab HS. (2003) Mechanism of chaperone function in small heat-shock proteins. Phosphorylation-induced activation of two-mode binding in alphaB-crystallin. *J. Biol. Chem.* **278**, 10361–10367.
36. Aquilina JA, Benesch JL, Ding LL, Yaron O, Horwitz J, Robinson CV. (2004) Phosphorylation of alphaB-crystallin alters chaperone function through loss of dimeric substructure. *J. Biol. Chem.* **279**, 28675–28680.
37. Rogalla, T., Ehrnsperger, M., Preville, X., Kotlyarov, A., Lutsch, G., Ducasse, C., Paul, C., Wieske, M., Arrigo, A.P., Buchner, J., and Gaestel, M. (1999) Regulation of Hsp27 oligomerization, chaperone function, and protective activity against oxidative stress/tumor necrosis factor alpha by phosphorylation. *J. Biol. Chem.* **274**, 18947–18956.
38. Zantema A, Vries MV-D, Maasdam D, Bol S, Van der Eb A. (1992) Heat shock protein 27 and aB-cristallin can form a complex, which dissociates by heat shock. *J. Biol. Chem.* **267**, 12936–12941.
39. Sun TX, Liang JJ. (1998) Intermolecular exchange and stabilization of recombinant human alphaA- and alphaB-crystallin. *J. Biol. Chem.* **273**, 286–290.
40. Saha S, Das KP. (2004) Relationship between chaperone activity and oligomeric size of

recombinant human alphaA- and alphaB-crystallin: a tryptic digestion study. *Proteins.* **57**, 610–617.

41. Paul C, Manero F, Gonin S, Kretz-Remy C, Virot S, Arrigo A-P. (2002) Hsp27 as a negative regulator of cytochrome C release. *Mol. Cell. Biol.* **22**, 816–834.
42. Mehlen P, Préville X, Chareyron P, Briolay J, Klemenz R, Arrigo A-P. (1995) Constitutive expression of human hsp27, Drosophila hsp27, or human alpha B-crystallin confers resistance to TNF- and oxidative stress-induced cytotoxicity in stably transfected murine L929 fibroblasts. *J. Immunol.* **154**, 363–374.
43. Brummelkamp TR, Bernards R, Agami R. (2002) A system for stable expression of short interfering RNAs in mammalian cells. *Science.* **296**, 550–553.
44. Mehlen P, Mehlen A, Guillet D, Préville X, Arrigo A.-P. (1995) Tumor necrosis factor-α induces changes in the phosphorylation, cellular localization, and oligomerization of human hsp27, a stress protein that confers cellular resistance to this cytokine. *J. Cell. Biochem.* **58**, 248–259.
45. Pichon S, Bryckaert M, Berrou E. (2004) Control of actin dynamics by p38 MAP kinase – Hsp27 distribution in the lamellipodium of smooth muscle cells. *J. Cell. Sci.* **117**, 2569–2577.
46. Diaz-Latoud C, Buache E, Javouhey E, Arrigo AP. (2005) Substitution of the unique cysteine residue of murine hsp25 interferes with the protective activity of this stress protein through inhibition of dimer formation. *Antioxid. Redox. Signal.* 7, 436–445.
47. Arrigo A-P, Welch W. (1987) Characterization and purification of the small 28,000-dalton mammalian heat shock protein. *J. Biol. Chem.* **262**, 15359–15369.
48. Merendino, A., Paul, C., Costa, M A., Melis, M., Chiappara, G., Izzo, V., Vignola, AM., Arrigo, A-P. (2002) Heat shock protein-27 protects human bronchial epithelial cells against oxidative stress-mediated apoptosis: possible implication in asthma. *Cell. Stress Chaperones.* 7, 269–280.

Chapter 10

Combined Lentiviral and RNAi Technologies for the Delivery and Permanent Silencing of the *hsp25* Gene

Punit Kaur, Ganachari M. Nagaraja, and Alexzander Asea

Abstract

Elevated heat shock protein 27 (Hsp27) expression has been found in a number of tumors, including breast, prostate, gastric, uterine, ovarian, head and neck, and tumor arising from the nervous system and urinary system, and determined to be a predictor of poor clinical outcome. Although the mechanism of action of Hsp27 has been well documented, there are currently no available inhibitors of Hsp27 in clinical trials. RNA interference (RNAi) has the potential to offer more specificity and flexibility than traditional drugs to silence gene expression. Not surprisingly, RNAi has become a major focus for biotechnology and pharmaceutical companies, which are now in the early stages of developing RNAi therapeutics, mostly based on short interfering RNA (siRNAs), to target viral infection, cancer, hypercholesterolemia, cardiovascular disease, macular degeneration, and neurodegenerative diseases. However, the critical issues associated with RNAi as a therapeutic are delivery, specificity, and stability of the RNAi reagents. To date, the delivery is currently considered the biggest hurdle, as the introduction of siRNAs systemically into body fluids can result in their degradation, off-target effects, and immune detection. In this chapter, we discuss a method of combined lentiviral and RNAi-based technology for the delivery and permanent silencing of the *hsp25* gene.

Key words: Gene silencing, Hsp25, Heat shock protein, Lentivirus, miRNA, siRNA, RNAi

1. Introduction

Heat shock proteins (HSP) are a group of physiologically essential, highly conserved proteins that are induced by heat shock as well as by other environmental and pathophysiological stresses (1). HSP play an essential role in protein folding, assembly, and transport and are considered to act as molecular chaperones. There is evidence that these molecules play a role in the regulation of cell growth and differentiation (2). In 1983, Ciocca and colleagues

Stuart K. Calderwood and Thomas L. Prince (eds.), *Molecular Chaperones: Methods and Protocols*,
Methods in Molecular Biology, vol. 787, DOI 10.1007/978-1-61779-295-3_10, © Springer Science+Business Media, LLC 2011

described the distribution of an estrogen-induced protein with a molecular weight of 24,000 in normal and malignant human tissues and cells (3), which later turned out to be Hsp27 (4, 5). Adams and McGuire later successfully measured Hsp27 in breast tumors using an enzyme-linked immunosorbent assay (ELISA) technique and concluded that breast tumors expressed elevated levels of Hsp27 (6). On the basis of this study, several research groups confirmed the overexpression of Hsp27 in breast cancer (3, 7, 8) and later also confirmed in a variety of cancers (9–12), which led to the overexpression of Hsp25 in cancers in turn promotes tumor cell survival. Cancer cells have distinct features as compared to normal cells due to number of specific genetic changes that cause their abnormal growth. As cancer develops, at least two cell processes are disrupted: cell growth is promoted and cell death is inhibited. Cancer cells have special features such as the ability to secrete unique proteins on the cell surface, and researchers have taken advantage of this feature to target these proteins (13, 14). The 4T1 cell line is a poorly immunogenic but highly metastatic breast adenocarcinoma cell line (15), which exhibits high baseline expression of Hsp25 (16).

With the advancement of molecular biology techniques, RNA interference (RNAi) has emerged as a novel pathway for the study of loss of function of vast number of genes in various model systems (17). The emergence of RNAi research has led to a better understanding of cancer biology. Fire and Mello first introduced this concept in 1998 and laid the foundation for this process by which cell systems have the inherent ability to alter gene expression (18) and then it was rapidly applied to wide range of organisms (19, 20). RNAi is a posttranscriptional gene silencing mechanism that can be triggered by small RNA molecules such as microRNA (miRNA) and short interfering RNA (siRNA). RNAi complexes are endogenously produced within cells to target host RNA sequences that result in gene silencing. In addition to regulation of normal biological processes in humans, miRNAs seem to serve an important role in human cancers. Recent studies show that components of RNAi pathways may serve as valuable diagnostic, prognostic, and therapeutic tools in cancer research (21–23). Briefly, long dsRNA or hairpin RNA is processed by the endonuclease Dicer into smaller RNA fragments. These siRNA are separated into single strands and then loaded into the multicomponent RNA-Induced Silencing Complex (RISC). The siRNA within the RISC complex serves as a template to guide the selective cleavage of the complementary mRNA. The cleaved mRNA, which has two unprotected ends, is highly susceptible to degradation by nucleases. A similar process takes place with microRNA (miRNA) that, like siRNA, is the product of dsRNA that has been cleaved by Dicer. A key difference is that microRNAs are typically endogenously produced and regulate the translation of mRNAs to which they have partial complementarity; by contrast, siRNAs, either directly introduced into the cell or generated from long dsRNA as products of Dicer, are fully complementary to the target mRNA and bring about complete mRNA degradation.

RNAi technology has demonstrated exceptional utility in understanding of gene function and biological process (24, 25). RNAi has the potential to offer more specificity and flexibility than traditional drugs to silence gene expression because any protein that causes or contributes to a disease is susceptible to RNAi previous disease targets considered "undruggable" is now accessible. Not surprisingly, RNAi has become a major focus for biotechnology and pharmaceutical companies, which are now in the early stages of developing RNAi therapeutics, mostly based on siRNAs, to target viral infection, cancer, hypercholesterolemia, cardiovascular disease, macular degeneration, and neurodegenerative diseases (26). However, the critical issues associated with RNAi as a therapeutic are delivery, specificity, and stability of the RNAi reagents. Till today, the delivery is currently considered the biggest hurdle, as the introduction of siRNAs systemically into body fluids can result in their degradation, off-target effects, and immune detection and subsequent reactions (27). Thus, many efforts are focused on developing ways to modify RNAi or attach them to delivery agents that will protect them until they reach their therapeutic destinations including delivery as particles or complexes using lipid nanoparticles that encapsulate the siRNA or combining siRNA molecules with peptide-based polymers. Additional advances in therapeutic applications are likely to come from chemical modifications or other approaches to improve the specificity and potency of RNAi reagents.

The lentiviral delivery system is one the powerful gene delivery technologies available today (28). Combining RNAi and lentiviral approach to silence the target gene is the most widely used approach for gene silencing and delivery into cells and organs (29–31). Lentiviral vector is derived from HIV and is one of the efficient gene delivery vehicles for gene therapy. Lentiviral vectors can target a wide range of targets (32, 33), primary cells (31), nondividing cells, neurons, and other cell types (34, 35). The preclinical studies showed that lentivirus that could deliver genes into multiple cell types may have broad utility in the treatment of multiple diseases (36), including genetic disorders and cancer (37).

We report that high expression of Hsp25 functions as a repressor of normal proteasome function, which in turn results in poor antigen presentation, inefficient $CD8^+$ T cell stimulation, and resultant tumor growth (38). Silencing Hsp25 using lentivirus-RNAi enhanced PA28α expression and proteasome activity, which induced the regression of established tumors by tumor-specific $CD8^+$ T lymphocytes (38). The adoptive transfer of reactive $CD8^+$ T lymphocytes from mice bearing Hsp25 silenced tumors efficiently induces the regression of established tumors in nontreated mice which normally succumb to tumor burden (38). In this chapter, we describe the methods of production using combined lentiviral and RNAi approach to permanently silence the Hsp25 gene in mouse highly metastatic breast carcinoma cells, which would be a step ahead in the understanding of the mechanism of RNAi as a therapeutic approach in the treatment of tumors expressing high levels of Hsp27.

2. Materials

2.1. Buffers and Regents

1. Hsp25shRNA/controlshRNA.
2. Enhanced green fluorescent protein (EGFP).
3. Topoisomerase I inhibitor hydroxycamptothecin (HCPT).
4. Dithiothreitol (DTT).
5. T4 DNA ligase buffer.
6. OptiMEM medium.
7. RNase.
8. KCl.
9. NH_4Cl.
10. Glycerol.
11. ATP.
12. $MgCl_2$.
13. *Mlu*I and *Cla*I enzymes.
14. SOC medium.
15. LB medium.
16. Ampicillin.
17. Polyethyleneglycol (PEG).

2.2. Cells and Vectors

1. Sf9 insect cells.
2. TOP10 *Escherichia coli* competent cells.
3. Max Stb12 competent cells.
4. pLVTHM plasmid.
5. psPAX2 and pMD2.G plasmid.
6. Herpes Simplex Virus-1 (HSV-1).
7. Simian virus 40 (SV40).
8. M13 virion.
9. Baculovirus expression system (BVS).
10. Lentivirus.

3. Methods

3.1. siRNA Delivery Systems

There are still many factors that must be considered when designing and using siRNA, such as target sequence, sequence length, overhangs, and secondary structure. siRNA effector molecules exist in several forms; siRNA can be delivered as single-stranded oligomers, double-stranded duplexes, RNA hairpins, or other forms (39). Short siRNAs induce interferon (IFN) responses through both sequence-dependent and sequence-independent mechanisms (40). TLR7 and 8 recognize specific sequences of both ssRNA and dsRNA, which ultimately leads to the secretion of a number of cytokines and IFN subtypes. However, not all cell types are activated to the same extent. RNA is highly susceptible to degradation; therefore, it is important to introduce chemical modifications by replacing phosphodiester linkages with phosphorothioates or boranophosphates to the siRNA structure to confer heightened stability to the siRNA.

3.1.1. Viral Vectors

Pseudovirions have the advantage of being virally derived and so often have efficient delivery with defined uptake mechanisms. Pseudovirions are generally easy to modify, and there are two general types of pseudoviral systems: (1) Transpackaging systems in which viral machinery and proteins have been repurposed to package genetic material other than the original viral genome including phagemid particles, Herpes Simplex Virus-1 (HSV-1) amplicons, and Simian virus 40 (SV40) in vitro packaged vectors. These vectors are distinguished from recombinant viral vectors in that they contain no coding viral sequences and at most carry the noncoding sequences necessary for packaging. (2) Viral envelope systems, in which native enveloped virus particles are inactivated and used to package other cargos.

Phagemid Particles

Phagemid particles are an attractive delivery vehicle because, unlike animal viruses, bacteriophages exhibit no natural tropism toward mammalian cells and, by virtue of their ligand display properties, have the potential to be developed as highly specific, targeted vectors. The ability to generate targeted ligands quickly through phage-display technology and directed evolution is well known and, while not always dramatic, results showing that phage-derived vectors can be targeted toward mammalian cells are promising (41). One of the minor coat proteins, pIII, makes up part of the cap on the end of the M13 virion and is the protein that has been most successfully modified to present recombinant proteins in phage-display applications. Uptake of phage in mammalian cells is achieved via receptor-mediated endocytosis through the use of targeted ligands displayed on the phage surface (42). Unlike conventional plasmid vectors, phagemids contain elements that cause them to be packaged into phage capsids as ssDNA when the phagemid-transformed bacteria are superinfected by helper phage, resulting in phagemid particles (41). Helper phages are bacteriophages whose genomes have been modified such that they provide all the proteins necessary to generate new phages, but are themselves very inefficiently packaged in the presence of phagemid vectors that carry optimal packaging signals (phage origin of replication). Jiang et al. used their phagemid encoding siRNA against the expression of enhanced green fluorescent protein (EGFP) to estimate the transduction efficiency of their system. Cells were transfected with EGFP and then 12 h later transduced with 10^{11} plaque-forming units/ml of phagemid particles. At 24 h after transduction, topoisomerase I inhibitor hydroxycamptothecin (HCPT) was added to a final concentration of 2.5 μM. When assayed 48 h after transduction, the percentage of EGFP-positive cells in H1299 cells treated with siEGFP phagemid particles and 2.5 μM HCPT was reduced to approximately 50% of the control cells that had been treated with phagemid particles carrying an empty pSilencer vector (43).

Herpes Simplex Virus Type-1

Herpes Simplex Virus-1 (HSV-1) is an enveloped virus whose attachment is mediated first by the interactions of envelope glycoproteins gB and gC with heparin sulfate, followed by binding of envelope glycoprotein gD with heparin sulfate, herpes virus entry mediator (HVEM), or nectins, which leads to fusion of the viral envelope with the cell membrane (44). The genome encodes numerous immediate-early, early, and late gene products including ICP0, ICP27, ICP4, and ICP22, which are involved in the regulation of viral gene expression (44). Amplicon plasmids contain two key elements, namely, the viral origin of replication (ori) and the DNA cleavage/packaging signal (pac), and comprise the base of the system. HSV amplicons are helper-dependent vectors. When helper functions are supplied in *trans* by one of the various helper systems, the amplicon plasmid is replicated and amplified as ~150-kb head-to-tail concatamers of linear DNA, which allows large fragments of foreign DNA, and is packaged into particles (45).

Simian Virus 40

Simian virus 40 (SV40) is a nonenveloped polyomavirus with a 5.2-kb circular ddDNA. The wild-type SV40 genome encodes two early genes: the large T antigen (T-ag), which is essential for viral genome replication and expression of the late genes, and the small t antigen (t-ag). The expression of T-ag and t-ag are driven by the SV40 early promoter. The late promoter, located on the opposite strand from the early promoter, is responsible for the transcription of the three capsid genes: VP1, VP2, and VP3. Human cells support viral replication in a nonefficient manner and, therefore, are categorized as "semipermissive," and other species, such as mice, are categorized as "nonpermissive." Maintenance of the wild-type virus within the host genome can occur either by integration or by the formation of a self-replicating mini-chromosome, a process that requires the expression of T-ag. Entry of the wild-type virus into cells differs from most other viral infection pathways in that it largely avoids a cytoplasmic phase and differs from enveloped viruses because of its inability to fuse with the cell membrane. Viral entry begins with the virus binding to its primary receptor, the major histocompatibility complex class I (MHC class I) (46). The virus then binds to GM1 gangliosides and travels with them to the endoplasmic reticulum (ER). The transport to the ER is mediated by caveolae and facilitated by ER chaperones. Following disassembly in the ER, nuclear pore complexes assist in delivery of the viral DNA to the nucleus for replication. The first step in the in vitro preparation of the SV40 particles is the production of VP1 in Sf9 insect cells using a baculovirus expression system. Sf9 nuclear extract containing VP1 is mixed in a tube with the nucleic acid cargo (either plasmid DNA or siRNA) around which VP1 assembles into pseudoviral particles in the presence of ATP and $MgCl_2$. After assembly, the pseudoviral particles are stabilized by the addition of $CaCl_2$. No encapsulation sequences are required, so the process does not incorporate any of the wild-type SV40 genetic material. Mukherjee and colleagues developed a slightly different methodology for encapsulation using the same VP1-containing nuclear extracts from Sf9 cells treated with a reducing agent dithiothreitol (DTT) and RNase. During the assembly step, their method calls for KCl, NH_4Cl, glycerol, and a pH of 7.9 in addition to ATP and $MgCl_2$ (47).

Lentivirus

Lentiviruses can infect nondividing cells, allowing stable gene transfer in postmitotic cells such as mature neurons. These vectors provide a unique tool to integrate siRNA expression constructs with the aim to locally knockdown expression of a specific gene, enabling to assess the function of a gene in a very specific neuronal pathway. There has been limited success in gene delivery with such viruses because for the gene to enter into their target cells, the cells must be dividing so that nuclear membranes are broken down, for the gene to enter and integrate into host chromosome (48). The members of the subfamily lentivirus derived from human immunodeficiency virus (HIV) would have same capability to transfer genes into the genomes of the cells, but could perform this with nondividing, dormant cells in vivo and growth-arrested cells in vitro (49). Lentiviral vectors are a type of retroviruses that can infect both dividing and nondividing target cells because they can enter through the intact membrane of the nucleus of the target cell. HIV is a very potential lentivirus vector because it has capability to infect and express its genes in human helper T cells and other macrophages. The only cells Lentiviruses cannot infect are quiescent cells (in the G_0 state) because their state blocks the reverse transcription step (49). Lentiviruses are a single kind of viruses that are diploid, have two strands of RNA that are approximately 10 kb long. The lentivirus vector ends are flanked with long terminal repeats (LTR). A psi-sequence present near the 5¢ end is necessary for packaging viral RNA into virus capsids to continue the infection of HIV in its host. Lentiviral vectors are replication-defective and are composed of hybrid viral particles and generated by the core proteins, enzymes of a lentivirus, and the envelope of a different virus. Mostly, the envelope derives from vesicular stomatitis virus (VSV) or the envelope from MLV (35). Pseudotyping the lentivirus with VSV envelope glycoprotein enhances the virus to target a broad range of tissues and cells and increases its therapeutic applications. Lentivirus genome is highly complex. Lentiviruses have common structural genes (gag, pol, and env) and also have essential regulatory elements (tat and rev) and other accessory virus genes required for modulation of lentiviral gene expression, and viral assembly. Lentiviruses replication depends mainly on *cis*-acting viral elements, which do not encode any proteins. Normally, viral *cis*-acting elements are needed for lentivirus functioning and become a part of lentiviral gene transfer construct (some portion of lentivirus that integrates in the host chromosome and carries the genes of interest). The lentiviral *trans*-acting elements encode three important proteins for lentivirus functioning: structural, regulatory, and accessory. HIV-derived lentiviral backbone vectors are replication defective, the initial steps of lentiviral life cycle, namely, attachment, entry, reverse transcription, nuclear transport, and integration, are functioning in lentiviruses. The early steps of lentivirus life cycle are not required for viral protein synthesis, and mostly, *trans*-acting elements must be removed from the lentiviral backbone that encodes only the gene of interest.

The method of combined lentiviral and RNAi-based technology for the delivery and permanent silencing of the *hsp25* gene is as follows (50). Briefly, design siRNA by using Web-based siRNA designer tools (Invitrogen) and convert siRNA into shRNA by using BD Web program. Digest the HIV-derived plasmid pLVTHM with *Mlu*I and *Cla*I and ligate to an oligonucleotide pair containing Hsp25 shRNA or control shRNA carrying restriction overhangs and transform into Max Stbl2 competent cells. The positive clones should be identified by digesting the control pLVTHM vector and the vector containing Hsp25 shRNA inserts. Double-digest the pLVTHM plasmid in a 20-μl final volume containing 5 μg pLVTHM vector, 2 μl of *Mlu*I and *Cla*I enzymes (NEB) and incubate the reaction mixture at 37°C for 4 h. After gel electrophoresis, excise the digested pLVTHM fragment using scalpel blade and transfer into a clean eppendorf tube and proceed for gel purification (Qiagen purification kit) and elute in 50 μl. For the shRNA oligo preparation, take 10 μl of each forward and reverse shRNA oligo primers (500 pmol/μl) and incubate at 95°C for 10 min, then slowly allow to cool down the annealed oligos to room temperature. After cooling, the oligos can be stored at −20°C. For ligation, prepare the ligation reaction containing 1 μl plVTHM digested and purified vector (80 ng), 2.5 μl annealed shRNA oligo (100 ng), 1 μl T4DNA ligase buffer (NEB), 1 μl T4DNA ligase enzyme (NEB) and incubate at 16°C for 16 h and uncut pLVTHM as a positive control. For transformation, add Stbl2-competent cells (Invitrogen) to 2 μl ligation mixtures, mix gently by tapping the tube, and incubate on ice for 30 min. Heat-shock at 42°C for 25 s and immediately place the tube on ice for 5 min. Add 250 μl room-temperature SOC medium (Invitrogen). Mix gently and incubate the tube at 37°C for 90 min at 225 rpm shaking speed. In the meantime, prepare the Luria–Bertani (LB) agar plates containing ampicillin (100 μg/ml). Plate the 200 μl mixture on LB–ampicillin plate and incubate at 37°C for overnight. Next day, inoculate multiple single positive colonies in 5 ml LB medium containing ampicillin and incubate at 37°C overnight in an incubator with 260 rpm shaking speed. Isolate plasmid DNA (Qiagen) according to the instructions. In order to identify Hsp25 shRNA/control shRNA inserts, digest the pLVTHM and pLVTHM plasmid containing Hsp25 shRNA/control shRNA clones with *Xba*I enzyme and then with *Mlu*I enzyme. After restriction digestion, pLVTHM plasmid vector should give a 280-bp fragment, and HSp25 shRNA/control shRNA containing pLVTHM clones should give a 348-bp fragment on agarose gel. Positive clones will also be identified by sequencing the pLVTHM plasmid containing Hsp25 shRNA/control shRNA inserts. For the lentivirus production and transduction, lentivirus transfection was carried out according to the standard protocol. Briefly, 4T1 cells were plated into six-well plates (3×10^4 cells/well) and 1 ml concentrated high-titer virus (5×10^8) was directly overlaid on the cells and polybrene was added at a final concentration of 8 μg/ml. Five days later, the cells were harvested and analyzed by fluorescence activated cell sorting (FACS).

For recombinant packaging and enveloping lentiviral plasmid preparation, transform psPAX2 and pMD2.G plasmids into TOP10 *E. coli* cells (Invitrogen) and grow in LB medium in presence of ampicillin. Similarly, grow Hsp25 shRNA/control shRNA positive clones in LB medium and isolate plasmids by using EndoFree maxiprep plasmid DNA isolation kit (Qiagen) to get high-quality endotoxin-free plasmid DNA. Preparation of lentiviral vectors describes the reagents required for 13 transfections (38). Seed 2.5×10^6 293FT cells (Invitrogen) in a 60-mm tissue culture dish containing 293FT complete medium without antibiotic (penstrep) and incubate at 37°C for overnight. Replace the medium with OptiMEM transfection medium (Invitrogen) 2 h before transfection. Prepare the plasmid DNA by mixing 8 μg pLVTHM vector containing Hsp25 shRNA/control shRNA, 6 μg core packaging plasmid (psPAX2), and 2 μg envelope plasmid. Make up the plasmid solution to a final volume of 6.5 ml with OptiMEM medium in a 15-ml polypropylene tube and incubate at room temperature for 10 min, then add 0.5 ml transfection solution dropwise over the cells and incubate at 37°C for 6 h. After 6 h transfection, remove the transfection medium and add 293FT complete growth medium containing antibiotic and incubate at 37°C. Collect 24 h virus supernatant and add fresh 293FT complete growth medium. Centrifuge the virus supernatant at 200 × *g* for 5 min at 5°C, pass through HVDF filter (Millipore), and store at 4°C. Similarly, collect 48 and 72 h virus supernatant and add fresh 293FT complete growth medium. Pool the 24, 48, and 72 h virus supernatants and concentrate by centrifuging at 1,000 × *g* for 2 h at 4°C. The virus titer was checked as per the manufacturer's instructions. Remove 4T1 growth medium and overlay 1 ml virus concentrate and add 1 μl polybrene (8 mg/ml), which enhances the transduction. Incubate the cells at 37°C for 24–48 h and then check the GFP fluorescence.

3.1.2. Nonviral Vectors

Nonviral vectors, such as liposomes, are usually nonimmunogenic and are easy to prepare and modify quickly. However, they often suffer from poor transfection efficiency. Therapeutic applications of siRNA require the development of carriers that will protect siRNA from degradation during circulation (51), delivering siRNA to the target cells, avoiding delivery to nontarget cell types, facilitating cellular uptake and endosomal escape, and releasing siRNA intracellularly so that it will be accessible to the cellular machinery (52). This can be achieved by introducing a hydrated steric barrier to surround the carrier using materials such as polyethyleneglycol (PEG) (53). When a carrier is injected into a peripheral vein, it enters the right side of the heart and is pumped out to the lungs; the lungs contain the first capillary beds and act as an initial mechanical filtration barrier (52). If small enough, the carriers leave the lungs and enter the left side of the heart and are pumped into the systemic circulation. Small interfering RNA is negatively charged and typically cannot cross the cell membrane by free diffusion. A number of approaches have been developed to facilitate siRNA uptake, including conjugating siRNA to a ligand, such as a cell-penetrating peptide or small molecule, to facilitate siRNA uptake, endocytosis of siRNA encapsulated within nanoparticles, fusion of the carrier with the cell membrane, thereby releasing the carriers' content into the cytoplasm. One study using siRNA lipoplexes generated from the commercially available cationic lipid DharmaFECT reported that 95% of the lipoplexes enter cells through endocytosis (54); 50% of endocytosis was clathrin-mediated (54). In the same study, about 20% of the remaining material delivered to the cytoplasm was internalized via lipid-raft/caveolin-mediated endocytosis (54). As siRNA carriers typically enter cells via endocytosis, a means of endosomal escape is necessary (54). Although the precise mechanisms of endosomal escape by siRNA delivery systems remain unclear, one hypothesis is that certain materials can facilitate endosomal escape via "the proton sponge effect" (55). The uptake of ions creates an osmotic imbalance; water enters the endosome to counter this effect, causing the endosome to inflate until it ruptures. Rupture of the endosome releases its contents to the cytoplasm (55). Another strategy to improve nucleic-acid delivery from lipid-based systems can induce phase or structural transformations that can promote unpackaging of siRNA from the complex used also to trigger drug release in tumors (56).

Naked siRNA

To increase the cellular permeability of naked siRNA, manual hydrodynamic injection (57) has been used. Generally, these methods are not ideal for in vivo applications, and in comparison, systemic delivery of siRNA nanoparticles, which has the ability to target and penetrate into desired cells and maintain siRNA integrity, is more favorable.

Neutral Liposomes

Neutral liposomes are unilammelar liposomes with hydrophilic cores and hydrophobic surfaces that effectively shield encapsulated siRNA (58). Recently, researchers have sandwiched siRNA between a cationic core composed of 1,2-dioleoyl-3-trimethylammoniumpropane (DOTAP) and an outer lipid bilayer of PEG-1,2-distearoyl-sn-glycero-3-phosphoethanolamine (PEG-DSPE) and egg phosphatidylcholine (egg-PC) and have showed that it remained in circulation even 20 h post injection (59). Stable nucleic-acid lipid particles (SNALPs) represented a major advancement in lipid-based siRNA delivery after injection of clinically relevant doses of ApoB-siRNA produced knockdown in the liver of nonhuman primates (27).

Cationic Lipids

The proposed mechanism of formation of lipoplexes is that negatively charged nucleic acids bind to positively charged lipid vesicles, which causes formation of a multilamellar structure of positively charged lipid bilayers (60) 3.7 nm thick, spaced 2 nm apart from each other by negatively charged nucleic acids. These liposomes can be downsized to small unilamellar vesicles (SUVs) <100 nm in diameter. Cholesterol as a component of certain DNA/RNA carriers has been reported to improve transfection in vivo, increase structural rigidity and stability of the carrier, and protect nucleic acids from extraliposomal degradative entities such as RNases (61). Cholesterol may play a dual role in the delivery of siRNA by facilitating endosomal internalization of the carrier and acting as a targeting entity. Increasing the charge ratio had little effect on the total knockdown, whereas it did, however, increase nonspecific knockdown. Interestingly, they found that multivalent lipids exhibited lower toxicity, higher total knockdown, and lower nonspecific knockdown in comparison with a similar charge ratio carried by several univalent cationic lipids (60). Although tumor uptake was not affected by particle charge, imaging analysis revealed that increasing positive charge on the particles did increase accumulation of the carriers in tumor vasculature.

Anionic Lipids

The negatively charged nucleic acid is complexed with an amine-based polypeptide poly-L-lysine), generating a particle with a net positive charge. The particles are then treated with negatively charged lipids to protect nucleic acids from deactivation by polyanions present in tumor ascitic fluid, such as hyaluronic acid (HA) (62). Avnir et al., using 80-nm PEGylated liposomes to treat an autoimmune arthritis model, showed that the higher levels of liposome accumulation in an arthritic joint than those found in the liver, kidney, or spleen due to increased accumulation are explained by the high permeability of vasculature surrounding the inflamed tissue (53). Worm-like micelles from degradable copolymers several microns long persisted in rodent circulation for up to 1 week after IV injection, ten times longer than spherical counterparticles, and were internalized by A549 human lung cancer cells (63).

Nanoparticles

Hyaluronic acid (HA) nanoparticles, also known as nanogels, with encapsulated siRNA were able to target HCT-116 cells in vitro, which overexpress HA-specific CD44 receptors. Depending on the engineered nanogel characteristics, the siRNA release rate was also regulated. siRNA has also been incorporated into poly-(D, L-lactic-co-glycolic) acid (PLGA) nanoparticles, and it produced significant levels of gene knockdown in vivo while causing less irritation and inflammation when compared to siRNA lipoplexes (64). Aggregation of siRNA nanoparticles typically occurs due to the surface charge of siRNA loaded nanoparticles (typically net positive), which is circumvented by PEG (65) or sugar molecules (cyclodextrin) (66). Targeting has been introduced to siRNA containing nanoparticles through the modification of the nanoparticles with ligands or antibodies recognizing cell surface receptors or antigen integrins, respectively. RGD peptides have been used to specifically target vascular endothelial growth factor (VEGF) overexpressing human umbilical vein endothelial cells (HUVECs) (67). Both cell penetrating peptides (CPPs) and cholesterol conjugation have been shown to enhance the efficiency of siRNA gene silencing in vivo (68). siRNA nanoparticles enter more than 90% of the cells through endocytosis; however, the release of the nanoparticles and/or siRNA to the cytosol occurs less readily. Following internalization of a nanoparticle, an early endosome is formed, which is accompanied by acidification of the vesicular lumen that continues into the late endosomal and lysosomal compartments, reaching a final pH of approximately 4.5 in the perinuclear lysosome. pH-responsive polymers include polymers that contain protonable amines that disrupt the endosomal membrane through the hypothesized proton sponge effect (69).

4. Notes

1. RNAi-based methods are providing unprecedented tools useful to address fundamental questions in the biology of living organisms by enhancing and/or replacing more classical genetic approaches and manipulations.
2. Since most organisms possess the cellular machinery for RNAi, this near universal approach makes loss-of-function studies approachable in organisms in which genetic tools do not exist.
3. With the growing appreciation for the fundamental potential of RNAi and a burgeoning collection of RNAi technologies and reagents, the diversity in in vivo applications to biology, medicine, and agriculture is seemingly limitless. RNAi applications have become one of the most powerful tools for biomedical research today and may be the novel application needed for advancing cancer therapy.
4. Recently, we have reported that RNAi machinery expression has a direct impact on survival in mice with metastatic breast cancer. In this study, we demonstrated that silencing the expression of the *hsp25* gene abrogates tumor growth, suppresses metastatic potential and enhances specific anti-tumor immune responses of a highly metastatic and poorly immunogenic murine breast cancer cell line (38).

Acknowledgments

This work was supported in part by Research Advancement Awards from Scott & White Memorial Hospital and Clinic (G.M.N. and P.K.), Institutional support from Scott & White Memorial Hospital and Clinic, Texas A&M Health Science Center College of Medicine, the Central Texas Veterans Health Administration, an Endowment from the Cain Foundation, and the US National Institutes of Health grant RO1CA91889 (A.A.).

References

1. Lindquist, S. (1986) The heat-shock response *Annu Rev Biochem* **55**, 1151–91.
2. Hightower, L. E. (1991) Heat shock, stress proteins, chaperones, and proteotoxicity *Cell* **66**, 191–7.
3. Ciocca, D. R., Adams, D. J., Edwards, D. P., Bjercke, R. J., and McGuire, W. L. (1983) Distribution of an estrogen-induced protein with a molecular weight of 24,000 in normal and malignant human tissues and cells *Cancer Res* **43**, 1204–10.
4. Ciocca, D. R., Adams, D. J., Edwards, D. P., Bjercke, R. J., and McGuire, W. L. (1984) Estrogen-induced 24K protein in MCF-7 breast cancer cells is localized in granules *Breast Cancer Res Treat* **4**, 261–8.
5. Ciocca, D. R., Asch, R. H., Adams, D. J., and McGuire, W. L. (1983) Evidence for modulation

of a 24K protein in human endometrium during the menstrual cycle *J Clin Endocrinol Metab* **57**, 496–9.

6. Adams, D. J., and McGuire, W. L. (1985) Quantitative enzyme-linked immunosorbent assay for the estrogen-regulated Mr 24,000 protein in human breast tumors: correlation with estrogen and progesterone receptors *Cancer Res* **45**, 2445–9.
7. Horne, G. M., Angus, B., Wright, C., Needham, G., Nicholson, S., Harris, A. L., Innes, B., and Horne, C. H. (1988) Relationships between oestrogen receptor, epidermal growth factor receptor, ER-D5, and P24 oestrogen regulated protein in human breast cancer *J Pathol* **155**, 143–50.
8. Ciocca, D. R., Stati, A. O., and Amprino de Castro, M. M. (1990) Colocalization of estrogen and progesterone receptors with an estrogen-regulated heat shock protein in paraffin sections of human breast and endometrial cancer tissue *Breast Cancer Res Treat* **16**, 243–51.
9. Ciocca, D. R., Puy, L. A., Edwards, D. P., Adams, D. J., and McGuire, W. L. (1985) The presence of an estrogen-regulated protein detected by monoclonal antibody in abnormal human endometrium *J Clin Endocrinol Metab* **60**, 137–43.
10. Langdon, S. P., Rabiasz, G. J., Hirst, G. L., King, R. J., Hawkins, R. A., Smyth, J. F., and Miller, W. R. (1995) Expression of the heat shock protein HSP27 in human ovarian cancer *Clin Cancer Res* **1**, 1603–9.
11. Ciocca, D. R., Jorge, A. D., Jorge, O., Milutin, C., Hosokawa, R., Diaz Lestren, M., Muzzio, E., Schulkin, S., and Schirbu, R. (1991) Estrogen receptors, progesterone receptors and heat-shock 27-kD protein in liver biopsy specimens from patients with hepatitis B virus infection *Hepatology* **13**, 838–44.
12. Malusecka, E., Zborek, A., Krzyzowska-Gruca, S., and Krawczyk, Z. (2001) Expression of heat shock proteins HSP70 and HSP27 in primary non-small cell lung carcinomas. An immunohistochemical study *Anticancer Res* **21**, 1015–21.
13. Raymond, E., Faivre, S., and Armand, J. P. (2000) Epidermal growth factor receptor tyrosine kinase as a target for anticancer therapy *Drugs* **60 Suppl 1**, 15–23; discussion 41–2.
14. Slamon, D. J., Leyland-Jones, B., Shak, S., Fuchs, H., Paton, V., Bajamonde, A., Fleming, T., Eiermann, W., Wolter, J., Pegram, M., Baselga, J., and Norton, L. (2001) Use of chemotherapy plus a monoclonal antibody against HER2 for metastatic breast cancer that overexpresses HER2 *N Engl J Med* **344**, 783–92.
15. Aslakson, C. J., and Miller, F. R. (1992) Selective events in the metastatic process defined by analysis of the sequential dissemination of subpopulations of a mouse mammary tumor *Cancer Res* **52**, 1399–405.
16. Bausero, M. A., Bharti, A., Page, D. T., Perez, K. D., Eng, J. W., Ordonez, S. L., Asea, E. E., Jantschitsch, C., Kindas-Muegge, I., Ciocca, D., and Asea, A. (2006) Silencing the hsp25 gene eliminates migration capability of the highly metastatic murine 4T1 breast adenocarcinoma cell *Tumour Biol* **27**, 17–26.
17. Elbashir, S. M., Harborth, J., Lendeckel, W., Yalcin, A., Weber, K., and Tuschl, T. (2001) Duplexes of 21-nucleotide RNAs mediate RNA interference in cultured mammalian cells *Nature* **411**, 494–8.
18. Fire, A., Xu, S., Montgomery, M. K., Kostas, S. A., Driver, S. E., and Mello, C. C. (1998) Potent and specific genetic interference by double-stranded RNA in Caenorhabditis elegans *Nature* **391**, 806–11.
19. Caplen, N. J., Parrish, S., Imani, F., Fire, A., and Morgan, R. A. (2001) Specific inhibition of gene expression by small double-stranded RNAs in invertebrate and vertebrate systems *Proc Natl Acad Sci USA* **98**, 9742–7.
20. Kuttenkeuler, D., and Boutros, M. (2004) Genome-wide RNAi as a route to gene function in Drosophila *Brief Funct Genomic Proteomic* **3**, 168–76.
21. Croce, C. M. (2009) Causes and consequences of microRNA dysregulation in cancer *Nat Rev Genet* **10**, 704–14.
22. Negrini, M., Nicoloso, M. S., and Calin, G. A. (2009) MicroRNAs and cancer--new paradigms in molecular oncology *Curr Opin Cell Biol* **21**, 470–9.
23. Shah, P. P., Hutchinson, L. E., and Kakar, S. S. (2009) Emerging role of microRNAs in diagnosis and treatment of various diseases including ovarian cancer *J Ovarian Res* **2**, 11.
24. Carpenter, A. E., and Sabatini, D. M. (2004) Systematic genome-wide screens of gene function *Nat Rev Genet* **5**, 11–22.
25. Chi, J. T., Chang, H. Y., Wang, N. N., Chang, D. S., Dunphy, N., and Brown, P. O. (2003) Genomewide view of gene silencing by small interfering RNAs *Proc Natl Acad Sci USA* **100**, 6343–6.
26. Shah, R. G., Ghodgaonkar, M. M., Affar el, B., and Shah, G. M. (2005) DNA vector-based RNAi approach for stable depletion of poly(ADP-ribose) polymerase-1 *Biochem Biophys Res Commun* **331**, 167–74.
27. Zimmermann, T. S., Lee, A. C., Akinc, A., Bramlage, B., Bumcrot, D., Fedoruk, M. N., Harborth, J., Heyes, J. A., Jeffs, L. B., John,

M., Judge, A. D., Lam, K., McClintock, K., Nechev, L. V., Palmer, L. R., Racie, T., Rohl, I., Seiffert, S., Shanmugam, S., Sood, V., Soutschek, J., Toudjarska, I., Wheat, A. J., Yaworski, E., Zedalis, W., Koteliansky, V., Manoharan, M., Vornlocher, H. P., and MacLachlan, I. (2006) RNAi-mediated gene silencing in non-human primates *Nature* **441**, 111–4.

28. Wiznerowicz, M., and Trono, D. (2005) Harnessing HIV for therapy, basic research and biotechnology *Trends Biotechnol* **23**, 42–7.
29. Zwierska, I., Walker, R. D., Choksy, S. A., Male, J. S., Pockley, A. G., and Saxton, J. M. (2006) Relative tolerance to upper- and lower-limb aerobic exercise in patients with peripheral arterial disease *European Journal of Vascular and Endovascular Surgery* **31**, 157–63.
30. Wiznerowicz, M., and Trono, D. (2003) Conditional suppression of cellular genes: lentivirus vector-mediated drug-inducible RNA interference *J Virol* 77, 8957–61.
31. Rubinson, D. A., Dillon, C. P., Kwiatkowski, A. V., Sievers, C., Yang, L., Kopinja, J., Rooney, D. L., Ihrig, M. M., McManus, M. T., Gertler, F. B., Scott, M. L., and Van Parijs, L. (2003) A lentivirus-based system to functionally silence genes in primary mammalian cells, stem cells and transgenic mice by RNA interference *Nat Genet* **33**, 401–6.
32. Abbas-Terki, T., Blanco-Bose, W., Deglon, N., Pralong, W., and Aebischer, P. (2002) Lentiviral-mediated RNA interference *Hum Gene Ther* **13**, 2197–201.
33. Gupta, S., Schoer, R. A., Egan, J. E., Hannon, G. J., and Mittal, V. (2004) Inducible, reversible, and stable RNA interference in mammalian cells *Proc Natl Acad Sci USA* **101**, 1927–32.
34. Lo Bianco, C., Ridet, J. L., Schneider, B. L., Deglon, N., and Aebischer, P. (2002) alpha -Synucleinopathy and selective dopaminergic neuron loss in a rat lentiviral-based model of Parkinson's disease *Proc Natl Acad Sci USA* **99**, 10813–8.
35. Naldini, L., Blomer, U., Gallay, P., Ory, D., Mulligan, R., Gage, F. H., Verma, I. M., and Trono, D. (1996) In vivo gene delivery and stable transduction of nondividing cells by a lentiviral vector *Science* **272**, 263–7.
36. Lu, X., Yu, Q., Binder, G. K., Chen, Z., Slepushkina, T., Rossi, J., and Dropulic, B. (2004) Antisense-mediated inhibition of human immunodeficiency virus (HIV) replication by use of an HIV type 1-based vector results in severely attenuated mutants incapable of developing resistance *J Virol* **78**, 7079–88.
37. Li, L., Lin, X., Staver, M., Shoemaker, A., Semizarov, D., Fesik, S. W., and Shen, Y. (2005) Evaluating hypoxia-inducible factor-1-alpha as a cancer therapeutic target via inducible RNA interference in vivo *Cancer Res* **65**, 7249–58.
38. Nagaraja, G. M., Kaur, P., Zheng, H., Bausero, M. A., Multhoff, G., and Asea, A. (2011) Silencing *hsp25* gene expression augments proteasome activity and increases CD8+ T cell-mediated memory and tumor killing. *Cancer Prevention Res (in press).*
39. Amarzguioui, M., Rossi, J. J., and Kim, D. (2005) Approaches for chemically synthesized siRNA and vector-mediated RNAi *FEBS Lett* **579**, 5974–81.
40. Judge, A., and MacLachlan, I. (2008) Overcoming the innate immune response to small interfering RNA *Hum Gene Ther* **19**, 111–24.
41. Kehoe, J. W., and Kay, B. K. (2005) Filamentous phage display in the new millennium *Chem Rev* **105**, 4056–72.
42. Monaci, P., Urbanelli, L., and Fontana, L. (2001) Phage as gene delivery vectors *Curr Opin Mol Ther* **3**, 159–69.
43. van Nimwegen, M. J., and van de Water, B. (2007) Focal adhesion kinase: a potential target in cancer therapy *Biochem Pharmacol* **73**, 597–609.
44. Lachmann, R. (2004) Herpes simplex virus-based vectors *Int J Exp Pathol* **85**, 177–90.
45. Wade-Martins, R., Smith, E. R., Tyminski, E., Chiocca, E. A., and Saeki, Y. (2001) An infectious transfer and expression system for genomic DNA loci in human and mouse cells *Nat Biotechnol* **19**, 1067–70.
46. Pelkmans, L., Kartenbeck, J., and Helenius, A. (2001) Caveolar endocytosis of simian virus 40 reveals a new two-step vesicular-transport pathway to the ER *Nat Cell Biol* **3**, 473–83.
47. Mukherjee, S., Abd-El-Latif, M., Bronstein, M., Ben-nun-Shaul, O., Kler, S., and Oppenheim, A. (2007) High cooperativity of the SV40 major capsid protein VP1 in virus assembly *PLoS One* **2**, e765.
48. Sikorski, R., and Peters, R. (1998) Gene therapy. Treating with HIV *Science* **282**, 1438.
49. Amado, R. G., and Chen, I. S. (1999) Lentiviral vectors – the promise of gene therapy within reach? *Science* **285**, 674–6.
50. Sinn, P. L., Sauter, S. L., and McCray, P. B., Jr. (2005) Gene therapy progress and prospects: development of improved lentiviral and retroviral vectors – design, biosafety, and production *Gene Ther* **12**, 1089–98.
51. Larson, S. D., Jackson, L. N., Chen, L. A., Rychahou, P. G., and Evers, B. M. (2007) Effectiveness of siRNA uptake in target tissues by various delivery methods *Surgery* **142**, 262–9.

52. Wolff, J. A., and Rozema, D. B. (2008) Breaking the bonds: non-viral vectors become chemically dynamic *Mol Ther* **16**, 8–15.
53. Avnir, Y., Ulmansky, R., Wasserman, V., Even-Chen, S., Broyer, M., Barenholz, Y., and Naparstek, Y. (2008) Amphipathic weak acid glucocorticoid prodrugs remote-loaded into sterically stabilized nanoliposomes evaluated in arthritic rats and in a Beagle dog: a novel approach to treating autoimmune arthritis *Arthritis Rheum* **58**, 119–29.
54. Lu, J. J., Langer, R., and Chen, J. (2009) A novel mechanism is involved in cationic lipid-mediated functional siRNA delivery *Mol Pharm* **6**, 763–71.
55. Akinc, A., Thomas, M., Klibanov, A. M., and Langer, R. (2005) Exploring polyethylenimine-mediated DNA transfection and the proton sponge hypothesis *J Gene Med* **7**, 657–63.
56. Simoes, S., Moreira, J. N., Fonseca, C., Duzgunes, N., and de Lima, M. C. (2004) On the formulation of pH-sensitive liposomes with long circulation times *Adv Drug Deliv Rev* **56**, 947–65.
57. Bradley, S. P., Rastellini, C., da Costa, M. A., Kowalik, T. F., Bloomenthal, A. B., Brown, M., Cicalese, L., Basadonna, G. P., and Uknis, M. E. (2005) Gene silencing in the endocrine pancreas mediated by short-interfering RNA *Pancreas* **31**, 373–9.
58. Peer, D., Park, E. J., Morishita, Y., Carman, C. V., and Shimaoka, M. (2008) Systemic leukocyte-directed siRNA delivery revealing cyclin D1 as an anti-inflammatory target *Science* **319**, 627–30.
59. Yagi, N., Manabe, I., Tottori, T., Ishihara, A., Ogata, F., Kim, J. H., Nishimura, S., Fujiu, K., Oishi, Y., Itaka, K., Kato, Y., Yamauchi, M., and Nagai, R. (2009) A nanoparticle system specifically designed to deliver short interfering RNA inhibits tumor growth in vivo *Cancer Res* **69**, 6531–8.
60. Bouxsein, N. F., McAllister, C. S., Ewert, K. K., Samuel, C. E., and Safinya, C. R. (2007) Structure and gene silencing activities of monovalent and pentavalent cationic lipid vectors complexed with siRNA *Biochemistry* **46**, 4785–92.
61. Liu, Y., Liggitt, D., Zhong, W., Tu, G., Gaensler, K., and Debs, R. (1995) Cationic liposome-mediated intravenous gene delivery *J Biol Chem* **270**, 24864–70.
62. Lee, R. J., and Huang, L. (1996) Folate-targeted, anionic liposome-entrapped polylysine-condensed DNA for tumor cell-specific gene transfer *J Biol Chem* **271**, 8481–7.
63. Srinivas, G., Discher, D. E., and Klein, M. L. (2005) Key roles for chain flexibility in block copolymer membranes that contain pores or make tubes *Nano Lett* **5**, 2343–9.
64. Woodrow, K. A., Cu, Y., Booth, C. J., Saucier-Sawyer, J. K., Wood, M. J., and Saltzman, W. M. (2009) Intravaginal gene silencing using biodegradable polymer nanoparticles densely loaded with small-interfering RNA *Nat Mater* **8**, 526–33.
65. Rozema, D. B., Lewis, D. L., Wakefield, D. H., Wong, S. C., Klein, J. J., Roesch, P. L., Bertin, S. L., Reppen, T. W., Chu, Q., Blokhin, A. V., Hagstrom, J. E., and Wolff, J. A. (2007) Dynamic PolyConjugates for targeted in vivo delivery of siRNA to hepatocytes *Proc Natl Acad Sci USA* **104**, 12982–7.
66. Heidel, J. D., Yu, Z., Liu, J. Y., Rele, S. M., Liang, Y., Zeidan, R. K., Kornbrust, D. J., and Davis, M. E. (2007) Administration in non-human primates of escalating intravenous doses of targeted nanoparticles containing ribonucleotide reductase subunit M2 siRNA *Proc Natl Acad Sci USA* **104**, 5715–21.
67. Pirollo, K. F., and Chang, E. H. (2008) Targeted delivery of small interfering RNA: approaching effective cancer therapies *Cancer Res* **68**, 1247–50.
68. Soutschek, J., Akinc, A., Bramlage, B., Charisse, K., Constien, R., Donoghue, M., Elbashir, S., Geick, A., Hadwiger, P., Harborth, J., John, M., Kesavan, V., Lavine, G., Pandey, R. K., Racie, T., Rajeev, K. G., Rohl, I., Toudjarska, I., Wang, G., Wuschko, S., Bumcrot, D., Koteliansky, V., Limmer, S., Manoharan, M., and Vornlocher, H. P. (2004) Therapeutic silencing of an endogenous gene by systemic administration of modified siRNAs *Nature* **432**, 173–8.
69. Reischl, D., and Zimmer, A. (2009) Drug delivery of siRNA therapeutics: potentials and limits of nanosystems *Nanomedicine* **5**, 8–20.

Chapter 11

Quantification of HSP27 and HSP70 Molecular Chaperone Activities

Jessica Gobbo, Caroline Gaucher-Di-Stasio, Stéphanie Weidmann, Jean Guzzo, and Carmen Garrido

Abstract

Stress-inducible heat-shock proteins (HSPs, like HSP70 and HSP27) are molecular chaperones that protect cells from stress damage by keeping cellular proteins in a folding competent state and preventing them from irreversible aggregation. HSP27 and HSP70 chaperone activities are useful indicators to test chemical products and physical stress impact on protein denaturation, to select HSP inhibitors, or to determine the implication of the chaperone function in other HSP activities, such as apoptosis. We have developed two simple and fast chaperone activity tests for HSP27 and HSP70 that we initially set up to test the effect of potential HSP inhibitors obtained after screening of chemical and small molecule libraries. These chaperone quantification tests are based on the capacity of HSP to counteract chemical or thermal protein aggregation.

Key words: HSP27, HSP70, Chaperone activity, Chemical and thermal aggregation, Screening

1. Introduction

Molecular chaperones were first defined in 1978 by Laskey (1) as proteins that can recognize misfolded or denaturated proteins. Heat-shock proteins (HSPs) are powerful chaperones and the expression of some of them is induced in response to a wide range of physiological and environmental stresses. HSP accumulation is necessary for cell survival after thermal, chemical, or physical insults. Hence, it is not surprising that molecular chaperones are implicated in the pathogenesis of many relevant diseases and could be regarded as potential pharmacological targets. HSP27 and HSP70 are the most stress-inducible HSPs. Inducible HSP70 (also called HSP72 or HSPA1) is an ATP-dependent molecular chaperone and

Stuart K. Calderwood and Thomas L. Prince (eds.), *Molecular Chaperones: Methods and Protocols*,
Methods in Molecular Biology, vol. 787, DOI 10.1007/978-1-61779-295-3_11, © Springer Science+Business Media, LLC 2011

is very low expressed under normal conditions. Like other members of the HSP70 family such as HSC70, HSP70 assists the folding of newly synthesized polypeptides, the assembly of multiprotein complexes, and the transport of proteins across cellular membranes (2–4). Under stressful conditions, elevated HSP70 levels are involved in the correct folding of misfolded or denatured proteins (5). HSP27 (also called HSPB1 and HSPB2) is ATP-independent molecular chaperone that inhibits protein aggregation (6). HSP27 can form oligomers of up to 1,000 kDa. The affinity of HSP27 for the proteins to be chaperoned is modulated by its oligomerization status, the multimer being the competent state (7). The range of oligomerization size and the magnitude of chaperone activity increase as the temperature increases (8, 9). Chaperone activities of HSP27 and HSP70 are complementary to counteract the aggregation or denaturation of proteins after a stress and to contribute to cell survival.

Here, we detailed two screening techniques to monitor eukaryotic HSP27 and HSP70 chaperone activities. These methods were adapted from techniques already described to measure bacterial HSP chaperone activity after a thermal or chemical stress.

Protein aggregation by chemical insult was induced by the reduction of disulfide bond of bovine serum albumin (BSA) using dithiotreitol (DTT). The resulting BSA aggregation is prevented by HSP27. A deficiency in HSP27 chaperone activity results in an increase of turbidity induced by BSA aggregation.

The thermal aggregation is based on protein defolding by temperature resulting in the exposition of charged and hydrophobic surfaces. These surfaces can be recognized by HSP70 that stops the nonspecific aggregation of defolded proteins.

Protein aggregation measurements after chemical stress are better suited for HSP27 while physical stress-induced protein aggregation (i.e., heat shock) seems best suited for HSP70. One possible explanation is that protein aggregation induced by a heat shock may be preceded by progressive proteins' misfolding/denaturation that could be correctly folded by HSP70, thereby preventing their aggregation. In our experimental conditions of chemical insult, protein aggregation might happen too fast for HSP70 to interfere with protein misfolding.

The protein aggregation assays described in this chapter were set up to quantify HSP27 and HSP70 chaperone inhibitors or activators obtained after high-throughput screening. The techniques are also useful to test HSP27 or HSP70 mutants in order to investigate the structure–function dynamism of HSP chaperone activity or to evaluate the implication of the chaperone activity in other HSP functions (i.e., apoptosis). Furthermore, in a more general basis, they could also be used to test the impact of chemical products and physical stress on protein denaturation.

2. Materials

2.1. Cell Culture (Mouse Embryonic Fibroblast HSF1$^{-/-}$, See Note 1)

1. 25-cm^2 culture flask.
2. Cell culture medium: DMEM (Lonza).
3. Fetal bovine serum (FBS) (Lonza).
4. DMEM w/FBS: DMEM supplemented with 10% FBS.
5. Cell culture grade trypsin solution (Lonza).
6. Ethylenediaminetetraacetic acid (EDTA): 1% (Sigma).
7. Sterile washing buffer: HBSS (Lonza).
8. Trypsin w/EDTA: 2.5% trypsin, 0.05% EDTA. Prepare in advance in washing buffer. Store at 4°C.

2.2. Cell Transfection

1. Transfection reagent: DreamFectTM Gold® (Oz biosciences).
2. DMEM w/FBS 10%.
3. Sterile washing buffer: HBSS (Lonza).
4. Opti-MEM medium (Gibco).
5. Human HSP27 and HSP70 plasmids: pcDNA 3.1(Invitrogen), HA hHSP27 or hHSP70 (Home-made).

2.3. Cell Lysis

1. Trypsin w/EDTA.
2. Sterile washing buffer: HBSS (Lonza).
3. DMEM w/FBS 10%.
4. Phosphate-buffered saline (PBS) 1× (Lonza).
5. Lysis buffer: 50 mM HEPES, 0.25 M EDTA, 4 M NaCl, 0.1% NP40. Prepare in advance. Store at 4°C up to 2 months.
6. Antiprotease: Prepare in advance in sterile water. Store at −20°C.
7. Lowry protein assay (Bio-Rad).
8. Sonicator.

2.4. Aggregation Tests

2.4.1. Chemical Aggregation Test

1. 96-well plate (flat bottom).
2. BSA: Prepare in advance at 100 g/L, store at −20°C.
3. DTT: Prepare in advance at 1 M, store at −20°C.
4. Sodium phosphate buffer at 50 mM, pH 7. Store at 4°C.
5. Hsp27 recombinant protein: Store separate aliquots at −80°C (see Note 2).
6. Cell lysates.
7. 96-well plate spectrophotometer reader (for 340 nm).

2.4.2. Thermal Aggregation Test

1. Microtubes.
2. Cell lysates.
3. Tris–HCl buffer at 20 mM, pH 7. Store at 4°C.
4. HSP70 recombinant protein.
5. Water bath.

3. Methods

3.1. Cell Culture

1. Seed 160,000 cells by flask in 7 mL DMEM w/FBS MEF (Mouse Embryonic Fibroblast) HSF1$^{-/-}$ (see Note 3).
2. Incubate at 37°C, 5% CO_2, for 24 h.

3.2. Cell Transfection (see Note 4)

1. Prepare one microtube for DNA and one microtube for transfection reagent per condition.
2. Add in DNA tube 3 μg of HSP plasmid and complete to 100 μL with opti-MEM.
3. Add in transfection tubes 12 μL of DreamFect Gold® solution and 90 μL of opti-MEM (see Note 5).
4. Put the DNA in the DreamFect tube. Mix gently the DNA and transfection reagent by carefully pipetting up and down two to three times. Incubate for 15–20 min at room temperature. Do not vortex or centrifuge (see Note 6).
5. Add the mix (dropwise) onto the cells. Homogenize by gently rocking the flask (for uniform distribution of the mixture) (see Note 7).
6. Carefully add 7 mL of DMEM w/FBS on the well wall close to the bottom.
7. Incubate at 37°C, 5% CO_2, for 24 h (see Note 8).

3.3. Cell Lysis

1. Remove DMEM w/FBS medium.
2. To avoid FBS interference with trypsin, rinse with 5 mL of washing buffer.
3. Add 1.5 mL of trypsin w/EDTA.
4. Incubate for 5 min at 37°C. Shake slightly the flask to detach cells.
5. Recover the cells with 7 mL DMEM w/FBS (see Note 9).
6. Pellet the cells by centrifugation at 200 × *g* for 5 min.
7. Discard the supernatant and resuspend the cells in 7 mL PBS 1×. Centrifuge at 200 × *g* for 5 min.
8. Discard the supernatant and keep the cells on ice (see Note 10).

9. For protease inhibition, add 300 μL of antiprotease to 10 mL lysis buffer. Keep the buffer on ice (see Note 11).
10. Add 300 μL of lysis buffer (see Note 12).
11. Transfer in microtube and vortex for 10 s. Keep on ice during 20 min for cell lysis.
12. Sonicate for 10 s, power 10%.
13. Centrifuge at 14,000 × *g* for 15 min at 4°C.
14. Transfer the supernatant in another tube.
15. To determine protein concentration, use Lowry protein assay. Adjust protein concentration of all samples at 2 mg/mL with sodium phosphate buffer or Tris–HCl buffer (see Note 13).
16. Samples can be stored at −80°C supplemented with 3% of glycerol before aggregation test (see Note 14).

3.4. Aggregation Test

3.4.1. Chemical Aggregation Test (Fig. 1a)

1. Defreeze BSA and DTT at room temperature.
2. Run a 2-h kinetic at 340 nm at 45°C, shake for 10 s before read, and read every 2 min.
3. In each sample, add successively 20 μL of BSA, 100 μL of cell lysates at 2 mg/mL of protein, complete to 197 μL with sodium phosphate buffer, and finally add 3 μL DTT (see Note 15).

3.4.2. Thermal Aggregation Test (Fig. 1b) (see Note 16)

1. Dilute the samples to a protein final concentration of 2 mg/mL.
2. Add 100 μL of Tris–HCl buffer to 200 μL of samples (see Note 17).
3. Incubate for 45–60 min at 55°C in a water bath.
4. Put samples on ice to stop protein aggregation.
5. Centrifuge for 10 min, 16,000 × *g*, at 4°C to pellet aggregated proteins.
6. Measure protein concentration in the supernatant with Lowry technique.
7. Subtract the final protein concentration measured in the supernatant from the initial concentration (2 mg/mL) = protein aggregation.

4. Notes

1. MEF HSF1$^{-/-}$ cells do not express HSP70 and HSP27.
2. HSP27 recombinant protein is used as protein aggregation negative control.

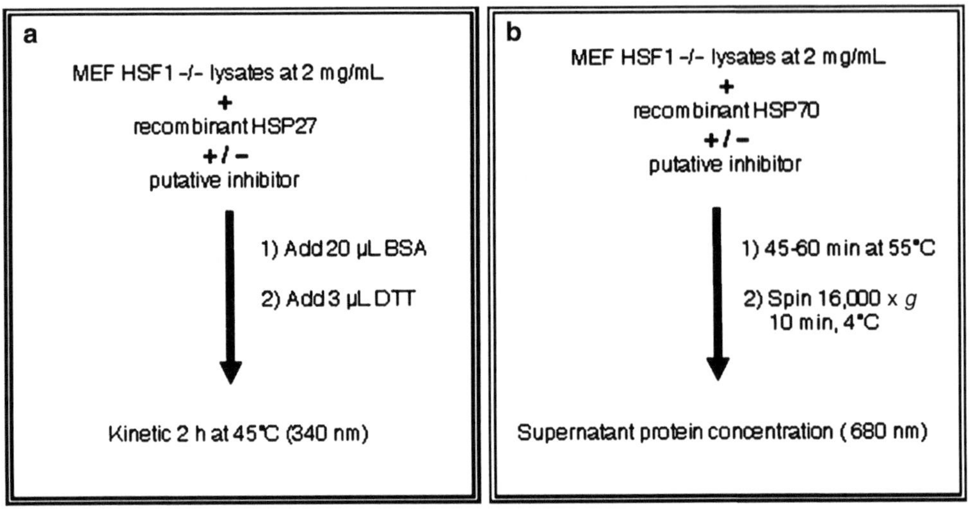

Fig. 1. Schematic representation of the two chaperone activity tests. (**a**) Chemical aggregation test for HSP27. *BSA* Bovine serum albumin, *DTT* Dithiothreitol. (**b**) Thermal aggregation test for HSP70.

3. Prepare two 25-cm² flasks by transfection conditions. Do not use 75-cm² flasks.
4. Alternatively to cell transfection, HSP recombinant proteins can be added to the cell lysates (see Fig. 1).
5. Always prepare the complexes in medium without serum or PBS because serum can interfere with vector assembly.
6. The diluted solutions should be combined within 5 min. It is important to put the DNA in the Dreamfect® never do the opposite. Because the transfection would be ineffective.
7. To avoid the dissociation of the transfection complexes, take slowly the mix with P1000 micropipet.
8. After transfection, do not change the culture medium.
9. To pellet the cells, pool two 25-cm² flasks in the same tube.
10. The cells can be kept at −20°C for 2 weeks. After centrifugation, discard the supernatant very well and put at −20°C.
11. The lysis buffer containing antiprotease cannot be stored for more than 24 h to keep antiprotease activity.
12. Use the smallest volume of lysis buffer possible to avoid sample dilution (final concentration of 2 mg/mL).
13. For chemical aggregation test, use sodium phosphate buffer, and Tris–HCl buffer for thermal aggregation test. Before aggregation test, control HSP expression by western blot.
14. Glycerol protects protein activity from freezing damage.

15. Add DTT at last. Use HSP27 as aggregation negative control. If you want to test HSP27 inhibitors, put inhibitors in these steps.
16. Keep the samples on ice during preparation.
17. Add HSP70 inhibitors in these step to test the effect of inhibitors of the protein.

References

1. Laskey, R. A., Honda, B. M., Mills, A. D., Morris, N. R., Wyllie, A. H., Mertz, J. E., De Roberts, E. M. and Gurdon, J. B. (1978). Chromatin assembly and transcription in eggs and oocytes of Xenopus laevis. *Cold Spring Harb Symp Quant Biol.* **42**, 171–178.
2. Beckmann, R.P., Mizzen, L.E., Welch, W. J. (1990) Interaction of Hsp 70 with newly synthesized proteins: implications for protein folding and assembly. *Science.* **248**, 850–854.
3. De Los Rios, P., Ben-Zvi, A., Slutsky, O., Azem, A., Goloubinoff, P. (2006) Hsp70 chaperones accelerate protein translocation and the unfolding of stable protein aggregates by entropic pulling. *Proc Natl Acad Sci. USA.* **103**, 6166–6171.
4. Shi, Y., Thomas, J. O. (1992) The transport of proteins into the nucleus requires the 70-kilodalton heat shock protein or its cytosolic cognate. *Mol Cell Biol.* **12**, 2186–2192.
5. Nollen, E. A., Brunsting, J. F., Roelofsen, H., Weber, L. A., Kampinga, H. H. (1999) In vivo chaperone activity of heat shock protein 70 and thermotolerance. *Mol Cell Biol.* **19**, 2069–2079.
6. Ehrnsperger, M., Gräber, S., Gaestel, M., Buchner, J. (1997) Binding of non-native protein to Hsp25 during heat shock creates a reservoir of folding intermediates for reactivation. *EMBO J.* **16**, 221–229.
7. Shashidharamurthy, R., Koteiche, H. A., Dong, J., McHaourab, H. S. (2005) Mechanism of chaperone function in small heat shock proteins: dissociation of the HSP27 oligomer is required for recognition and binding of destabilized T4 lysozyme. *J Biol Chem.* **280,** 5281–5289.
8. Bruey, J. M., Paul, C., Fromentin, A., Hilpert, S., Arrigo, A. P., Solary, E., Garrido, C. (2000) Differential regulation of HSP27 oligomerization in tumor cells grown in vitro and in vivo. *Oncogene.* **19**, 4855–4863.
9. Lelj-Garolla, B., Mauk, A. G. (2006) Self-association and chaperone activity of Hsp27 are thermally activated. *J Biol Chem.* **281**, 8169–8174.

Chapter 12

Measuring Hsp72 (HSPA1A) by Indirect Sandwich ELISA

H. Elyse Ireland and John H.H. Williams

Abstract

The enzyme-linked immunosorbent assay (ELISA) is an immunological technique which is used to determine the presence or quantity of an antigen within a sample. ELISAs rely on the use of at least one antibody (Ab) specific for the antigen being measured. This antibody is covalently linked to an enzyme which is detected through the use of an enzymatic substrate, which can be colorimetric, fluorogenic, or chemiluminescent.

The ELISA for Hsp72 described here is a typical indirect sandwich ELISA, which can be used for measuring Hsp72 from cellular/tissue extracts, tissue culture supernatant, and serum. Typically, a 96-well ELISA plate is coated with a specific antibody which captures Hsp72 from the sample, and another antibody specific for a different Hsp72 epitope is used to detect Hsp72. An enzyme-labelled species-specific antibody conjugate is then applied which is consequently detected using a colorimetric enzyme substrate. The quantity of Hsp72 present in the samples is interpolated using a standard curve of known amounts of pure Hsp72.

Key words: Hsp70, Hsp72, HSPA1A, Heat-shock protein, ELISA, Antibody, Quantification, Cell extracts, Supernatant

1. Introduction

Hsp72 has been shown in vitro to be released by passive or active mechanisms and demonstrated to stimulate immune responses, such as cytokine release, and elevated levels of Hsp72 in vivo are related to certain inflammatory conditions (1). The basal and elevated levels of Hsp72 reported by these studies vary widely depending upon the method used (1–6). Quantification of Hsp72 from such samples has generally utilised enzyme-linked immunosorbent assays (ELISAs) which reportedly accurately measure Hsp72 protein levels in a variety of matrices (3, 5–9). ELISAs are robust, quick, and a relatively cheap method for the detection and quantification

Stuart K. Calderwood and Thomas L. Prince (eds.), *Molecular Chaperones: Methods and Protocols*,
Methods in Molecular Biology, vol. 787, DOI 10.1007/978-1-61779-295-3_12, © Springer Science+Business Media, LLC 2011

of antigen within a sample. The type of ELISA typically used for Hsp72 quantification is an indirect sandwich ELISA which allows the detection of Hsp72 within complex matrices, such as cell extracts, serum, and cell culture supernatants. This is achieved through the use of matched antibody (Ab) pairs to Hsp72 which recognise specific, non-overlapping regions of Hsp72. Generally, a capture Ab is immobilised to the plastic surface of a treated ELISA plate overnight, and then a non-specific blocker, such as bovine serum albumin (BSA), is added to prevent non-specific protein/Ab binding. Following washing, antigen (Hsp72) in the form of standards and unknown samples are then applied to the plate, and any Hsp72 present binds to the capture Ab. The bound Hsp72 is then recognised by a detector Ab, which is subsequently bound by a species-specific enzyme-conjugated secondary Ab. The addition of an enzyme substrate enables the colorimetric detection of bound Hsp72 (5, 6, 9) (see Fig. 1).

Most Hsp72 studies have employed commercial ELISAs or their own in-house ELISAs, and the reported levels of extracellular Hsp72 vary widely (3, 7, 10–14), with the lower levels being obtained with the commercial ELISA (13, 14) (see Table 1). Intriguingly, the studies reporting high levels of Hsp72 in serum all perform ELISAs using the same antibody pair utilised by the commercial ELISA (3, 7, 10–12). Also, most cell culture supernatant measurements do not include serum within the test media which may skew the treatment effects (15). Therefore, it cannot be assumed that measurements of Hsp72 in serum and cell culture supernatants which have been reported are accurate. Measurements should be confirmed through the use of a secondary technique, such as Western blot, which although not quantitative gives qualitative data, and/or mass spectroscopy for identification.

Matrix interference is known to occur in many sample types, including serum, caused by abundant proteins and lipids, which can interfere with binding to the target analyte (16). To reduce the possibility of inaccurate results, protein standards used in ELISAs should be diluted in a solution identical to that of unknown samples, such as cell culture supernatant.

This chapter describes an indirect sandwich ELISA for the detection and quantification of Hsp72 from cell/tissue extracts and cell culture supernatants which has been validated to quantify Hsp72 within cell/tissue extracts and cell culture supernatant, being able to detect as little as 0.39 ng/mL Hsp72 within a complex matrix, such as RPMI media supplemented with 10% serum.

Intra-assay variability for this particular ELISA was determined to be: CVs <5% ($r^2 = 0.996$ for cell/tissue extracts; $r^2 = 0.999$ for cell culture supernatant; for both $n = 6$), and inter-assay variability was: CVs <11% ($r = 0.948$ for cell/tissue extracts; $r = 0.998$ for cell culture supernatant; for both $n = 6$). Recovery rates, derived from

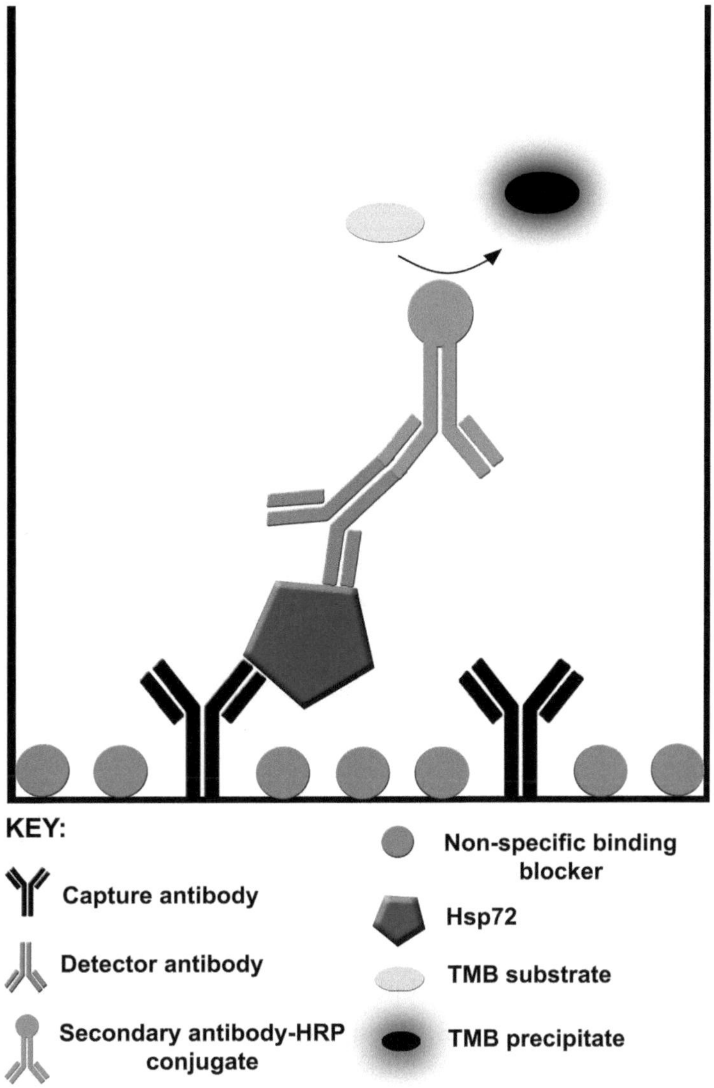

Fig. 1. Illustration of a sandwich ELISA for the detection of Hsp72. Plates are coated with capture Ab at 4°C overnight, followed by blocking of unbound sites with BSA. Standards, samples, and controls are added to the plate and any Hsp72 present binds to the capture Ab. Unbound protein is then washed off and the detector Ab is added to the plate. This binds to any Hsp72 bound. Following washing of excess detector Ab, the secondary HRP-labelled Ab is applied which binds to the Fc region of the detector Ab. Following washing, substrate is then added to the plate for the time specified and the reaction is stopped using stop solution. The ODs are read at 450 nm on a plate reader. The signal is directly proportional to the amount of Hsp72 present and unknown samples can be interpolated through use of the standard curve ODs.

control or treated samples which were spiked with a known quantity of Hsp72, were: 89–113% for cell/tissue extracts and 102–111% for cell culture supernatants within the standard acceptable limits of 80–120%.

Table 1
Reported Hsp72 serum levels using commercial or in-house ELISAs

Reference	Hsp72 range (ng/mL)	Antibodies utilised
(Njemini et al. (3); Terry et al. (9))	10–20,000	Capture – Stressgen SPA-810 Detector – Stressgen SPA-812
(Pockley et al. (10))	20–18,500	Capture – Stressgen SPA-810 Detector – Stressgen SPA-812
(Njemini et al. (7))	235–6,000	Capture – Stressgen SPA-810 Detector – Stressgen SPA-812
(Rea et al. (12))	20–1,500	Capture – Stressgen SPA-810 Detector – Stressgen SPA-812
(Fukushima et al. (13))	2–130	Stressgen Kit
(Zhu et al. (14))	0.2–30	Stressgen Kit

2. Materials

2.1. General Buffers

1. Reagents for the ELISA assay chosen by the investigator (capture Ab see Note 1), Pure Hsp72 (SPR-103; Stressmarq, Victoria, Canada), detector Ab (see Note 1), and a secondary Ab–horseradish peroxidise (HRP) conjugate (see Note 2).
2. Plate coating buffer: 0.1 M sodium carbonate–bicarbonate buffer, pH 9.6. Store at 4°C for up to 1 month.
3. Phosphate-buffered saline (PBS): 0.01 M sodium phosphate buffer, pH 7.2, 0.15 M NaCl.
4. Wash buffer: PBS plus 0.05% (v/v) Tween®20 and 0.01% (w/v) thimerosal.
5. Plate blocking buffer: PBS plus 0.5% (w/v) BSA. Prepare fresh prior to use.
6. Ab buffer: Wash buffer plus 0.5% (w/v) BSA. Prepare fresh prior to use.
7. HRP enzyme substrate: Tetramethyl benzidine (TMB) solution (UP664781; Interchim SA, Montluçon, France) (see Note 3).
8. Stop solution: 1.0 M phosphoric acid.

2.2. Preparation of Cell/Tissue Extracts

1. Prepare extraction buffer: 0.02 M Tris base, 0.1 M EDTA, 0.1 mM DL-dithiothreitol, pH 7.2, 0.002 M phenylmethylsulfonyl fluoride, 0.005 M ε-amino-*n*-caproic acid, 0.001 M benzamidine, plus 0.1% (v/v) Triton X-100. Store at 4°C for up to 1 month.

2. Prepare cell extracts by re-suspending cells/tissue in extraction buffer (~1×10^6 cells/mL) using reflux pipetting.
3. Transfer cell suspensions to polypropylene micro-centrifuge tubes and centrifuge at $13,500 \times g$ for 20 min at 25°C.
4. Transfer supernatant to clean polypropylene micro-centrifuge tubes and store at −80°C until required (see Note 4).

2.3. Preparation of Tissue Culture Supernatants

1. Remove supernatants from cell cultures, transfer to polypropylene micro-centrifuge tubes, and centrifuge at $400 \times g$ for 5 min.
2. Transfer supernatants to clean polypropylene micro-centrifuge tubes and store at −80°C until required (see Note 4).

2.4. Preparation of Hsp72 Standards

1. Prepare a stock solution of the highest concentration of pure Hsp72 required in the appropriate diluent: either extraction buffer for quantifying Hsp72 in cell/tissue extracts or appropriate cell culture media for quantifying Hsp72 in tissue culture supernatants.
2. Prepare up to ten known concentrations of Hsp72 standards from the stock solution using doubling dilutions. Also prepare a zero standard of buffer or media only (see Note 5).
3. Prepare standards up to 2 h prior to application to the ELISA plate. Keep at 4°C until ready to be applied to the ELISA plate.

3. Methods

1. Prepare capture Ab at 2 μg/mL in plate coating buffer. One plate requires 10 mL (see Notes 1 and 6). Prepare fresh.
2. Add 100 μL of the diluted capture Ab to each well of a high-binding 96-well ELISA plate.
3. Seal plate with micro-plate sealing film and incubate overnight at 4°C.
4. Wash the plate three times with 350 μL per well of wash buffer and blot dry.
5. Block non-specific protein absorption by adding 300 μL per well of blocking solution, seal plate, and incubate for 1 h at 25°C.
6. Wash the plate three times with 350 μL per well of wash buffer and blot dry.
7. Transfer 100 μL per well of known standards, samples, and controls to be tested, in triplicate, to the ELISA plate (see Notes 7–8).
8. Seal plate with micro-plate sealing film and incubate for 2 h at 37°C.
9. Wash the plate five times with 350 μL per well of wash buffer, allowing to soak for 30 s between washes, and blot dry.

10. Prepare the detector Ab by diluting in Ab buffer (see Notes 1 and 6).
11. Add 100 μL of detector Ab to each well of the ELISA plate, except from the appropriate control wells, then seal plate with micro-plate sealing film, and incubate for 1 h at 37°C.
12. Wash the plate five times with 350 μL per well of wash buffer, allowing to soak for 30 s between washes, and blot dry.
13. Prepare the secondary Ab–HRP conjugate by diluting in Ab buffer (see Notes 2 and 6).
14. Add 100 μL of the secondary Ab–HRP conjugate to each well of the ELISA plate, except from the appropriate control wells, then seal plate with micro-plate sealing film, and incubate for 1 h at 37°C.
15. Wash the plate five times with 350 μL per well of wash buffer, allowing to soak for 30 s between washes, and blot dry.
16. Add 100 μL HRP enzyme substrate (TMB) to each well of the ELISA plate and incubate for 45 min at 25°C in the dark on a plate shaker (~400 revolutions per minute) (see Note 9).
17. Stop the reaction by adding 100 μL of stop solution to each well and place back on the plate shaker for 15 s.
18. Read the OD of each well at 450 nm in a plate reader.
19. Determine the quantity of Hsp72 per sample by data interpolation using the standard curve derived from the known concentrations of Hsp72 (see Fig. 2).

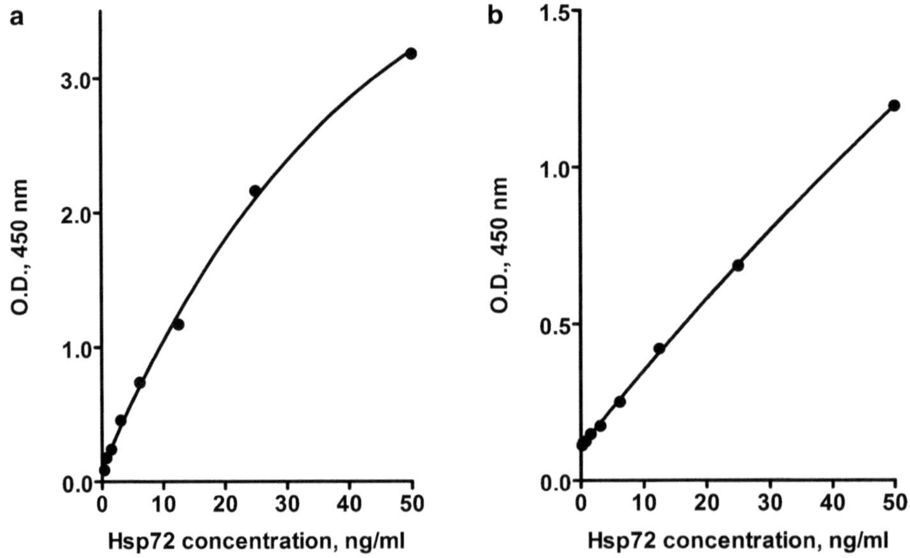

Fig. 2. Typical standard curves obtained from optimised ELISAs for measuring Hsp72 in cell/tissue extracts and tissue culture supernatants. For cell/tissue extracts (**a**), detector Ab was diluted at 1/5,000 and secondary Ab diluted at 1/2,500. For tissue culture supernatants (**b**), detector Ab was diluted at 1/1,000 and secondary Ab diluted at 1/1,250.

4. Notes

1. It is essential to choose the right Ab pairs for use in a sandwich ELISA. These can be commercially available or can be produced by carefully selecting two distinct non-overlapping sequences from Hsp72 so that when Hsp72 is bound by the capture Ab the epitope recognised by the detector Ab is not inaccessible or hidden. The matched Ab pair used here was developed in-house: the capture being raised in sheep against a short peptide sequence from Hsp72 followed affinity purification and the detector being a mouse monoclonal Ab raised against bovine Hsp70.
2. The secondary Ab used in the method described here is conjugated to HRP. Other enzyme labels are available, such as alkaline phosphatase.
3. The substrate used here is for use with HRP conjugates and colorimetric detection. Other substrates can be used depending on the sensitivity required and instrumentation used for detection, such as spectrophotometer, fluorometer, or luminometer. Colorimetric substrates, although not as sensitive as other substrates, also allow direct visualization and, if required, kinetic studies.
4. It is possible to store samples to be tested at −20°C for approximately 1 month. Following this time, some deterioration is likely to occur. Avoid repeated freeze thaw cycles as this also leads to sample deterioration.
5. Careful consideration should be given to the standard curve range used. The range of standards required is dependent upon the concentrations likely to be obtained from the samples tested. For example, tissue samples are likely to have higher levels of Hsp72 than cultured cell samples, tissue culture supernatants, and serum. Dilution of samples may be required for optimization of the standard curve range.
6. All ELISA conditions must be optimized to ensure that the highest signal-to-noise (S/N) ratio is achieved. The S/N ratio should be at least 5, preferably 10. The noise (std 0) should be as close to zero (0.0 OD) as possible and no higher than 0.2 OD. Maximal signal (i.e. the highest std concentration) should be between 1.0 and 3.0 OD. Particularly important are the optimal Ab concentrations to be used. Optimisation is achieved through the sequential use of checkerboard ELISAs for each Ab required (17). As a guide, the concentrations of detector and secondary antibodies used for the method described here are: for cell/tissue extracts, 200 ng/mL detector and 1,000 ng/mL secondary; for cell culture supernatant, 400 ng/mL detector and 800 ng/mL secondary.

7. Control samples to be included are: detector Ab only; detector and secondary Ab only; secondary Ab only; substrate only. These determine any non-specific signal which is not due to antigen presence or maybe masking low antigen values. Optimisation of all experimental parameters should overcome any non-specific signal. Commercial kits have already been optimised, but non-specific signals may occur due to experimenter error.
8. Samples, especially tissue extracts, may require dilution before application to the ELISA plate due to potentially high levels of Hsp72 present. Extracts should be diluted in extraction buffer. Samples may also require dilution due to potential matrix interference. The use of commercial kits may become costly as samples may or may not require dilution before application to the ELISA (read the manufacturer's instructions very carefully before performing the ELISA).
9. The incubation time of the enzymatic substrate requires optimisation depending on whether it is colorimetric, fluorogenic, or chemiluminescent and the concentration of enzyme-labelled Ab used.

References

1. Ireland, H. E., Leoni, F., Altale, O., Birch, C. S., Coleman, R. C., Hunter-Lavin, C., and Williams, J. H. H. (2007) Measuring the secretion of heat shock proteins from cells, *Methods* **43**, 176–183.
2. Njemini, R., Lambert, M., Demanet, C., and Mets, T. (2003) Elevated serum heat-shock protein 70 levels in patients with acute infection: use of an optimized enzyme-linked immunosorbent assay, *Scand J Immunol* **58**, 664–669.
3. Njemini, R., Demanet, C., and Mets, T. (2005) Comparison of two ELISAs for the determination of Hsp70 in serum, *J Immunol Methods* **306**, 176–182.
4. Fehrenbach, E., Passek, F., Niess, A. M., Pohla, H., Weinstock, C., Dickhuth, H. H., and Northoff, H. (2000) HSP expression in human leukocytes is modulated by endurance exercise, *Med Sci Sport Exer* **32**, 592–600.
5. Walsh, R. C., Koukoulas, I., Garnham, A., Moseley, P. L., Hargreaves, M., and Febbraio, M. A. (2001) Exercise increases serum Hsp72 in humans, *Cell Stress Chaperon* **6**, 386–393.
6. Dempsey, N. C., Leoni, F., Ireland, H. E., Hoyle, C., and Williams, J. H. H. (2010) Differential heat shock protein localization in chronic lymphocytic leukemia., *J Leukocyte Biol* **87**, 467–476.
7. Njemini, R., Demanet, C., and Mets, T. (2003) Determination of intracellular heat shock protein 70 using a newly developed cell lysate immunometric assay, *J Immunol Methods* **274**, 271–279.
8. Wright, B. H., Corton, J. M., El-Nahas, A. M., Wood, R. F. M., and Pockley, A. G. (2000) Elevated levels of circulating heat shock protein 70 (hsp70) in peripheral and renal vascular disease, *Heart Vessels* **15**, 18–22.
9. Dempsey, N. C., Ireland, H. E., Smith, C. M., Hoyle, C. F., and Williams, J. H. H. (2010) Heat Shock Protein translocation induced by membrane fluidization increases tumor-cell sensitivity to chemotherapeutic drugs, *Cancer Lett* doi:10.1016/j.canlet.2010.04.016 DOI:dx.doi.org .
10. Terry, D. F., Wyszynski, D. F., Nolan, V. G., Atzmon, G., Schoenhofen, E. A., Pennington, J. Y., Anderson, S. L., Wilcox, M. A., Farrer, L. A., Barzilai, N., Baldwin, C. T., and Asea, A. (2006) Serum heat shock protein 70 level as a biomarker of exceptional longevity, *Mech Aging Dev* **127**, 862–868.
11. Pockley, A. G., Shepherd, J., and Corton, J. M. (1998) Detection of heat shock protein 70 (hsp70) and anti-hsp70 antibodies in the serum of normal individuals, *Immunol Invest* **27**, 367–377.

12. Rea, I. M., McNerlan, S., and Pockley, A. G. (2001) Serum heat shock protein and anti-heat shock protein antibody levels in aging, *Exp Gerontol* **36**, 341–352.

13. Fukushima, A., Kawahara, H., Isurugi, C., Syoji, T., Oyama, R., Sugiyama, T., and Horiuchi, S. (2005) Changes in serum levels of heat shock protein 70 in preterm delivery and pre-eclampsia, *J Obstet and Gynaecol Res* **31**, 72–77.

14. Zhu, J. H., Quyyumi, A. A., Wu, H. S., Csako, G., Rott, D., Zalles-Ganley, A., Ogunmakinwa, J., Halcox, J., and Epstein, S. E. (2003) Increased serum levels of heat shock protein 70 are associated with low risk of coronary artery disease, *Arterioscl Throm Vas* **23**, 1055–1059.

15. Mambula, S. S., and Calderwood, S. K. (2006) Heat induced release of hsp70 from prostate carcinoma cells involves both active secretion and passive release from necrotic cells, *Int J Hyperther* **22**, 575–585.

16. Murray, K. S., Rouse, J. C., Tangarone, B. S., Peterson, K. A., and Van Cleave, V. H. (2001) Identification of human serum interferants in the recombinant P-selectin glycoprotein ligand-1 clinical ELISA using MALDI MS and RP-HPLC, *J Immunol Methods* **255**, 41–56.

17. Nemzek, J. A., Siddiqui, J., and Remick, D. G. (2001) Development and optimization of cytokine ELISAs using commercial antibody pairs, *J Immunol Methods* **255**, 149–157.

Chapter 13

Analysis of Heat-Shock Protein Localisation Using Flow Cytometry

Nina C. Dempsey, Francesca Leoni, Christine Hoyle, and John H.H. Williams

Abstract

The use of flow cytometry in heat-shock protein (HSP) research is increasing rapidly due to the high sensitivity and versatility of the technique. The method allows the simultaneous analysis of multiple proteins within numerous cell types in a heterogeneous sample, providing advantages over alternative techniques, such as ELISA and Western blotting. As a result, flow cytometry is becoming the leading technique used in this area of research. The current chapter describes the methodology for preparing samples for this technique and outlines two protocols for the analysis of surface- and intracellular-localised HSPs.

Key words: Flow cytometry, Heat-shock proteins, PBMCs

1. Introduction

Over the last decade, the use of flow cytometry for both diagnostic and research purposes has increased dramatically. Due to the versatility and high sensitivity of this technique, it is employed to measure a vast array of parameters in many different sample types, including cells, nuclei, and chromosomes. As a result, flow cytometry has become an indispensable tool in the fields of immunology, cell biology, and haematology.

1.1. The Flow Cytometer

Flow cytometry employs the principles of light scattering, light excitation, and light emission as a sample of interest is streamed passed multiple beams of light of single wavelength. Critically, the cells in the sample are streamed in single file so that each cell is analysed individually. As the cells within the sample intercept the

Stuart K. Calderwood and Thomas L. Prince (eds.), *Molecular Chaperones: Methods and Protocols*, Methods in Molecular Biology, vol. 787, DOI 10.1007/978-1-61779-295-3_13, © Springer Science+Business Media, LLC 2011

light source, the light is deflected or "scattered". Forward-scattered (FSC) light is proportional to the size or surface area of the corresponding cell while side-scattered (SSC) light is proportional to the granularity of the cell. The light emitted following interaction with the sample is detected using photodetectors (PMTs). An electrical pulse is generated from these photodetectors which is then processed and analysed in real time. Samples may be stained with fluorochrome-conjugated antibodies prior to analysis and so, in addition to FSC and SSC light, the fluorescent signal intensity of the cell is also analysed.

Many cell characteristics, including the presence of cell surface receptors and intracellular proteins, can be analysed simultaneously by the use of various combinations of fluorochrome-conjugated antibodies. Using flow cytometry, homogeneous sub-populations within a heterogeneous cell population can be detected and "gated" so that further analysis can be made on specific populations (1, 2). This has made flow cytometry the favoured technique over other methods, such as ELISA, Western blot, and spectrofluorometry. Confirmation of flow cytometry results is typically done through the use of fluorescence microscopy.

1.2. Flow Cytometry in HSP Research

Flow cytometry is fast becoming the method of choice in the area of heat-shock protein (HSP) research (3–11). The method allows rapid, quantitative analysis of HSP levels within many different cell types within a single sample. Some flow cytometers can analyse samples at rates of approximately 100,000 cells per second while measuring between 10 and 20 parameters for each cell (1). Figure 1 shows the levels of intracellular Hsp90 and Hsp27 in neutrophils, monocytes, and lymphocytes analysed simultaneously from a single blood sample. In addition to allowing the analysis of HSPs on different cell types, flow cytometry also allows the cellular localisation of these HSPs to be determined by labelling with fluorochrome-conjugated

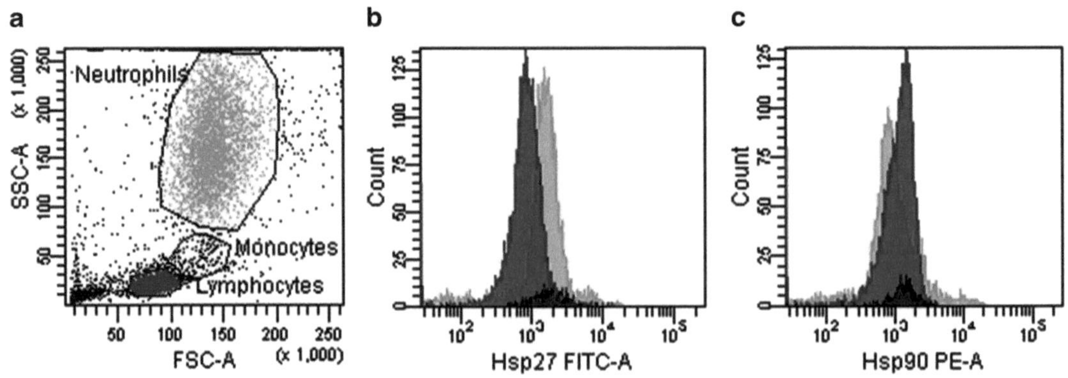

Fig. 1. (**a**) Neutrophils (*light grey*), monocytes (*black*), and lymphocytes (*dark grey*) in lysed whole blood distinguished using FSC and SSC light. (**b**) Intracellular Hsp27 and (**c**) intracellular Hsp90 levels in the three cell types.

antibodies before or after cell permeabilisation (3, 4). The following methods describe the analysis of surface- and intracellular-localised Hsp72 in lysed whole blood (LWB) or peripheral blood mononuclear cells (PBMCs). The methods may be adapted to incorporate alternative HSP or other cellular protein antibodies.

2. Materials

2.1. Isolation of Leukocytes Using the Lysed Whole Blood Method

1. 7-ml EDTA tubes.
2. 15-ml centrifuge tubes.
3. DPBS. Store at RT.
4. Lysing Buffer (BD Biosciences; UK, 349202). Store at 4°C.
5. Trypan blue. Store at RT. Toxic. Wear suitable protective clothing.
6. Haemocytometer.

2.2. Isolation of Peripheral Blood Mononuclear Cells Using Density Gradient Centrifugation

1. 7-ml EDTA tubes.
2. 15-ml centrifuge tubes.
3. DPBS.
4. Histopaque 1077. Store at 4°C.
5. Trypan blue. Toxic. Wear suitable protective clothing.
6. Haemocytometer.

2.3. Detection of Surface Heat-Shock Proteins

1. Binding buffer; 5% FBS in DPBS. Make up fresh each time.
2. Antibodies:

   Anti-human CD5-APC (BD Biosciences; 555355).

   Store at 4°C. Protect from the light.

   Anti-human CD19-PE (BD Biosciences; 555413).

   Store at 4°C. Protect from the light.

   Anti-human Hsp72-FITC (Multimmune Ltd., Germany; cmHsp70.1).

   Store at 4°C. Protect from the light.
3. DPBS.
4. 12 × 75-mm plastic flow cytometry tubes.
5. Flow cytometer (FACScanto, Becton Dickinson, San Diego, CA, USA).

2.4. Detection of Intracellular Heat-Shock Proteins

1. Binding buffer; 5% FBS in DPBS. Make up fresh each time.
2. Antibodies:

   Anti-human CD5-APC (BD Biosciences; 555355).

Store at 4°C. Protect from the light.

Anti-human CD19-PE (BD Biosciences; 555413).

Store at 4°C. Protect from the light.

Anti-human Hsp72-FITC (Assay Designs; SPA-810, clone number C92F3A-5). Store at 4°C. Protect from the light.

3. Cell Cytofix/Cytoperm Solution (BD Biosciences; 554722). Store at 4°C. Toxic. Wear suitable protective clothing.
4. DPBS.
5. 12 × 75-mm plastic flow cytometry tubes.
6. Flow cytometer (FACScanto, Becton Dickinson, San Diego, CA, USA).

3. Methods

3.1. Isolation of Leukocytes Using the LWB Method

1. Collect blood by venipuncture into 7-ml EDTA tubes.
2. Transfer the whole blood into a 15-ml centrifuge tube and wash the sample by adding two volumes of DPBS.
3. Centrifuge the sample at 500 × *g* for 5 min at RT and discard the supernatant.
4. Lyse the red blood cells by adding 5 ml of lysing buffer to the cell pellet. HINT: The sample should become opaque as the red blood cells lyse. This usually takes approximately 3–5 min.
5. Centrifuge the sample at 500 × *g* for 5 min at RT and discard the supernatant.
6. Wash the leukocyte cell pellet by adding 5 ml of DPBS and centrifuge at 500 × *g* for 5 min at RT.
7. After discarding the supernatant, resuspend the cell pellet in 1 ml of DPBS.
8. Take two 100-μl aliquots of the sample into microcentrifuge tubes and add 100 μl of trypan blue in order to distinguish any non-viable cells.
9. Count the cells by applying the aliquots to a haemocytometer.

   HINT: Subsequent experiments should be performed only when the percentage of dead cells is <1%.
10. Aliquot volumes of sample containing 1×10^6 cells/ml into 1.5-ml microcentrifuge tubes (one aliquot per test).
11. Centrifuge the samples at 500 × *g* for 5 min at RT.
12. Discard the supernatant and proceed to Subheading 3.3 for surface staining of HSPs.

3.2. Isolation of PBMCs Using Density Gradient Centrifugation

1. Collect blood by venipuncture into 7-ml EDTA tubes.
2. Dilute the blood 1:1 with DPBS in a 15-ml centrifuge tube.
3. In two separate 15-ml centrifuge tubes, add 3 ml of Histopaque 1077.
4. Carefully layer 6 ml of the blood/DPBS solution over the top of the Histopaque in each tube, being very careful not to disturb the top of the Histopaque layer.
5. Centrifuge the samples at 400 × *g* for 30 min at RT.
6. After centrifugation, discard the plasma at the top of each sample and transfer the buffy coats containing the mononuclear cells from each sample into one clean 15-ml centrifuge tube.

   HINT: Be careful not to transfer any of the Histopaque with the buffy coat.
7. Add 10 ml of DPBS and centrifuge at 250 × *g* for 10 min at RT.
8. Repeat step 7.
9. After discarding the supernatant, resuspend the cell pellet in 1 ml of DPBS.
10. Take two 100-μl aliquots of the sample into microcentrifuge tubes and add 100 μl of trypan blue in order to distinguish any non-viable cells.
11. Count the cells by applying the aliquots to a haemocytometer.

    HINT: Subsequent experiments should be performed only when the percentage of dead cells is <1%.
12. Aliquot volumes of sample containing 1×10^6 cells/ml into microcentrifuge tubes (one aliquot per test).
13. Centrifuge the samples at 500 × *g* for 5 min at RT.
14. Discard the supernatant and proceed to Subheading 3.3 for surface staining of HSPs.

3.3. Detection of Surface Heat-Shock Proteins

Surface proteins are detected on cells using fluorochrome-conjugated antibodies to specific surface-localised antigens. Each antibody is conjugated to a different fluorochrome to ensure that the distinction can be made between each antibody. The following method describes the detection of membrane-embedded Hsp72 on CD19+/CD5+ co-expressing CLL cells using an FITC-conjugated anti-human Hsp72, a PE-conjugated anti-human CD19, and an APC-conjugated anti-human CD5 (see Note 1). The presence of membrane-embedded Hsp72 is detected using a FITC-labelled anti-Hsp72, termed cmHsp70.1 which specifically recognises the TKD peptide, TKDNNLLGRFELSG, in the N-terminal region of the Hsp72 protein, which is presented outside the cell membrane

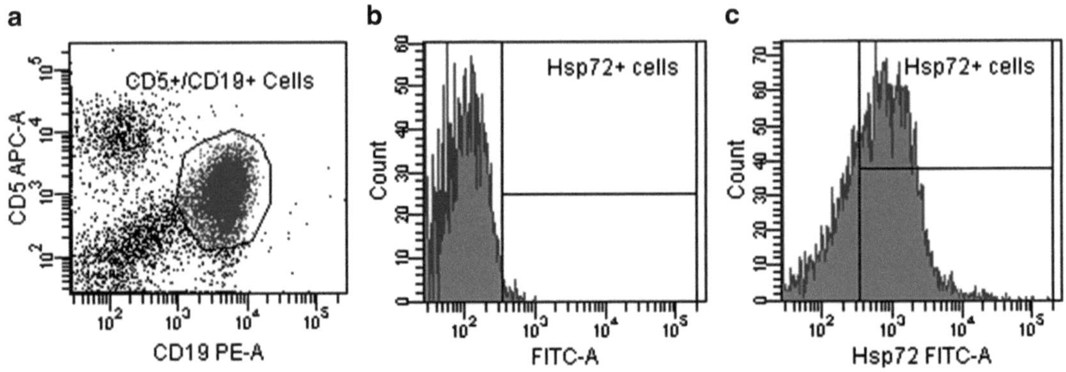

Fig. 2. Triple staining of LWB with APC-conjugated anti-human CD5, PE-conjugated anti-human CD19, and FITC-labelled anti-human Hsp72 (cmHsp70.1). (**a**) CD5+/CD19+ CLL cells are distinguished by double fluorescence and "gated" to distinguish them from the remaining cell population on the histograms. (**b**) A sample of the LWB is analysed without prior staining to evaluate background fluorescence. (**c**) Histogram shows CD5+/CD19+ cells expressing membrane-embedded Hsp72.

when the protein is embedded (12). The same method may be followed to analyse the levels of other HSPs on the cell surface. Figure 2 shows the surface Hsp72 levels in CD5+/CD19+ cells from a CLL patient.

1. Wash the cells by adding 100 μl binding buffer to each sample and centrifuge at 500×*g* for 5 min. This limits non-specific binding of antibodies in subsequent steps. Remove the supernatant.
2. Make up a cocktail of each "panel" of antibodies to be used by multiplying the number of samples by the volume of antibody required; 20 μl of anti-human CD5-APC, 20 μl of anti-human CD19-PE, and 1 μl of anti-human Hsp72-FITC are required per sample. Therefore, for example, for four CLL samples, mix 80 μl of anti-human CD5-APC, 80 μl of anti-human CD19-PE, and 4 μl of anti-human Hsp72-FITC in a foil-wrapped microcentrifuge tube.
3. Add 41 μl of antibody cocktail to each cell pellet and mix (see Notes 2 and 3 for information on including control samples).
4. Incubate at 4°C for 30 min in the dark.
5. Add 100 μl binding buffer to each sample to wash any unbound antibodies and centrifuge at 500×*g* for 5 min at RT.
6. Remove the supernatant and resuspend each cell pellet in 100 μl DPBS.
7. Transfer the cell suspension to 12×75-mm tubes ready for immediate analysis on the flow cytometer (see Note 4).
8. For triple staining protocols, it is necessary to run compensation experiments prior to the main analysis (see Note 3).

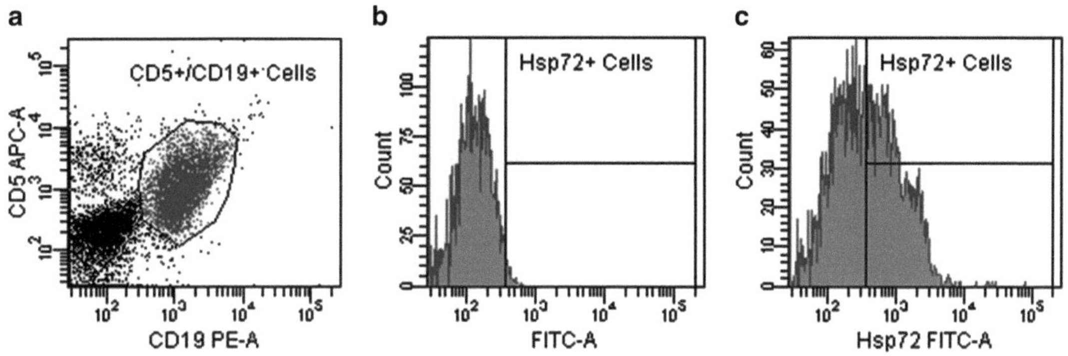

Fig. 3. Double staining of LWB with APC-conjugated anti-human CD5 and PE-conjugated anti-human CD19 followed by permeabilisation and labelling with FITC-labelled anti-human Hsp72. (**a**) CD5+/CD19+ CLL cells are distinguished by double fluorescence and "gated" to distinguish them from the remaining cell population on the histograms. (**b**) A sample of the LWB is analysed without prior staining to evaluate background fluorescence. (**c**) Histogram shows CD5+/CD19+ cells expressing intracellular Hsp72.

3.4. Detection of Intracellular Heat-Shock Proteins

Intracellular proteins are detected on cells using fluorochrome-conjugated antibodies to specific antigens following permeabilisation of the cells using a saponin-based permeabilisation solution. Each antibody is conjugated to a different fluorochrome to ensure that the distinction can be made between each antibody. The following method describes the detection of intracellular Hsp72 on CD19+/CD5+ co-expressing CLL cells using a FITC-conjugated anti-human Hsp72, a PE-conjugated anti-human CD19, and an APC-conjugated anti-human CD5. In the following method, the presence of intracellular Hsp72 is detected using a FITC-conjugated anti-Hsp72 (Assay Designs). This antibody differs from the antibody used in the surface staining method (cmHsp70.1) in that it recognises a different epitope of the same protein. Figure 3 shows the levels of intracellular Hsp72 in CD5+/CD19+ cells from a CLL patient.

1. Wash the cells by adding 100 μl binding buffer to each sample and centrifuge at 500×*g* for 5 min. This limits non-specific binding of antibodies in subsequent steps. Remove the supernatant.
2. Make up the cocktail of each "panel" of antibodies to be used by multiplying the number of samples by the volume of antibody required; 20 μl of anti-human CD5-APC and 20 μl of anti-human CD19-PE are required per sample. Therefore, for example, for four CLL samples, mix 80 μl of anti-human CD5-APC and 80 μl of anti-human CD19-PE in a foil-wrapped microcentrifuge tube.
3. Add 40 μl of antibody cocktail to each cell pellet and mix.
4. Incubate at 4°C for 30 min in the dark.

5. Add 100 μl binding buffer to each sample to wash any unbound antibodies and centrifuge at 500 × *g* for 5 min at RT.
6. Fix and permeabilise the cells by adding 100 μl/sample of cytofix/cytoperm solution and incubate for 20 min at 4°C. This allows any subsequent antibodies to diffuse into the cytosol of the cells and allows intracellular detection (see Note 5).
7. Add 100 μl/well binding buffer to each sample to wash the cells and centrifuge at 500 × *g* for 5 min at RT.
8. After removing the supernatant, add 50 μl/sample of a 1:50 dilution of the FITC-conjugated anti-Hsp72 for 60 min at 4°C in the dark.
9. Add 100 μl binding buffer to each sample and leave for 5 min to allow any unbound antibodies to diffuse out of the cells.
10. Centrifuge at 500 × *g* for 5 min at RT.
11. Remove the supernatant and resuspend the cell pellet in 100 μl DPBS.
12. Transfer the cell suspension to 12 × 75-mm tubes ready for analysis on the flow cytometer.

4. Notes

1. The HSP detection methods described above may also be followed to analyse HSP levels in tissue culture samples (4). However, adherent cell lines require trypsinization prior to staining in order to get the cells into suspension. HSP levels in primary tissue samples may also be analysed using these methods, but must be in single cell suspension prior to staining. This can be achieved by prior treatment of the sample with an enzymatic solution, such as trypsin, collagenase, or elastase, followed by filtration of the sample through a nylon mesh.
2. A number of control samples must be analysed alongside the test samples in order to ensure accurate results.
   (a) A no-stain control sample which has not been probed with any antibodies is used as a negative control. This sample is analysed prior to the stained samples in order to adjust the PMT voltages and to place the negative population within the first log decade of the histogram (Figs. 2b and 3b).
   (b) Isotype control samples must also be analysed. Isotype control antibodies have no relevant specificity and help to distinguish background fluorescence. Isotype controls must be matched to the specific test primary antibody by

both species and isotype. For example, a mouse IgG1–FITC can be used as an isotype control against the mouse monoclonal anti-human Hsp72-FITC.

3. Compensation experiments must be performed prior to running the test samples when more than one fluorochrome-conjugated antibody is added to a single sample. If the wavelengths of two fluorochromes overlap, the fluorescent signal detected by the flow cytometer may not be the actual signal emitted by the cell. This results in the cell appearing to possess a phenotype that it does not actually have. This overlap can be corrected by running a number of samples stained with a single antibody from each of the "panel" of antibodies used in the experiment (1). The need to run compensation experiments can be avoided by carefully selecting fluorochromes that do not have spectral overlap.
4. Samples can be stained in microcentrifuge tubes and transferred to 12 × 75-mm flow cytometry tubes before analysis. Alternatively, the cells may be transferred to 96-well v-bottom plates prior to surface staining so that all subsequent steps and flow cytometry are performed in the wells. The analysis of samples in the plate requires the use of a high-throughput sampler (HTS) attachment on the flow cytometer. Staining in plates not only saves time at washing stages, but also allows the use of three times lower concentrations of antibody as cell concentrations can be adjusted to 3×10^5 cells/well.
5. The use of a fixative reagent in the intracellular HSP detection method prevents any further biochemical reactions occurring within the cells. Immediate analysis on the flow cytometer following staining is, therefore, not required and the cells can be stored in DPBS at 4°C protected from the light for up to 2 weeks before analysis. For surface HSP analysis, cells may be resuspended in 100 μl of 4% paraformaldehyde (PFA) following staining and stored at 4°C, protected from the light for up to 2 weeks before analysis. Samples should be centrifuged at 500 × *g* for 5 min to remove the PFA and resuspended in 100 μl of DPBS before analysis.

References

1. Robinson, P.J. (2004). Flow Cytometry. *In* G. Wnek & G.L. Bowlin. *Encyclopedia of Biomaterials and Biomedical Engineering*. New York: Informa Healthcare
2. Bachelet, M., Mariethoz, E., Banzet, N., Souil, E., Pinot, F., Polla, C.Z., Durand, P., Bouchaert, I. and Polla, B.S. (1998). Flow cytometry is a rapid and reliable method for evaluating heat shock protein 70 expression in human monocytes. *Cell Stress Chaperones* **3**, 168–76
3. Dempsey, N.C., Leoni, F., Ireland, H.E., Hoyle, C. and Williams, J.H.H. (2010). Differential Heat Shock Protein Localization in Chronic Lymphocytic Leukemia. *J. Leukoc. Biol* **87**, 467–76
4. Dempsey, N.C., Ireland, H.E., Smith, C.M., Hoyle, C. and Williams, J.H.H. (2010). Heat Shock Protein translocation induced by membrane fluidization increases tumor-cell sensitivity to chemotherapeutic drugs. *Cancer Lett.* doi:10.1016/j.canlet.2010.04.016

5. Schilling, D., Gehrmann, M., Steinem, C., De Maio, A., Pockley, A.G., Abend, M. Molls, M. and Multhoff, G. (2007). Binding of heat shock protein 70 to extracellular phosphatidylserine promotes killing of normoxic and hypoxic tumor cells. *FASEB Journal* **23**, 2467–77
6. Gehrmann, M., Liebisch, G., Schmitz, G., Anderson, R., Steinem, C., De Maio, A., Pockley, G. and Multhoff, G. (2008). Tumor-Specific Hsp70 Plasma Membrane Localization Is Enabled by the Glycosphingolipid Gb3. *PlosOne* **3**, 1–9
7. Hromadnikova, I . and Sedlackova, L., (2008). Analysis of cell surface and relative gene expression of heat shock protein 70 in human leukemia cell lines. *Leukaemia Lymphoma* **49**, 570–76
8. Lin, L., Kim, S.C., Wang, Y., Gupta, S., Davis, B., Simon, S.I., Torre-Amione, G. and Knowlton, A.A. (2007a). HSP60 in heart failure: abnormal distribution and role in cardiac myocyte apoptosis. *Am. J. Physiol* **293**, 2238–47
9. Pfister, K., Radons, J., Busch, R., Tidball, J.G., Pfeifer, M., Freitag, L., Feldmann, H.J., Milani, V., Issels, R. and Multhoff, G. (2007). Patient survival by Hsp70 membrane phenotype. *Cancer* **110**, 926–35
10. Thomas, X., Campos, L., Mounier, C., Cornillon, J., Flandrin, P., Le, Q.H., Piselli, S. and Guyotat, D. (2005). Expression of heat-shock proteins is associated with major adverse prognostic factors in acute myeloid leukemia. *Leukaemia Res* **29**, 1049–58
11. Becker, B., Multhoff, G., Farkas, B., Wild, P.J., Landthaler, M., Stolz, W. and Vogt, T. (2004). Induction of Hsp90 protein expression in malignant melanomas and melanoma metastases. *Exp Dermatol* **13**, 27–32
12. Multhoff, G., Pfister, K., Gehrmann, M., Hantschel, M., Gross, C., Hafner, M. And Hiddemann, W. (2001). A 14-mer Hsp70 peptide stimulates natural killer (NK) cell activity. *Cell Stress Chaperones* **6**, 337–344

Chapter 14

Quantitation of Heat-Shock Proteins in Clinical Samples Using Mass Spectrometry

Punit Kaur and Alexzander Asea

Abstract

Mass spectrometry (MS) is a powerful analytical tool for proteomics research and drug and biomarker discovery. MS enables identification and quantification of known and unknown compounds by revealing their structural and chemical properties. Proper sample preparation for MS-based analysis is a critical step in the proteomics workflow because the quality and reproducibility of sample extraction and preparation for downstream analysis significantly impact the separation and identification capabilities of mass spectrometers. The highly expressed proteins represent potential biomarkers that could aid in diagnosis, therapy, or drug development. Because the proteome is so complex, there is no one standard method for preparing protein samples for MS analysis. Protocols differ depending on the type of sample, source, experiment, and method of analysis. Molecular chaperones play significant roles in almost all biological functions due to their capacity for detecting intracellular denatured/unfolded proteins, initiating refolding or denaturation of such malfolded protein sequences and more recently for their role in the extracellular milieu as chaperokines. In this chapter, we describe the latest techniques for quantitating the expression of molecular chaperones in human clinical samples.

Key words: Heat-shock proteins, Mass spectrometry, LC-MS/MS, Proteomics

1. Introduction

In the postgenomic era, researchers have realized the importance of proteomic approaches for clinical and diagnostic research of disease. Detection of these differences in protein expression levels, as a function of disease progression, is an important area of research. This quantitative regulation approach to proteomic research monitors the expression of large numbers of proteins to observe how expression patterns change quantitatively with disease progression, environmental changes, or exposure to external chemical agents (1).

Stuart K. Calderwood and Thomas L. Prince (eds.), *Molecular Chaperones: Methods and Protocols*,
Methods in Molecular Biology, vol. 787, DOI 10.1007/978-1-61779-295-3_14, © Springer Science+Business Media, LLC 2011

Extensive changes can occur to proteins after translation so that multiple protein forms or products result from a single gene. Consequently, proteomic approaches are essential to obtain detailed information about posttranslational modifications and protein expression quantitation. Differences in protein p*I*s, M_r, or expression levels between samples can be visualized, and disease biomarkers can be identified as potential targets for diagnosis or drug development. Using 2-dimensional gel electrophoresis (2-DE) to perform protein expression mapping combined with mass spectrometry (MS) to identify and characterize proteins has become the basis of proteomic-based research. The separation can be directly coupled to an electrospray mass spectrometer for intact protein M_r analysis. Liquid chromatography-mass spectrometry (LC-MS) is, thus, the most promising technique for the measurement of bioactive compounds and their metabolites in blood/plasma due to the inherent high selectivity afforded through both of the hyphenated techniques.

A handful of biomarkers, described by Adkins et al. as "proteins that undergo a change in concentration or state in association with a biological process or disease," are currently used routinely for population screening, disease diagnosis, prognosis, monitoring of therapy, and prediction of therapeutic response (2). Unfortunately, most of these biomarkers suffer from low sensitivity, specificity, and predictive value, particularly when applied to rare diseases in population screening programs (3). Serum and plasma potentially contain elements of all proteins produced in the body (4). The technical challenge in the analysis of the serum/plasma proteome is that the serum/plasma proteins are present at unequal concentrations. A few are so dominant, such as serum albumin and immunoglobulins (almost 90% of total serum/plasma protein by weight), that they mask the detection of other proteins, especially low-abundance proteins. Not surprisingly, sample preparation strategies, including preliminary procedures for sampling, handling, and storage, methods for removal of high-abundance proteins, and fractionation approaches, constitute an important field of study for scientists involved in biomarker-related projects.

Dried blood spot (DBS) offers a number of advantages over conventional whole blood, plasma, or serum sample collection in therapeutic drug monitoring as it requires a less-invasive sampling method (finger or heel prick, rather than conventional venous cannula), simpler storage, and easier transfer (5). There is no requirement for freezers or dry ice, requires a smaller blood volume (<100 μl, compared with more than 0.5 ml blood) (6). The analysis of DBS samples using MS could be traced back to the early 1990s when Millington and Chace et al. used the precursor ion scan of *m/z* 99 for the detection of acylcarnitines and the constant neutral loss scan of 102 Da and product ion scan for the analysis of phenylalanine and tyrosine in newborn screening (7). In addition to the above MS/MS detection modes where (a) all precursor ions that produce a single product ion are detected in the precursor ion scan, (b) all precursor ions that undergo the loss of a common fragment are detected in the constant neutral loss scan, and (c) all product ions that are produced from a single precursor ion are detected in the product ion scan, MS/MS offers some sensitive detection methods, like selected reaction monitoring (SRM) and multiple reaction monitoring (MRM). In SRM or MRM, only the selected precursor ion(s) and the selected product ion(s) are detected. However, the specificity of MS/MS assay alone (infusion or flow injection) is often hampered by interferences of the matrix components, especially those sharing the same MS/MS transitions with the target analytes (8).

LC column/mobile-phase optimization is needed to achieve appropriate peak shape, sensitivity, and the separation of the analyte from interfering metabolites and phospholipids. Use of quality methods developed in this manner reduces bioanalytical risk due to ionization suppression/enhancement (matrix effect) caused by plasma phospholipids (9). The significance of using incurred sample during method development is essential in order to ensure that metabolites and other components present only in the incurred sample, and not in quality control (QC) samples, do not interfere with the accurate quantitation of the drug or any particular metabolite of interest (9).

Urinalysis carried out in accredited laboratories is generally considered to be a standard technique for detecting drug doping. The major disadvantage of urinalysis is that it provides only short-term information of the individual's drug consumption and does not determine sustained drug abuse (10). Hair testing is noninvasive, tamper resistant, and the most convenient technique for the detection and control of drug doping in comparison to urinalysis and blood tests. Although hair analysis may not completely replace urinalysis, it can be used in special cases to provide information complementary to that obtained from urine and blood tests (11). It can be used to confirm repetitive intake of drugs by directly detecting the parent drugs. The extended lower limit of detection and lower limit of quantification are capable of detecting these steroids even at very low concentrations, hence reducing the amount of hair required to 14–15 hairs affording sample collection without leaving noticeable bald patch.

A highly selective, sensitive, and high-throughput LC-MS/MS method was developed by employing a "surrogate analyte" strategy for simultaneous quantitation of endogenous ethanolamides (fatty acid amides), including arachidonyl ethanolamide (AEA), oleoyl ethanolamide (OEA), and palmitoyl ethanolamide (PEA), are substrates of fatty acid amide hydrolase (FAAH) which are important in biomarker discovery in human plasma (12). LC-MS/MS is a sensitive and selective analysis technique which is suitable to quantify low-concentrated target analytes in microdialysates (13).

2. Materials

2.1. Samples

1. Human plasma.
2. Human serum.
3. Human urine.
4. Human hair.
5. Tumor samples from breast, prostate, and cancer patients.

2.2. Buffers and Reagents

1. Anticoagulants (citrate or heparin).
2. Ethylenediaminetetraacetic acid (EDTA).
3. Acetonitrile.
4. Sodium acetate buffer.
5. Methanol.
6. 50 mM ammonium biocarbonate.
7. Coomassie blue stain.
8. Tris (2-carboxyethyl) phosphine hydrochloride (TCEP).
9. Iodoacetamide (IAA).
10. Trifluoroacetic acid (TFA).
11. Bovine serum albumin (BSA).
12. Sodium dodecyl sulfate (SDS).
13. Dithiothreitol (DTT).
14. 3-(*N*-morpholino) propane sulfonic acid (MOPS).
15. 1 M hydrochloric acid.
16. 0.2 M phosphate buffer.
17. Liquid–liquid extraction (LLE) buffer.
18. Vacutainer® SST serum separator tube.
19. DBS card/paper.
20. Solid-phase extraction (SPE) buffer.
21. C18 columns.
22. Molecular weight cutoff (MWCO) membranes.
23. Multiple-affinity removal system (MARS).
24. Syringe filter.

25. Concentrator.
26. Zip-tip.
27. Microdialysis (MD).

2.3. Equipment and Software

1. Ingenuity software.
2. Spectrum mill software.
3. 1100 Series LC/MSD Trap XCT Ultra ion Chip (Agilent Technologies).

3. Methods

3.1. Sample Collection and Preparation

Plasma is the largest single component of blood, comprising about 55% of total blood volume. To prevent clotting, an anticoagulant, such as citrate or heparin can be added to the blood immediately and centrifuged to separate plasma. Serum is obtained by letting a blood specimen clot prior to centrifugation in a serum-separating tube (5 ml; Vacutainer® SST serum separator tube, Greiner bio-one, UK). Animal blood is collected from an ear vein prior to immunization and at fortnightly intervals during the experimental period into heparinized containers and plasma obtained by centrifugation at 1,500 × *g* for 10 min at 4°C. In order to prevent ex vivo change of endogenous fatty acid amides, like ethanolamide levels in the plasma, it is critical to separate the plasma from blood cells as soon as possible so that the enzymes/transporters involved in the biosynthesis and catabolism of ethanolamides are removed to make ethanolamides stable in plasma (14). Although serum is preferred for many tests because the anticoagulants in plasma can sometimes interfere with the results, plasma seems to be more stable than serum (15). Commercially available blood collection tubes contain multiple components, like silicones, rubber stoppers, and the plastics that may appear as interfering or confounding peaks during the MS analysis (15). At this time, insufficient information is available to decide whether serum or plasma should be preferred in proteomic studies for biomarker discovery.

EDTA-treated blood is only slightly less stable, but over longer periods of time, marked changes appear as elapsed time before centrifugation increases because the activity of many proteases requires metals and the chelating action of EDTA may help prevent coagulation (16). The effects of the clotting time on protein profiles obtained from serum and the incubation time before separation of plasma from blood cells have been extensively studied by Hsieh et al. (15). The authors attributed the changes observed to the continuous metabolism of blood cells, the alterations of cell membrane integrity resulting in continuous release of metabolites, and/or the release of degraded products from the clot (15). Changes in protein profiles obtained from serum samples were also observed for the different clotting times at both 4°C and room temperature (15). The degradation of plasma peptides or the formation/accumulation of new peptides during and after the clotting process showed changes in the peptide peak intensities. For serum and plasma stored up to 24 h at 4°C, the profiles were quite similar. For long-term storage of plasma/serum samples, no major differences have been observed at −20, −80°C, or using liquid nitrogen (17). After repeated freezing at −80°C and thawing on ice of serum/plasma samples up to ten times, no detectable changes were observed on the basis of whole spectra (15).

Recently, a new method for blood collection is provided by Edelbroek et al. by pricking the finger, heel, or toe with a lancet, and the blood drops are then spotted onto preprinted circles on specially manufactured DBS card/paper (5). After the pricks, the blood drop is directly applied onto the sampling paper/card within a premarked circle, ideally one drop per spot. It is very important to dry blood spots completely before storage or transportation. In general, a minimum of 3 h drying in an open space at room temperature (15–22°C) is recommended (5). After drying, the DBS samples should be protected against humidity and moisture by covering them with a paper overlay and packing them in low gas-permeable zip-closure bags with desiccant packages and stored at a lower temperature, for example 2–8°C, ≤ −15°C, or ≤ −60°C in order to enhance stability (18). The DBS paper/card might be further treated via impregnation for an improved extraction recovery (19). The "whole blood" is then enriched with ^{125}I-l-thyroxine, and a 100 μl volume of the blood is applied onto the paper lots to be tested.

3.1.1. Off-Line Preparation

Small volume of plasma or urine (20 μl–4 ml) was added to an equivalent volume of hydrochloric acid (4–6.8 M) and was heated for 30 min at 80°C to hydrolyse the flavonoid glycosides. Studies by Mauri et al., where terpene lactones were analyzed in plasma, an LLE of those compounds (ginkgolides A, B and bilobalide) was performed using an ethyl acetate volume that was equivalent to the plasma volume (300 μl) (20). Following vortexing, the resulting upper organic layer or amount thereof would then be transferred to another centrifuge vial and evaporated to dryness using a stream of nitrogen. Evaporated samples would then be reconstituted in 10% methanol90% water (20), mobile phase (21), or acetonitrile (ACN) (22) with an approximate volume of 100 μl and then centrifuged at high speed. For quantitative analysis, one or more DBS disks are punched from DBS card/paper into assay tubes or the appropriate wells of 96-well microtiter plate for extraction with methanol, ACN, or a mixture of water/organic containing the internal standard(s). The extraction solvent must be strong enough to interrupt the binding of analyte to protein in the matrix or the paper material. The analyte of interest is then extracted with gentle shaking or vortexing, and sonication may be necessary for better extraction efficiency (19). After centrifugation, the resulting extracts are transferred manually or by an automated liquid handler to new tubes or microtiter plates. The extracts can be directly injected onto the LC-MS/MS system for analysis or dried for reconstitution using an MS-friendly solvent prior to analysis (23).

3.1.2. Online Preparation

The online SPE technique generally had higher recoveries than the off-line mode. Ding et al. illustrated the use of a novel online extraction configuration for solution of human urine, sodium acetate buffer and hydrolysis enzymes were injected into a mobile phase of 90%, 0.1% acetic acid in water – 10%, 50:50 methanol:ACN (24) by compound trapping and band focusing on a capillary trapping column. A change of the valve configuration and a change to high-organic mobile-phase conditions then allowed the elution of compounds onto the analytical capillary column. Chromatographic analysis from plasma using LC-MS has been carried out using reversed-phase small-diameter C18 columns (with geometries 5.0 × 2.0 to 150 × 2.0-mm) with flow rates 0.3 ml/min in negative ion mode electrospray ionization (ESI) (22) and atmospheric chemical ionization (APCI) (25). Zhao et al. also indicated that the addition of very low concentrations of mobile-phase additives appeared to enhance the sensitivity of the ginkgo flavonoids and terpenes (22). Another sample preparation procedure for the analysis of metabolites involves SPE. Briefly, hydrophobic SPE cartridges were preconditioned using a number of volumes of methanol and then a number of volumes of water. Aliquots of 100 μl–1 ml of plasma or urine and/or plasma/internal standard were introduced onto the SPE cartridges along with a volume (3 ml) of water in order to capture ginsenoside metabolites onto the cartridges. The ginsenoside components from plasma were then eluted from the SPE cartridges with a number of volumes of either methanol or 95% methanol–water. Extracts were then either injected (26) or evaporated under a stream of nitrogen and reconstituted in 100 μl of mobile phase (27).

3.1.3. Immunodepletion

Centrifugal ultrafiltration has been extensively used for the removal of high-molecular-weight species for serum/plasma-based biomarker discovery for clinical diagnosis of several diseases. Besides use of different MWCO membranes and various centrifugation speeds between 3,000 and 4,000 × *g*, addition of ACN to diluted serum or plasma to a final concentration of 25% (v/v) has been shown to have a drastic positive effect on the enrichment of low-molecular-weight (LMW) protein/peptides without affecting the ability of the procedure to remove larger proteins (28). Low-speed centrifugation, use of diluted serum or plasma, and use of denaturing conditions are key variables for successfully performing plasma/serum centrifugal ultrafiltration for removal of highly abundant proteins together with the enrichment of LMW protein/peptides. The selective removal of large abundant from serum is performed by precipitating LMW proteins using organic solvents. The precipitation with organic solvents in the presence of ion-pairing agents dissociates peptides and smaller proteins from large abundant proteins, thereby facilitating their extraction. It has been shown that two volumes of ACN added to serum/plasma samples efficiently precipitate large abundant proteins, such as albumin, while smaller proteins and peptides stay in solution and subsequently can be analyzed by MS (29).

SPE columns are probably the approach most widely used for depletion of high-abundance proteins in serum/plasma and they have been widely used as a first step in biomarker-related proteomic studies. This MARS commercialized by Agilent Technologies containing 14 polyclonal immunoglobulin yolk (IgY) antibodies, the SepproTM mixed 12 spin column (Genway Biotech), may be a promising step for depleting (95%) the most abundant proteins from serum/plasma (30, 31). Recently, we found that SERPINE1 protein might be a linking factor between obesity and diabetes after depletion of high-abundant proteins in plasma and urine samples (32). The major disadvantages of antibody-based depletion columns are those features inherent to working with antibodies, namely, relatively high cost and low sample capacity.

For the immunodepletion of blood samples, dilute human plasma or serum sample (10 μl) and for mouse sample (30 μl) with buffer and syringe filter (0.22 μm). Add sample to depleted column (Agilent Technologies) and spin at 100 × *g* for 1 min and save flow through in tube F1 (33). Let the column rest for 5 min at RT, add more buffer to the column, again spin at 100 × *g* for 2.5 min, and collect the flow through in the same tube F1. Add another buffer to elute bound proteins into new collection tube. This column could be used for multiple times. Fill concentrator with F1 fractions and centrifuge at 7,500 × *g*. Add 50 mM of ammonium bicarbonate (ambic) and fill the concentrator and centrifuge at same 7,500 × *g* (6,700 rpm) for three times for desalting. Digest the sample with trypsin without zip-tip (see below) and run on a mass spectrometer (32).

3.1.4. In-Gel Digestion

Many staining methods are compatible with subsequent tryptic digestion and mass spectrometric analysis. Staining by either Coomassie blue stain (Bio-Safe, Bio-Rad, Cat # 161-0786) or silver staining is recommended (34). Although 200 fM may be enough for a protein identification, however, to achieve high accuracy, 1 pmol (100 ng of 100 kDa protein) is required. However, for a Coomassie blue stained SDS-gel, at least 100 ng is required. After band/spot cutting, vortex tubes for 15 min to remove excess water from the gel pieces. Wash gel pieces with ddw and ACN (1:1), ACN, 50 mM ambic, and ACN sequentially for 15 min each with volume 2× of the gel pieces and speed vac the gel pieces. Add TCEP (1 mg/ml) in 50 mM ambic for reduction for 1 h at 65°C. Add IAA (10 mg/ml) with volume 2× of the gel pieces in 50 mM ambic and incubate at RT for 1 h in dark. Finally, add 250 ng of trypsin diluted with 200 μl of 50 mM ambic and incubate overnight at 37°C. Remove supernatant and concentrate sample in speed vac. The peptides are cleaned with zip-tip as standard protocol. Briefly, add 20 μl of 0.1% TFA to each sample tube. Pipet up and down the sample with zip-tip for 20 times in the sample tube to make sample retain in the tip. Pipet up and down 60% ACN in a 1.5-ml tube six times to elute sample. Dry down the sample in autosample tube in speed vac. Peptides can be analyzed after zip-tip, and resuspend in 50 μl of 0.1% TFA, transfer the supernatant to autosampler tube, and run on mass spectrometer (32).

Peptides can have the tendency to stick to pipette tips, test and storage tubes, tubing connections, injection syringes, or to chromatographic columns. The addition of BSA to the perfusion medium is efficient to reduce sticking of several peptides in the microdialysis tubing or onto the membrane (35). Some reports describe precautions, such as removing aliquots under laminar flow hood, filter sterilization of samples, flushing tubing and syringes with ethanol, and even the addition of peptidase inhibitors to the perfusion fluid to minimize the presence of peptidases, produced by bacteria and capable of degrading the already low levels of peptides in vivo samples (35).

3.2. Liquid Chromatography-Mass Spectrometry

LC-MS is an analytical chemistry technique that combines the physical separation capabilities of liquid chromatography (HPLC) with the mass analysis capabilities of mass spectrometry. LC-MS is a powerful technique used for many applications which has very high sensitivity and specificity. Generally, its application is oriented toward the specific detection and potential identification of chemicals in the presence of other chemicals (in a complex mixture). Before discussing the details of different methods for MS analysis, there are few important points that need to be addressed regarding a mass spectrometer.

3.2.1. Standard Curves and Quality Control Standards

Standard curve ranges for pharmacokinetic studies are highly dependent upon the dosing route and the type of subjects being tested. For example, people tend to use higher doses in animals than in human subjects. As a result, the concentrations of the target analytes in plasma or urine may be much higher (high ng/ml range) in animal studies than for human studies. Different administration routes also significantly impact on the concentrations of the target analytes in the biological fluids, for example oral administration generally leads to lower target compound concentrations than intravenous administration due to the absorption and elimination processes. The use of surrogate standards and internal standards for the quantitation of analytes in biological fluids is generally considered to be important in order to compensate for errors associated with sample preparation and the analytical instrument, respectively. Internal standard methods are often used to improve the precision and accuracy of assay measurements, and hence may be an important inclusion for quantitative measurements of biological fluids. Normally, the choice of an internal standard is driven by compound similarity, like digoxin and dioscin, respectively, which share structural similarities to the ginsenoside compounds measured (36).

Another issue that may arise, especially where solvent composition changes during a chromatographic run, is that if the surrogate or internal standard elutes at significantly different times the droplet properties in the ion source may be different for the analyte and for the internal standard which leads to inequality with signal responses. In addition, matrix effects that were discussed above may again affect different regions of the chromatogram and hence cause significant errors with surrogate or internal standard use, since this could lead to either overestimation or underestimation of the target analyte concentration in biological fluids (37). An alternative approach is to use isotopically labeled internal standards since no significant difference in chromatographic selectivity should exist (38). Ideally, this would decrease errors associated with different elution conditions and also ionization matrix effects.

3.2.2. Selectivity

Selectivity essentially concerns the ability to isolate and/or resolve analytical signals so that specific signals do not impact on the measurement of each other. When the selectivity factor is great enough (complete resolution), accurate measurement of pure analytical signals can be achieved. In this respect, a combination of a highly selective liquid chromatographic separation with a selective detection system is needed with a single mass filter like quadrupole ion trap and time of flight. The second is a three-stage mass filter design and can include quadrupoles, quadrupoles and ion traps, and quadrupoles and time-of-flight devices (39). This configuration, with the exception of quadrupole/ion trap configurations, is often operated, whereby the first mass filter M1 is set to monitor a single *m/z* for a precursor ion. The precursor ions passing through M1 enter the second mass filter M2, where collision-induced dissociation (CID) of the precursor ions occurs as a consequence of the introduction of a collision gas into M2. The third mass filter is then used for tandem MS to monitor a single (product) ion formed from CID within M2 SRM or MRM or every ion formed from the fragmentation process (product ion scanning) (40). The limited blood volume taken from a punch, a larger punch, or multiple punches makes the lower concentration of analyte measurable (23). On the other hand, a larger punch from a small spot might result in a better reproducibility and a small punch from a large area might be one of the sources of poor assay precision due to the chromatographic effect (41). The selectivity of the DBS assay should be established by analyzing blank DBS samples prepared from six individual lots of blood with and without internal standard. Ideally, the six lots of blood should have different hematocrits for the target population. In the case where multianalytes need to be quantified, a cross-interference check should be conducted.

3.2.3. Matrix Effects

Interferences in mass spectrometry are more insidious in nature in that they cannot be directly observed and are described as an alteration in the ionization efficiency as a consequence of co-eluting compounds that can have a negative impact on accuracy and precision (42). The use of internal standards needed to be examined closely since the ionization efficiency of the target analytes in plasma and the internal standards were not the same (43). By increasing retention using a "more selective" extraction procedure, Matuszewski et al. observed a significant change in the slope of the reference curves (44). Also, BSA can hamper the LC-MS/MS assay performance and therefore, protein precipitation with methanol or ACN, followed by centrifugation, is an essential sample cleanup step. Not only matrix components are responsible for ion suppression, but also the mobile-phase solvents can deteriorate sensitivity. Therefore, they chose to use TFA as the mobile-phase additive and to minimize ion suppression effects, postcolumn addition of propionic acid was successfully utilized (13). Because of the nature of the mass spectrometric detection and unlike more nonspecific detectors, interferences in the detection of compounds are rare due to the detection and overlapping of peaks within the same mass channel and are less likely when using SRM mode with tandem mass spectrometers (45). Tong et al. conducted a study on signal interference caused by the dosing excipients using reversed-phase LC in conjunction with ESI (46). Changing the mobile phase to increase retention further enhanced the recoveries of compounds A4216–A4217 to 100 and 108%, respectively (47). It was observed that more polar compounds using a fast gradient were subject to greater ionization suppression. Similar to the finding by Tong et al., an increase in chromatographic retention appeared to decrease the matrix suppression effects (46).

Use of water or water/ethanol as perfusion medium instead of Ringer's solution is recommended to reduce the salt content in the ion source and to avoid the need of sample cleanup (48). Gao et al. added an ice-cold 2 M K_3PO_4 solution to the liver dialysis samples to salt the HIV protease inhibitors out, and afterward an ether extraction was performed (49). Igarashi et al. added methanol to the dialysates to extract opioidmimetics from the matrix (50). Sekar et al. performed SPE for the analysis of vesnarinone and its metabolites in human urine and dialysate fluid (51). Pickl et al. coupled a short polymeric SPE column online with the MS detector to desalt the microdialysates for the analysis of ketoprofen with run times of only 1.5 min (52). Online desalting and sample stacking on the RP analytical column by using water as mobile phase during the first minutes of the chromatographic run is another frequently used desalting step (53). Finally, different groups applied the column switching technique, which is an elegant solution for desalting dialysis samples (52). For example, pretreatment of the paper/card with pasteurized plasma proteins has been found useful in deactivating the sites of active adsorption on the paper and making desorption of analytes easier (19). As part of assay validation, the matrix effect in MS/MS detection and assay recovery should be evaluated by preparing three extracts or solutions: (a) DBS QC (these are the controls which are run in triplicates and contain low, medium and high sample extract values); (b) blank DBS sample extracts (at least in three replicates) postfortified with both the analyte and the internal standard with the concentrations the same as in (a); and (c) neat solutions in at least three replicates with the concentrations of the analytes and the internal standard the same as in (a) and (b). All the above extracts or solutions are assayed using LC-MS/MS with a set of calibration standards. The matrix effect and recovery could be assessed, respectively, by comparing the response (peak area) of (b) and (c), and (a) and (b), with CV% of ≤15% from replicate measurements. A precise estimate of the amount (ng/μg) of the analyte from a single DBS punch (3.2 mm) is necessary for the evaluation, or accurately pipetting the prepared QC blood onto the card/paper and cutting the whole spot for the above assessment should be considered (41). As long as there is no significant decrease or increase of the instrument (MS/MS) response in the retention time region of the target analyte from the blank DBS sample injection, one could assume that there is no apparent ion suppression or enhancement (19).

3.2.4. HPLC-Chip/Mass Spectrometry Analysis

During sample preparation, always wear gloves (powderless, rinsed with water and ethanol before use) to eliminate contamination, including keratins. The most frequent contaminants are BSA and human keratin. Use clean dishes for gel casting as well as staining. Always use siliconized polypropylene tubes as well as low-retention tips to minimize protein loss by adsorption to tube walls. Free acrylamide may react with the amino groups on proteins during polyacrylamide gel electrophoresis, therefore precast gels are good. Excise the bands; be sure to use extremely clean surfaces and new razor blades or scalpels. Ideally, this should be done in a laminar flow hood (tissue culture type) to minimize the possibility of any dust, hair, flakes of skin, or other forms of dirt. Samples must be very "clean" and must not contain any salts, positive ions (Na^+, K^+), negative ions (Cl^-, SO_4^{2-}), which can form ion pairs with basic sites and neutralize positive charge in the analyte(s). Furthermore, positive salt ions can be reduced in the mass spectrometer and plate out on the ion optics, damaging the instrument. The sample must not contain any organic buffer salts, like Tris or MOPS, reducing agents, like DTT, stabilizers (glycerol) and detergents, like SDS. The sample should never be prepared in colored Eppendorf tubes containing residual quantities of dyes, which can swamp out the signal. Solvent components for direct injection are (50% H_2O-distilled and deionized water) and 50% HPLC-grade organic solvent, like ACN, methanol, and isopropanol to increase the solubility of the analyte(s) and to promote desolvation of the charged droplets formed by the electrospray process. Mass spectrometer is sensitive to salts and buffers resulting in signal suppression because the salts normally form adduct peaks which compete with the molecular ion peaks and broaden the overall signal (especially for protein analysis).

A 1 μl of peptide was injected onto an LC/MS system consisting of a 1100 Series liquid chromatography, HPLC-Chip Cube MS interface, and 1100 Series LC/MSD Trap XCT Ultra ion trap mass spectrometer (all Agilent Technologies). The system was equipped with an HPLC-Chip (Agilent Technologies) that incorporated a 40-nl enrichment column and a 43-mm × 75-μm analytical column packed with Zorbax 300SB-C18 5 μm particles. Peptides were loaded onto the enrichment column with 97% solvent A (water with 0.1% formic acid). They were then eluted with a gradient from 3% B (ACN with 0.1% formic acid) to 45% B in 25 min, followed by a steep gradient to 90% B in 5 min at a flow rate of 0.3 μl/min. The total runtime, including column reconditioning, was 35 min. The column effluent was directly coupled to an LC/MSD Trap XCT Ultra ion trap mass spectrometer (Agilent Technologies) via an HPLC-Chip Cube nanospray source operated at ~1,900 V in ultra–ultra mode. The gain control (ICC) was set to 500,000 with a maximum accumulation time of 150 ms. CID was triggered on the six most abundant, not singly charged peptide ions in the *m/z* range of 450–1,500 Da. Precursors were set in an exclusion list for 1 min after two MS/MS spectra (32).

Spectrum Mill for MassHunter Workstation quickly identifies proteins and peptides *via* fast database searches with automatic and manual match validation and unique algorithms that minimize false positives. These also offer de novo spectral interpretation for proteins not found in any database. Spectrum Mill software can identify relative abundance differences of twofold or greater without complicated isotope labeling and summarizes and correlates results in ways that provide maximum insight and convenience. CID data was searched against the SwissProt all species database using the Agilent Spectrum Mill Server software (Rev A.03.03.) installed on an HP Intel® Xeon (TM) dual-processor server. Peak lists were created with the Spectrum Mill Data Extractor program with the following attributed: scans with the same precursor ±1.4 m/z were merged within a time frame of ±15 s. Precursor ions needed to have a minimum signal-to-noise value of 25. Charges up to a maximum of 7 were assigned to the precursor ion, and the 12C peak was determined by the Data Extractor. The SwissProt database was searched for tryptic peptides with a mass tolerance of ±2.5 Da for the precursor ions and a tolerance of ±0.7 Da for the fragment ions. Two missed cleavages were allowed. A Spectrum Mill autovalidation was performed first in the protein details followed by peptide mode using default values (minimum scores, minimum scored peak intensity (SPI), forward minus reversed score threshold, and rank 1 minus rank 2 score threshold). All protein hits found in a distinct database search by Spectrum Mill are nonredundant (32).

The Mascot search engine (http://www.matrixscience.com) was used to obtain peptide identifications. The product ion masses were used as input into the MS/MS search form from Mascot with search parameters as species, allowed number of missed cleavages, possible posttranslational modifications, peptide and MS/MS tolerance, and peptide charge. Mascot scores provide an objective measure of the significance of the search results, where an event, that is, the peptide matching a specific protein, is considered significant if it would be expected to occur at random with a frequency of less than 5%. Since Mascot uses a probability-based implementation of the molecular weight search (MOWSE) scoring algorithm, empirically determined factors are used to assign statistical weights to each individual peptide. Therefore, the value of a significant Mascot score varies depending upon the peptide and the size of the database being searched. For MS/MS search results, true confidence that a protein has been identified comes largely from obtaining matches to multiple peptides from the same protein. It is generally accepted that matching two peptides above the significance level equates to confident protein identification.

Identities obtained from mass spectrometry are sometimes difficult to assign to specific bands from the mass maps because M_r entries in databases do not always correspond closely to experimentally determined M_r. These proteins were either not expressed at sufficient cellular levels to be observed or due to posttranslational modifications that cause an increase in M_r, such as acetylation, phosphorylation, or modifications that decrease the M_r, such as posttranslational cleavage resulting in truncated protein forms. Many theoretical protein M_r in databases are calculated from amino acid sequences translated directly from gene sequences. MS-Fit and Mascot searches may result in protein identifications with extremely high confidence; however, unless an M_r parallels a band in the mass map and the chromatographic retention times match well, it is difficult to determine which M_r or isoform reported in the database is the correct one. By injecting protein standards of known M_r from 5 to 70 kDa into the mass spectrometer, it was determined that the deconvoluted M_r consistently fell within 200 ppm of the known mass.

Phospholipids

Association of plasma phospholipids with ionization suppression or enhancement (matrix effect) is a major source of bioanalytical risk. The total concentration of all phospholipids in human plasma is about 3 mg/ml. The phosphatidylcholine (PC) class of phospholipids accounts for about 60% of the total phospholipids found in plasma, which includes sphingomyelins (SM), phosphatidylethanolamine (PE), phosphatidylinositol (PI), lysophospholipids-PC, phosphatidylserine (PS), and phosphatidylglycerol (PG) listed in the order of decreasing concentration (54). Recently evaluated three tandem mass spectrometry techniques to monitor phospholipids using positive and negative ESI are (55): positive precursor ion scan of *m/z* 184 for the detection of all the PC, lyso-PC, and SM phospholipids; positive neutral loss scan of 141 Da for the detection of lysoPE and PE; and negative precursor ion scan of *m/z* 153 for the detection of the remaining phospholipids. The most effective organic eluent to remove phospholipids from the reversed-phase column used was isopropyl alcohol, followed by ACN and methanol, where methanol was indicated to be a stronger eluent than ACN (56). Under the reversed-phase chromatographic conditions used for LC-MS/MS bioanalytical methods, typically 0.6 ml/min flow rate for a 2 × 50-mm column with a mobile phase consisting of less than 80% methanol or ACN, lyso-PC, elutes near the analytes (drugs, metabolites, or biomarkers) while the phospholipids, such as PC, elute much later. It is highly desirable that the phospholipids elute after the analyte of interest since this would allow, if so desired, changing the chromatographic conditions to rapidly elute the phospholipids off the column after the analyte elution. One approach would be to completely elute all the phospholipids off the column before the next injection of a sequence consisting of a large batch of samples. Following the elution of the analyte, this can be accomplished in a rapid manner by increasing the flow rate and/or raising the organic component of the mobile phase to 100% or by introducing another organic solvent of greater eluting strength, such as isopropanol. The technique of LLE for the extraction of small-molecule drugs and metabolites from plasma samples is preferred over the other techniques of sample extraction by SPE, protein precipitation, and online extraction.

Cell Lysate

The cell lysates were obtained from patients with breast cancer, prostate cancer, and pancreatic cancer. Cells were lysed in a urea buffer system, and soluble proteins are separated by their p*I* using liquid-phase isoelectric focusing (IEF) in the first dimension. The second dimension employs nonporous silica reverse-phase-HPLC (RP-HPLC) to separate proteins according to their hydrophobicity. Eluent from the RP-HPLC step can be analyzed directly by electrospray ionization-quadrupole time-of-flight (ESI-TOF) MS or collected as fractions for off-line analysis by Matrix-assisted laser desorption/ionization-quadrupole time of flight (MALDI-QTOF) MS/MS. In order to make quantitative comparisons between samples from different cells, it is important to accurately measure the entire amount of protein present in each sample. This is achieved by creating a selected ion chromatogram (SIC) for each detected protein in LC/MS. The data collected from online LC/MS was initially viewed as a total ion chromatogram for multiply charged protein spectra, which were then integrated to provide a single combined spectrum for every protein present. A maximum entropy deconvolution algorithm (MaxEnt 1) (Waters, MA) was used to deconvolute multiply charged spectra and produce molecular mass spectra. Often, more than one peak from the protein umbrella must be integrated due to the complexity and number of proteins in the sample, where a given peak in a protein umbrella could overlap with an umbrella spectrum from another protein. The mass of each peptide was taken as the centroid mass of the isotopic envelope with the program MagTran (57). Masses and intensities from the molecular spectra were then converted to a single text file and imported into ProteoVue, a mass mapping program (58). Protein fractions collected off-line undergo tryptic digestion and SPE prior to analysis by either of the MALDI techniques. MALDI-QTOF MS is performed to obtain a profile of all tryptic peptides present in a fraction. Results from both MALDI techniques are compared against established protein databases to identify proteins, p*I*, mass, and relative abundance data.

Small Molecules in Dried Blood Spot

Unlike aqueous samples, including whole blood, plasma, or serum, for which sample homogeneity for quantitative analysis could be readily ensured by completely thawing and thoroughly vortex mixing the content in the sample vials/tubes, DBS is a special matrix in which the analyte of interest distributes after spotting. Although each individual DBS punch should theoretically represent a specific homogeneous blood volume (15 μl), the DBS sample quality might vary from spot to spot and from the spots of standard/QC samples to those of study samples after spots are made and dried (5). Unfortunately, remixing the dried blood prior to analysis is not possible and therefore, during the development of an LC-MS/MS assay method, a considerable effort is needed. According to the current industry practice, certain lots of blank matrix should be screened to ensure no interference from the matrix before its use for the preparation of calibration standards and QC samples and for the assessment of assay recovery and matrix effects.

For new chemical entities, like new drug candidates, there might be less difficulty in identifying the "clean" matrix to be used for preparation of calibration standard and QC samples. However, for small biomarker molecules, it might be difficult because these molecules are endogenous with varying concentrations. A common approach to overcoming this is to mix the washed blood cells with artificial serum or plasma to make artificial "whole blood," as demonstrated in several applications for the simultaneous analysis of multiendogenous compounds (23).

Microdialysates

The combination of brain dialysis with microelectrospray tandem MS for the analysis of neuropeptides was first described by Andren and Caprioli for $[Met]^5$-enkephalin (Met-enk) and neurotensin (NT) (59). A 10 μl of the dialysis samples were preconcentrated and desalted on a nanoflow SPE column containing C18 RP resin particles packed in the tip of a 50-μm i.d. fused silica capillary micro-ESI needle. With this device, peptides were concentrated and adsorbed to the column, which was rinsed to desalt the sample using a mobile phase containing 2% methanol and 0.25% acetic acid. The peptides of interest were subsequently eluted with a solvent containing 50% methanol and 0.25% acetic acid and introduced into the triple quadrupole MS. Nyitrai et al. used LC-ESI-MS to measure increases in the extracellular level of somatostatin in the hippocampus of the rat after local administration of a $GABA_B$ receptor antagonist (60). Pettersson et al. published a promising study of solid-supported enhanced MD for peptides (61).

Steroids in Human Hair

To overcome ELISA screening for false results, recently a new method capable of detecting 0.5 pg stanozolol and 3 pg nandrolone per mg of hair was developed by Deshmukh et al. when approximately 20 mg of hair were processed (11). Briefly, 20 mg of hair segments were incubated with 1 ml 1 M sodium hydroxide at 95°C for 15 min in the presence of stanozolol D3, which was used as an internal standard. After cooling, the homogenate was neutralized with 1 M hydrochloric acid followed by addition of 2 ml of 0.2 M phosphate buffer (pH 7.0). The homogenate was purified by LLE using pentane. After vortex mixing and centrifugation (4 min at 1,257 × *g*), the organic layer was transferred into a fresh glass tube after filtering through a hydrophilic (Teflon®) polytetrafluoroethylene (PTFE) membrane (0.45 μm). The organic layer was then dried under a gentle stream of nitrogen gas at 50°C. The extracted residue was reconstituted with 100 μl ACN. A 4-μl aliquot of the solution was injected into the LC-MS/MS system. An Agilent ZORBAX SB-C18 column (2.1 × 50 mm, 1.8 μm) was used for analysis. The column oven temperature was maintained at 60°C. The two mobile phases used composed of ACN (solvent A) and 0.1% formic acid in water (solvent B). The total flow rate through the column was 100 μl/min. The mass spectrometer was operated in the positive electrospray ionisation mode at a spray voltage of 4,500 V and capillary temperature of 350°C. The protonated molecules of nandrolone, stanozolol, and stanozolol D3 generated act as precursor ions for CID for MS-MS analysis. SRM was used for the confirmation of steroids. Calibration curves and quality controls at low, medium, and high concentration levels were prepared by spiking negative hair control with known amounts of steroids and internal standard (11).

Ethanolamides

The identification of biomarkers is one of the very important goals of researchers. Proteins that are differentially expressed (upregulated or downregulated) between normal and diseased cells at high frequency represent potential breast cancer biomarkers. These markers can be targets for drug development and therapeutics or serve as prognostic indicators. In order to compare quantitation level differences, the total peak areas of detected proteins are compared. The method employed for quantitation of three fatty acids in human plasma are: D_4-AEA, D_4-OEA, and $^{13}C_2$-PEA as "surrogate analytes" to establish the concentration–mass response relationship that is a regression equation by quantitation of AEA, OEA, and PEA, the stable isotope-labeled compounds D_4-AEA, D_4-OEA, and $^{13}C_2$-PEA (12). The concentrations of AEA, OEA, and PEA were calculated based on the regression equations derived from the surrogate analytes. This approach made it possible to prepare calibration standard and QC samples in plasma devoid of interferences from the endogenous analytes, and thus there is no true "blank" plasma available (62). The stable isotope-labeled ethanolamides are not present in human plasma and, therefore, provide response–concentration equations free of interference from the endogenous analytes. Both D_4-AEA and D_8-AEA are commercially available. The analytical methodology required 150 µl of human plasma that was processed via LLE using a 96-well plate format.

Chromatographic separation was achieved with an HPLC column using gradient elution, and the run time was 3 min. Surrogate analyte approach ensures that the SRM channels of the authentic analytes, surrogate analytes, and internal standards do not interfere with each other. Since endogenous PEA concentration is low (maximum concentration determined in the current study is 2.48 ng/ml), the contribution from endogenous PEA to $^{13}C_2$-PEA signal would be negligible. In addition, with careful selection of internal standard concentrations, the impact from D_2-OEA to D_4-OEA was minimized, so was that from $^{13}C_2$-PEA to D_4-PEA. The isotopic purity was determined for all three surrogate analytes and three internal standards (12).

Heat-Shock Proteins

Heat shock proteins (HSPs) are a highly conserved family of proteins also referred to as molecular chaperones. Their expression is induced when a cell undergoes various types of environmental stress, and their function is being explored in the development of cancer vaccines (63). Elevated levels of HSPs have been reported in breast cancer cells (64). To map the peptides produced by digestion of HSPs and malate dehydrogenase (MDH) with pepsin, 20 μM concentration of protein is taken. The sample was applied to a column of immobilized pepsin (2 × 50 mm) using water and 0.05% TFA as the mobile phase at a flow rate of 50 μl/min (65). The protein digest was collected by a micropeptide trap (Michrome BioResources, Auburn, CA) and washed for 2 min at a flow rate of 300 μl/min. Peptides in the trap were then eluted onto a microbore C18 HPLC column (Micro-Tech Scientific, Vista, CA) coupled to a Waters micro Q-TOF (Milford, MA) with a typical ESI voltage of 3 kV for accurate parent mass measurements (66). Peptides were eluted from the column over 12 min using a gradient of 15–45% ACN at a flow rate of 50 μl/min. The micropeptide trap and HPLC column were immersed in ice water during the entire process. The typical ESI voltage used was 4.5 kV. Both parent mass and tandem mass spectra were used for peptide identification. The Waters micro Q-TOF was used for solution-phase hydrogen/deuterium exchange (HX) coupled with mass spectrometry.

3.3. Bioinformatics

The bioinformatics analysis was performed with Ingenuity Pathway Analysis (IPA) as previously reported (32). IPA is all-in-one software application to identify the biological mechanisms, pathways, and functions most relevant to its experimental datasets or genes of interest, and thus is very effective for understanding protein–protein interactions within the context of metabolic or signaling pathways, understanding how proteins operate and form pathways by manually abstracting and curating a large fraction of the biomedical literature according to a very strict curation process, followed by storing the data in a highly structured manner.

4. Notes

1. The paradigm shift from DNA sequencing and mRNA profiling to proteomics is beginning to be realized. Proteins perform at the functional level in the cell and proteomic-based approaches are essential in the attempt to elucidate the cellular changes that occur during disease progression. Methodology must be capable of not only identifying proteins, but also detecting those that are differentially expressed in normal *versus* diseased organs, tissue and cells. These proteins, potential biomarkers, provide researchers with targets that can be used in diagnostics, drug development, and therapeutics. Two-dimensional electrophoresis with peptide mass fingerprinting by MALDI MS has

been virtually the only method possessing these capabilities, despite its inherent problems. Complementary and/or alternative techniques described herein, provides a desirable alternative to 2-DE especially due to its capabilities to interface with mass spectrometric analysis. Preparative liquid IEF followed by RP-HPLC provides the means for accurate interlysate comparisons based upon the high resolution and mass accuracy of M_r.

2. The use of LC/MS appears to be the method of choice for the measurement of target analytes of herbal medicine in biological fluids, such as blood, plasma, and urine. This is because of selectivity and specificity gains that can be achieved by both the chromatographic and the mass detection systems. The good sensitivity gained through LC/MS is through a combination of both increasing the analyte signal and reducing noise. Method development for the determination of analytes from biological fluids highlights the importance of minimizing the on-column dilution effect that can be achieved by using small-diameter traditional HPLC columns (2 mm diameter) or increasing signal intensity through the use of new particle size technologies for high-throughput chromatography. The "unseen" ionization suppression and enhancement effects must be evaluated and if ignored can severely impact on both the accuracy and precision of quantitative measurements for bioavailability and pharmacokinetic determinations.

3. DBS-LC-MS/MS is expected to play an increasingly important role in the quantitative analysis of drugs and metabolites in blood. Unfortunately, the metabolite concentration is usually very low compared with the parent compound, and assay sensitivity is always a challenge for the determination of some analytes at low circulating levels. Unlike aqueous matrix, such as whole blood, plasma, or serum, for which sample homogeneity could be easily ensured during sample collection and prior to analysis, the quality of DBS samples is often a concern for absolute quantification due to the possible DBS card/paper quality issue, possible noncompliance in sample collection (blood volume on the spot, degree of saturation, and degree of hemolysis), drying, storage, and transportation. Any of the above uncertainties could lead to a higher degree of assay errors using this nonvolumetrically (size) based approach for quantitative analysis than the conventional methods. More importantly, whereas most analytes may be stable on DBS after drying, some enzyme labile compounds, especially prodrugs, are unstable even with the pretreated card/paper. Hair specimens are collected without invasion of privacy and are easy to store for the determination of stanozolol and nandrolone,

which facilitate improved doping testing regimes that are negligible risk of infection and can be used in combination with urinalysis or blood test.

Acknowledgments

This work was supported in part by a Research Advancement Award from Scott & White Memorial Hospital and Clinic (P.K.), Institutional support from Scott & White Memorial Hospital and Clinic, Texas A&M Health Science Center College of Medicine, the Central Texas Veterans Health Administration, an Endowment from the Cain Foundation, and the US National Institutes of Health grant RO1CA91889 (A.A.).

References

1. Anderson, N. L., Matheson, A. D., and Steiner, S. (2000) Proteomics: applications in basic and applied biology *Curr Opin Biotechnol* **11**, 408–12.
2. Adkins, J. N., Varnum, S. M., Auberry, K. J., Moore, R. J., Angell, N. H., Smith, R. D., Springer, D. L., and Pounds, J. G. (2002) Toward a human blood serum proteome: analysis by multidimensional separation coupled with mass spectrometry *Mol Cell Proteomics* **1**, 947–55.
3. Diamandis, E. P. (2004) How are we going to discover new cancer biomarkers? A proteomic approach for bladder cancer *Clin Chem* **50**, 793–5.
4. Lathrop, J. T., Hayes, T. K., Carrick, K., and Hammond, D. J. (2005) Rarity gives a charm: evaluation of trace proteins in plasma and serum *Expert Rev Proteomics* **2**, 393–406.
5. Edelbroek, P. M., van der Heijden, J., and Stolk, L. M. (2009) Dried blood spot methods in therapeutic drug monitoring: methods, assays, and pitfalls *Ther Drug Monit* **31**, 327–36.
6. Garcia Boy, R., Henseler, J., Mattern, R., and Skopp, G. (2008) Determination of morphine and 6-acetylmorphine in blood with use of dried blood spots *Ther Drug Monit* **30**, 733–9.
7. Chace, D. H., Millington, D. S., Terada, N., Kahler, S. G., Roe, C. R., and Hofman, L. F. (1993) Rapid diagnosis of phenylketonuria by quantitative analysis for phenylalanine and tyrosine in neonatal blood spots by tandem mass spectrometry *Clin Chem* **39**, 66–71.
8. Shigematsu, Y., Hata, I., and Tanaka, Y. (2007) Stable-isotope dilution measurement of isovalerylglycine by tandem mass spectrometry in newborn screening for isovaleric acidemia *Clin Chim Acta* **386**, 82–6.
9. Jemal, M., Ouyang, Z., and Xia, Y. Q. Systematic LC-MS/MS bioanalytical method development that incorporates plasma phospholipids risk avoidance, usage of incurred sample and well thought-out chromatography *Biomed Chromatogr* **24**, 2–19.
10. Kintz, P., Cirimele, V., Sachs, H., Jeanneau, T., and Ludes, B. (1999) Testing for anabolic steroids in hair from two bodybuilders *Forensic Sci Int* **101**, 209–16.
11. Deshmukh, N., Hussain, I., Barker, J., Petroczi, A., and Naughton, D. P. (2010) Analysis of anabolic steroids in human hair using LC-MS/MS *Steroids* **75**, 710–14.
12. Jian, W., Edom, R., Weng, N., Zannikos, P., Zhang, Z., and Wang, H. Validation and application of an LC-MS/MS method for quantitation of three fatty acid ethanolamides as biomarkers for fatty acid hydrolase inhibition in human plasma *J Chromatogr B Analyt Technol Biomed Life Sci* **878**, 1687–99.
13. Lanckmans, K., Sarre, S., Smolders, I., and Michotte, Y. (2008) Quantitative liquid chromatography/mass spectrometry for the analysis of microdialysates *Talanta* **74**, 458–69.
14. McKinney, M. K., and Cravatt, B. F. (2005) Structure and function of fatty acid amide hydrolase *Annu Rev Biochem* **74**, 411–32.
15. Hsieh, S. Y., Chen, R. K., Pan, Y. H., and Lee, H. L. (2006) Systematical evaluation of the

effects of sample collection procedures on low-molecular-weight serum/plasma proteome profiling *Proteomics* **6**, 3189–98.

16. Banks, R. E., Stanley, A. J., Cairns, D. A., Barrett, J. H., Clarke, P., Thompson, D., and Selby, P. J. (2005) Influences of blood sample processing on low-molecular-weight proteome identified by surface-enhanced laser desorption/ionization mass spectrometry *Clin Chem* **51**, 1637–49.
17. West-Nielsen, M., Hogdall, E. V., Marchiori, E., Hogdall, C. K., Schou, C., and Heegaard, N. H. (2005) Sample handling for mass spectrometric proteomic investigations of human sera *Anal Chem* **77**, 5114–23.
18. Fingerhut, R. (2009) False positive rate in newborn screening for congenital adrenal hyperplasia (CAH)-ether extraction reveals two distinct reasons for elevated 17alpha-hydroxyprogesterone (17-OHP) values *Steroids* **74**, 662–5.
19. van der Heijden, J., de Beer, Y., Hoogtanders, K., Christiaans, M., de Jong, G. J., Neef, C., and Stolk, L. (2009) Therapeutic drug monitoring of everolimus using the dried blood spot method in combination with liquid chromatography-mass spectrometry *J Pharm Biomed Anal* **50**, 664–70.
20. Mauri, P., De Palma, A., Pozzi, F., Basilico, F., Riva, A., Morazzoni, P., Bombardelli, E., and Rossoni, G. (2006) LC-MS characterization of terpene lactones in plasma of experimental animals treated with Ginkgo biloba extracts Correlation with pharmacological activity *J Pharm Biomed Anal* **40**, 763–8.
21. Hua, L., Guangji, W., Hao, L., Minwen, H., Haitang, X., Chenrong, H., Jianguo, S., and Tian, L. (2006) Sensitive and selective liquid chromatography-electrospray ionization mass spectrometry analysis of ginkgolide B in dog plasma *J Pharm Biomed Anal* **40**, 88–94.
22. Zhao, Y., Wang, L., Bao, Y., and Li, C. (2007) A sensitive method for the detection and quantification of ginkgo flavonols from plasma *Rapid Commun Mass Spectrom* **21**, 971–81.
23. Higashi, T., Nishio, T., Uchida, S., Shimada, K., Fukushi, M., and Maeda, M. (2008) Simultaneous determination of 17alpha-hydroxypregnenolone and 17alpha-hydroxyprogesterone in dried blood spots from low birth weight infants using LC-MS/MS *J Pharm Biomed Anal* **48**, 177–82.
24. Ding, S., Dudley, E., Chen, L., Plummer, S., Tang, J., Newton, R. P., and Brenton, A. G. (2006) Determination of active components of Ginkgo biloba in human urine by capillary high-performance liquid chromatography/mass spectrometry with on-line column-switching purification *Rapid Commun Mass Spectrom* **20**, 3619–24.
25. Xie, J., Ding, C., Ge, Q., Zhou, Z., and Zhi, X. (2008) Simultaneous determination of ginkgolides A, B, C and bilobalide in plasma by LC-MS/MS and its application to the pharmacokinetic study of Ginkgo biloba extract in rats *J Chromatogr B Analyt Technol Biomed Life Sci* **864**, 87–94.
26. Wang, X., Zhao, T., Gao, X., Dan, M., Zhou, M., and Jia, W. (2007) Simultaneous determination of 17 ginsenosides in rat urine by ultra performance liquid chromatography-mass spectrometry with solid-phase extraction *Anal Chim Acta* **594**, 265–73.
27. Yu, K., Ma, Y., Shao, Q., Qu, H., and Cheng, Y. (2007) Simultaneously determination of five ginsenosides in rabbit plasma using solid-phase extraction and HPLC/MS technique after intravenous administration of 'SHENMAI' injection *J Pharm Biomed Anal* **44**, 532–9.
28. Orvisky, E., Drake, S. K., Martin, B. M., Abdel-Hamid, M., Ressom, H. W., Varghese, R. S., An, Y., Saha, D., Hortin, G. L., Loffredo, C. A., and Goldman, R. (2006) Enrichment of low molecular weight fraction of serum for MS analysis of peptides associated with hepatocellular carcinoma *Proteomics* **6**, 2895–902.
29. Chertov, O., Biragyn, A., Kwak, L. W., Simpson, J. T., Boronina, T., Hoang, V. M., Prieto, D. A., Conrads, T. P., Veenstra, T. D., and Fisher, R. J. (2004) Organic solvent extraction of proteins and peptides from serum as an effective sample preparation for detection and identification of biomarkers by mass spectrometry *Proteomics* **4**, 1195–203.
30. Echan, L. A., Tang, H. Y., Ali-Khan, N., Lee, K., and Speicher, D. W. (2005) Depletion of multiple high-abundance proteins improves protein profiling capacities of human serum and plasma *Proteomics* **5**, 3292–303.
31. Huang, L., Harvie, G., Feitelson, J. S., Gramatikoff, K., Herold, D. A., Allen, D. L., Amunngama, R., Hagler, R. A., Pisano, M. R., Zhang, W. W., and Fang, X. (2005) Immunoaffinity separation of plasma proteins by IgY microbeads: meeting the needs of proteomic sample preparation and analysis *Proteomics* **5**, 3314–28.
32. Kaur, P., Reis, M. D., Couchman, G. R., Forjuoh, S. N., Greene Jr, J. F., and Asea, A. (2010) SERPINE1 links obesity and diabetes: a pilot study *J Proteomics Bioinform* **3**, 191–199.
33. Brand, J., Haslberger, T., Zolg, W., Pestlin, G., and Palme, S. (2006) Depletion efficiency and recovery of trace markers from a multiparameter immunodepletion column *Proteomics* **6**, 3236–42.

34. Wilm, M., Shevchenko, A., Houthaeve, T., Breit, S., Schweigerer, L., Fotsis, T., and Mann, M. (1996) Femtomole sequencing of proteins from polyacrylamide gels by nano-electrospray mass spectrometry *Nature* **379**, 466–9.
35. Maidment, N. T., Brumbaugh, D. R., Rudolph, V. D., Erdelyi, E., and Evans, C. J. (1989) Microdialysis of extracellular endogenous opioid peptides from rat brain in vivo *Neuroscience* **33**, 549–57.
36. Li, X., Sun, J., Wang, G., Hao, H., Liang, Y., Zheng, Y., Yan, B., and Sheng, L. (2007) Simultaneous determination of panax notoginsenoside Rl, ginsenoside Rgl, Rd, Re and Rbl in rat plasma by HPLC/ESI/MS: platform for the pharmacokinetic evaluation of total panax notoginsenoside, a typical kind of multiple constituent traditional Chinese medicine *Biomed Chromatogr* **21**, 735–46.
37. Grova, N., Monteau, F., Le Bizec, B., Feidt, C., Andre, F., and Rychen, G. (2005) Determination of phenanthrene and hydroxyphenanthrenes in various biological matrices at trace levels using gas chromatography-mass spectrometry *J Anal Toxicol* **29**, 175–81.
38. Stokvis, E., Rosing, H., and Beijnen, J. H. (2005) Stable isotopically labeled internal standards in quantitative bioanalysis using liquid chromatography/mass spectrometry: necessity or not? *Rapid Commun Mass Spectrom* **19**, 401–7.
39. Chan, E. C., Yap, S. L., Lau, A. J., Leow, P. C., Toh, D. F., and Koh, H. L. (2007) Ultra-performance liquid chromatography/time-of-flight mass spectrometry based metabolomics of raw and steamed Panax notoginseng *Rapid Commun Mass Spectrom* **21**, 519–28.
40. Dooley, K. C. (2003) Tandem mass spectrometry in the clinical chemistry laboratory *Clin Biochem* **36**, 471–81.
41. Hoogtanders, K., van der Heijden, J., Christiaans, M., Edelbroek, P., van Hooff, J. P., and Stolk, L. M. (2007) Therapeutic drug monitoring of tacrolimus with the dried blood spot method *J Pharm Biomed Anal* **44**, 658–64.
42. Taylor, P. J. (2005) Matrix effects: the Achilles heel of quantitative high-performance liquid chromatography-electrospray-tandem mass spectrometry *Clin Biochem* **38**, 328–34.
43. Rogatsky, E., and Stein, D. (2005) Evaluation of matrix effect and chromatography efficiency: new parameters for validation of method development *J Am Soc Mass Spectrom* **16**, 1757–9.
44. Matuszewski, B. K., Constanzer, M. L., and Chavez-Eng, C. M. (1998) Matrix effect in quantitative LC/MS/MS analyses of biological fluids: a method for determination of finasteride in human plasma at picogram per milliliter concentrations *Anal Chem* **70**, 882–9.
45. Bonfiglio, R., King, R. C., Olah, T. V., and Merkle, K. (1999) The effects of sample preparation methods on the variability of the electrospray ionization response for model drug compounds *Rapid Commun Mass Spectrom* **13**, 1175–1185.
46. Tong, X. S., Wang, J., Zheng, S., Pivnichny, J. V., Griffin, P. R., Shen, X., Donnelly, M., Vakerich, K., Nunes, C., and Fenyk-Melody, J. (2002) Effect of signal interference from dosing excipients on pharmacokinetic screening of drug candidates by liquid chromatography/mass spectrometry *Anal Chem* **74**, 6305–13.
47. Gray, M. J., Chang, D., Zhang, Y., Liu, J., and Bensoussan, A. Development of liquid chromatography/mass spectrometry methods for the quantitative analysis of herbal medicine in biological fluids: a review *Biomed Chromatogr* **24**, 91–103.
48. Kobayashi, N., Kazui, M., and Ikeda, T. (2000) Rapid, real-time sampling of R-84760 in blood by in vivo microdialysis with tandem mass spectrometry *J Pharm Biomed Anal* **21**, 1233–42.
49. Gao, W., Kishida, T., Kimura, K., Kageyama, M., Sumi, M., Yoshikawa, Y., Shibata, N., and Takada, K. (2002) Sensitive and simultaneous determination of HIV protease inhibitors in rat biological samples by liquid chromatography-mass spectrometry *Biomed Chromatogr* **16**, 267–73.
50. Igarashi, K., Murabayashi, Y., Hotta, K., Kitamura, Y., Kasuya, F., Shiotani, K., Tingyou, L., Miyazaki, A., Tsuda, Y., Okada, Y., and Fukushima, S. (2004) Application of liquid chromatography-tandem mass spectrometry for the determination of opioidmimetics in the brain dialysates from rats treated with opioidmimetics intraperitoneally *J Chromatogr B Analyt Technol Biomed Life Sci* **806**, 53–7.
51. Sekar, K. S., and Bramer, S. L. (2003) Extended application of an LC-MS/MS method for the analysis of vesnarinone and its metabolites in human urine and dialysate fluid *J Pharm Biomed Anal* **33**, 711–7.
52. Pickl, K. E., Magnes, C., Bodenlenz, M., Pieber, T. R., and Sinner, F. M. (2007) Rapid online-SPE-MS/MS method for ketoprofen determination in dermal interstitial fluid samples from rats obtained by microdialysis or open-flow microperfusion *J Chromatogr B Analyt Technol Biomed Life Sci* **850**, 432–9.
53. Prokai, L., Zharikova, A. D., Janaky, T., and Prokai-Tatrai, K. (2000) Exploratory pharmacokinetics and brain distribution study of a neuropeptide FF antagonist by liquid

chromatography/atmospheric pressure ionization tandem mass spectrometry *Rapid Commun Mass Spectrom* **14**, 2412–8.

54. Pang, L. Q., Liang, Q. L., Wang, Y. M., Ping, L., and Luo, G. A. (2008) Simultaneous determination and quantification of seven major phospholipid classes in human blood using normal-phase liquid chromatography coupled with electrospray mass spectrometry and the application in diabetes nephropathy *J Chromatogr B Analyt Technol Biomed Life Sci* **869**, 118–25.
55. Xia, Y. Q., and Jemal, M. (2009) Phospholipids in liquid chromatography/mass spectrometry bioanalysis: comparison of three tandem mass spectrometric techniques for monitoring plasma phospholipids, the effect of mobile phase composition on phospholipids elution and the association of phospholipids with matrix effects *Rapid Commun Mass Spectrom* **23**, 2125–38.
56. Little, J. L., Wempe, M. F., and Buchanan, C. M. (2006) Liquid chromatography-mass spectrometry/mass spectrometry method development for drug metabolism studies: Examining lipid matrix ionization effects in plasma *J Chromatogr B Analyt Technol Biomed Life Sci* **833**, 219–30.
57. Zhang, Z., and Marshall, A. G. (1998) A universal algorithm for fast and automated charge state deconvolution of electrospray mass-to-charge ratio spectra *J Am Soc Mass Spectrom* **9**, 225–33.
58. Chong, B. E., Hamler, R. L., Lubman, D. M., Ethier, S. P., Rosenspire, A. J., and Miller, F. R. (2001) Differential screening and mass mapping of proteins from premalignant and cancer cell lines using nonporous reversed-phase HPLC coupled with mass spectrometric analysis *Anal Chem* **73**, 1219–27.
59. Emmett, M. R., Andren, P. E., and Caprioli, R. M. (1995) Specific molecular mass detection of endogenously released neuropeptides using in vivo microdialysis/mass spectrometry *J Neurosci Methods* **62**, 141–7.
60. Nyitrai, G., Kekesi, K. A., Emri, Z., Szarics, E., Juhasz, G., and Kardos, J. (2003) GABA(B) receptor antagonist CGP-36742 enhances somatostatin release in the rat hippocampus in vivo and in vitro *Eur J Pharmacol* **478**, 111–9.
61. Pettersson, A., Amirkhani, A., Arvidsson, B., Markides, K., and Bergquist, J. (2004) A feasibility study of solid supported enhanced microdialysis *Anal Chem* **76**, 1678–82.
62. Jemal, M., Schuster, A., and Whigan, D. B. (2003) Liquid chromatography/tandem mass spectrometry methods for quantitation of mevalonic acid in human plasma and urine: method validation, demonstration of using a surrogate analyte, and demonstration of unacceptable matrix effect in spite of use of a stable isotope analog internal standard *Rapid Commun Mass Spectrom* **17**, 1723–34.
63. Liu, B., DeFilippo, A. M., and Li, Z. (2002) Overcoming immune tolerance to cancer by heat shock protein vaccines *Mol Cancer Ther* **1**, 1147–51.
64. Franzen, B., Linder, S., Alaiya, A. A., Eriksson, E., Uruy, K., Hirano, T., Okuzawa, K., and Auer, G. (1996) Analysis of polypeptide expression in benign and malignant human breast lesions: down-regulation of cytokeratins *Br J Cancer* **74**, 1632–8.
65. Wang, L., Pan, H., and Smith, D. L. (2002) Hydrogen exchange-mass spectrometry: optimization of digestion conditions *Mol Cell Proteomics* **1**, 132–8.
66. Cheng, G., Cusanovich, M. A., and Wysocki, V. H. (2006) Properties of the dark and signaling states of photoactive yellow protein probed by solution phase hydrogen/deuterium exchange and mass spectrometry *Biochemistry* **45**, 11744–51.

Chapter 15

Bioinformatic Approach to Identify Chaperone Pathway Relationship from Large-Scale Interaction Networks

Yunchen Gong, Zhaolei Zhang, and Walid A. Houry

Abstract

We describe a computational protocol to identify functional modules and pathway relationship of chaperones based on physical interaction data derived from high-throughput proteomic experiments. The protocol first identifies interacting proteins shared by the different chaperone systems to organize the chaperones into functional modules. The chaperone functional modules represent groups of chaperones that are involved in mediating the folding of the shared interacting proteins. Either the chaperones in a module can function along a single folding pathway of a given substrate protein or the substrate protein might have two or more different folding pathways that the chaperones act on independently. As described in our computational protocol, probabilities of these pathway relationships between two chaperones in a two-component chaperone module can be determined using whole-genome expression and cellular pathways as reference. This protocol is potentially useful for identifying functional modules and pathway relationships in other biological systems that involve multiple proteins with many identified interactions.

Key words: Chaperone, Protein interaction network, Functional module, Pathway relationship

1. Introduction

Molecular chaperones represent a large and diverse group of proteins whose general function is to maintain protein homeostasis in the cell (1, 2). Consequently, molecular chaperones play a wide range of cellular roles including protein folding and unfolding, protein disassembly and disaggregation, protein degradation, protein translocation, endoplasmic reticulum associated protein degradation (ERAD), and ribosomal RNA processing among many other functions. In the well-studied model organism *Saccharomyces cerevisiae* (budding yeast), there are 7 small heat shock proteins, 3 chaperones of the AAA+ family, 8 of the CCT/TRiC complex,

Stuart K. Calderwood and Thomas L. Prince (eds.), *Molecular Chaperones: Methods and Protocols*, Methods in Molecular Biology, vol. 787, DOI 10.1007/978-1-61779-295-3_15, © Springer Science+Business Media, LLC 2011

6 of the prefoldin/GimC complex, 22 Hsp40s, 1 Hsp60 (& 1 Hsp10), 14 Hsp70s, and 2 Hsp90s (3). These 63 chaperones are localized in the cytoplasm/nucleus, mitochondria, and the endoplasmic reticulum.

Recently, we have identified the TAP-tag based interactors for all of these chaperones (4). A total of 21,687 unique pairs of interactions were identified with high confidence. These interactions are between the 63 chaperones and a total of 4,340 other proteins; in addition, there are 259 chaperone–chaperone interactions. All of our data is deposited in a publicly database that we created and termed ChaperoneDB (http://chaperonedb.ccbr.utoronto.ca/).

Two chaperones interacting with a given protein might functionally collaborate to assist in the folding of that protein or one chaperone might be redundant with the other. Both of these scenarios had been experimentally observed (see e.g. refs. 5, 6). In the former case, the chaperone-assisted folding of the substrate protein is along a single pathway, while in the latter case, the chaperone-assisted folding can proceed along alternate multiple parallel pathways (Fig. 1). Analysis of the large-scale chaperone interaction data that we have recently published (4) allowed us to identify "chaperone modules," herein defined as a group of chaperones interacting with a common set of proteins. It also allowed us to determine whether chaperones in a two-component module act along single or multiple folding pathways for a given protein substrate (Fig. 1). In this chapter, we present the details of the protocol applied in that previous work and focus on the algorithmic and programming aspects.

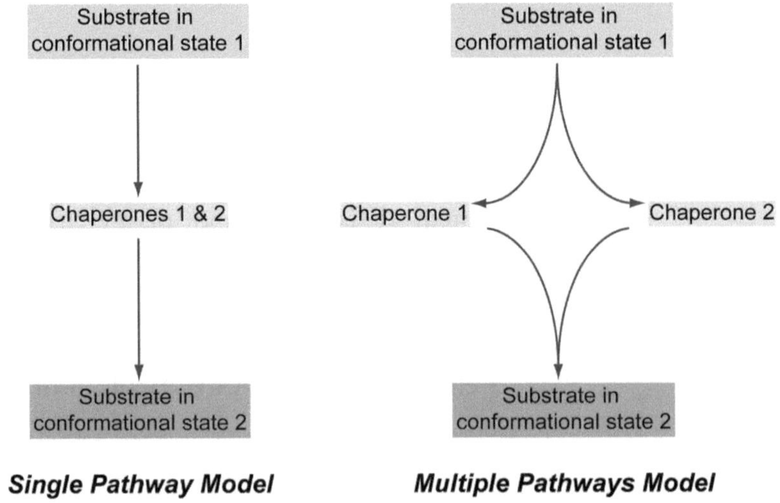

Fig. 1. The schematic depicts chaperone-mediated single pathway and multiple pathways folding models.

Briefly, our protocol consists of four steps: (1) Raw interactions are first filtered based on their experimental scores and demonstrated interactions in public interaction databases. (2) Chaperone functional modules are then inferred based on the numbers of shared interaction partners using a *Z*-score criteria. (3) To reveal the pathway relationships between the chaperones in a two-component functional module, genome-scale gene expression data are analyzed for all protein pairs located in the same and/or different pathways. (4) Finally, a statistical integration approach is applied for calculating the probability of the pathway relationship of the two chaperones.

2. Programs and Data Sources

This bioinformatic protocol relies heavily on implementations of a variety of algorithms for data processing and analysis. Some of them are simply implemented in a programming language of user's choice, while others are found in existing software packages. A scripting language such as PHP (http://www.php.net) and Perl (http://www.perl.org) is good enough for routine data processing such as custom data sorting, simple calculations, and visualization. The statistical package R (http://www.r-project.org/) is used to calculate enrichment of the documented protein–protein interactions in the target dataset. This is useful for data filtering. The network visualization package CytoScape (7) is used for visualization of complex networks. When colors of nodes and edges are varied, it is much easier to create graph files with a custom script than to manually input and modify the network components in CytoScape (see Note 1).

This protocol also needs several external low- or high-throughput datasets for data filtering and establishment of pathway relationships between the pair of chaperones in a two-component functional module. The interaction database BioGRID (8) provides manually curated known interactions from both low- and high-throughput experiments. The MIPS proteins complex database (9) provides yet another source of documented interactions. Our protocol uses BioGRID interactions for calculation of enrichment of known interactions in the target dataset, and MIPS interactions are used as criteria for reliable interactions, which are typically obtained from low-throughput methods, such as immunoprecipitation. Microarray datasets and KEGG pathway database (10) are used to establish a correlation between coexpression and pathway relationship. This correlation is the basis for inferring the pathway relationship between two chaperones in a functional module, as described below.

3. Methods

3.1. Data Filtering

Proteomics data are known to be noisy, with many false-positive interactions. It is, therefore, important to filter out these false positive interactions as much as possible. While proteomic interactions are often scored and ranked based on mass spectrometry database searches, comparison of the raw interactions obtained in proteomic studies with validated interactions from small-scale biochemical studies is also very important for data filtering. In our protocol, two datasets of such reliable interactions are used for this purpose.

3.1.1. Filtering Interactions Using MIPS Complexes

The MIPS database (9) contains 215 well-established complexes curated from numerous biochemical publications. Although this dataset has not been updated for a few years (latest complexes were derived from a publication in the year 2004), it is still frequently used as a gold standard for protein–protein interactions. To use this database in data filtering, complexes are downloaded from the MIPS Web site, and protein–protein interactions are assigned to each pair of proteins found in a complex. This interaction list comprises a reference interaction set. The raw interactions from the TAP-tag pulldown experiments in which the interactors are identified by mass spectrometry are then screened for their existence in the reference interaction set from MIPS. The raw interactions are further grouped into bins of scores derived from the mass spectrometry experiments. Histograms of the frequency distribution of the bins of both the raw interactions and those found in the reference interaction list are plotted. One can determine the score cutoff for reliable interactions in the raw data based on the score distribution for the reference interactions.

3.1.2. Enrichment of Known Interactions Using BioGRID

To further confirm the selection of score cutoff in the above procedure, the interactions selected from the above protocol can be compared to those from a curated interaction database. We use BioGRID (8) for this purpose. BioGRID documents published interactions from both high- and low-throughput experiments. As with MIPS, we use only the low-throughput interactions, as they are deemed more reliable. BioGRID contains many more interactions than MIPS, since it is continuously being updated. The principle of testing for enrichment for known interactions in the chaperone interaction data is as follows. Given all interactors of the chaperones and the interactors for each chaperone from the filtered data, find the interactors already documented in BioGrid for a particular chaperone, and the interactors found in both filtered chaperone interactors and BioGrid. We will thus have four subsets:

1. Filtered interactors of all chaperones (N).
2. Filtered interactors of one chaperone (m).

3. BioGrid interactors present in filtered interactors of all chaperones (n).
4. BioGrid interactors present in filtered interactors of one chaperone (k).

The fold of enrichment of BioGrid interactors that are also found in the filtered data is then:

$$\text{Enrichment\_fold} = \frac{(k / m)}{(n / N)}. \quad (1)$$

To test the significance of this enrichment, we assume a hypergeometric distribution:

$$f(k, N, m, n) = \frac{\binom{m}{k}\binom{N-m}{n-k}}{\binom{N}{n}}. \quad (2)$$

The p-value is calculated as the sum of f for all $i \geq k$, where i is the number of chaperone interactors that are found in BioGrid when m interactors are randomly drawn. A significant enrichment is when $p < 0.05$. Implementation of this algorithm can be found in the statistical package R using the function *phyper*, which is called with command line, *phyper(k,n,N−n,m,lower.tail=FALSE)*. In plain language, it refers, for example, to the probability of getting k or more red balls by chance, when drawing m balls from a mix of n red balls and $N-n$ blue balls.

If $p < 0.05$, we consider that the known interactors of a chaperone are enriched in the filtered dataset. Enrichment of known interactors for most of the chaperones would indicate the reliability of the filtered dataset.

3.2. Identification of Functional Chaperone Modules

3.2.1. Determination of the Number of Shared Interactors Among a Group of Chaperones

Two or more chaperones that interact with a common protein either do so to work together to promote the proper folding of that protein or the common protein has several alternative (parallel) folding pathways that the chaperones act on independently. We call the former scenario, the "single pathway model," and we call the latter the "multiple pathways model" (Fig. 1). In either case, the chaperones are considered to form a functional module. We find that the number of chaperones in the module typically ranges from two to five (4).

As a first step to identify the functional chaperone modules, we count the shared protein interactors for each group of chaperones with varied sizes. The following pseudo-code finds modules of arbitrary size:

```
Given n chaperones with their interactor list
Given the size of the functional module m < n
For i1=1 to n-(m-1)
  For i2=2 to n-(m-2)
      .....
          For im=m to n
              count the interactors shared by chaperone i1, i2, ..., im
          Endfor
      .....
  Endfor
Endfor
```

The numbers of shared interactors are stored in an array or written to a file for use in the next step.

3.2.2. Using Z-Score for Chaperone Module Identification

Given the number of shared interactors for a group of chaperones of a specified size, the Z-score is calculated for each group as follows:

$$Z = \frac{x - \overline{x}}{\sigma}, \quad (3)$$

where x is number of shared interactors for that group of chaperones, $\overline{x}$ is the average number of interactors for all chaperone groups, and σ is the standard deviation of the number of interactors, calculated as:

$$\sigma = \sqrt{\frac{1}{n}\left(\sum_{i=1}^{N} x_i^2 - n\overline{x}^2\right)}. \quad (4)$$

Here, x and $\overline{x}$ are defined as above, and n is the total number of chaperone groups. If there are a total of N chaperones, and if there are m chaperones in a given group, the calculation of n is as follows:

$$n = C_N^m = \frac{N!}{(N-m)!m!}. \quad (5)$$

Hence, the *Z*-score is calculated for each chaperone group. Chaperone modules are identified as those with *Z*-score ≥ 2.

3.2.3. Further Validation of Chaperone Modules: Retrieving Functional Modules from a Consensus Chaperone Network

It is likely that for some well-studied chaperones, models of their networks have been previously proposed. For example, Young et al. (11) compiled a chaperone network with the components HSP40, HSP70, HSP90, PFD, and CCT. In these networks, the nodes represent chaperones or chaperone complexes and the edges are flow of the protein substrates as they change conformation from newly translated peptides to the folded native state. Chaperone modules can be identified in these networks based on the above definition. Identifying chaperone modules from an expert annotated chaperone network is useful in validating modules obtained from the experimental high-throughput data.

We identify functional modules from the chaperone network models in three steps. The first step is to identify all chaperones on pathways starting from the nascent peptide to the folded protein. The second step is to group the pathways, with each group containing a specified number of pathways. In the third step, each pathway group is "cleaned" so that each chaperone appears only once.

Step 1: Searching all chaperone pathways from the network

There are two algorithms in graph theory for a search of all pathways from a network, namely, breadth-first and depth-first search, respectively (12). Here, we list the pseudo-code for breadth-first algorithm as an example, starting from the list of edges in the model network, which is a directed acyclic graph (DAG). Note that there is a function expanding_path, which could be called iteratively.

```
Given the edges of the network
For each edge
    Put the nodes in a two-dimensional array as edges[start] ≥
array(ends)
Endfor
Define candidate_pathway array with the nascent polypeptide as both the
key (candidate pathway) and value (end node of the candidate pathway)
Define all_pathway as an empty array
all_pathway = expanding_path(edges,all_pathway,candidate_pathway)
```

```
Function expanding_path(edges,all_pathway,candidate_pathway)
    Define candidate_pathway_temp array for storing unfinished pathways
    For each candidate pathway and its end node from candidate_pathway
        For each child node of the end node from edges
            If the child node is the native polypeptide
                Put the pathway to all_pathway
            Else
                Concatenating the child node to candidate pathway and
store it in candidate_pathway_temp as key and the child node as value
            Endif
        Endfor
    Endfor
    candidate_pathway=candidate_pathway_temp
    If candidate_pathway is not empty
        call                                                    function
expanding_path(edges,all_pathway,candidate_pathway)
    Else
        return all_pathway
    Endif
Endfunction
```

Step 2: Grouping pathways

The algorithm and pseudo-code is very much the same as that used for identifying shared interactors among the chaperones. The *m* is the size of pathway group. For different *m*, the number of nested For-Endfor loops will be different.

```
Given n pathways from the model as identified in step 1
For m=1 to n
    For i1=1 to n-(m-1)
        For i2=2 to n-(m-2)
            .....
                For im=m to n
                    combine chaperones in pathways 1 to m
                Endfor
            .....
        Endfor
    Endfor
Endfor
```

Step 3: Listing the chaperone modules
In this step, chaperone modules are listed by finding unique set of chaperones in each group from step 2. If a module occurs multiple times, only one is listed. This step is trivial, for example, a single function array_unique() in the programming language PHP is sufficient for this task.

3.2.4. Comparing Inferred Chaperone Modules with Consensus Modules

The purpose of comparing chaperone functional modules obtained from consensus models with those obtained from the new proteomics data is to confirm known modules and predict new ones. The comparison is straightforward in terms of programming. In PHP, one can use the functions array_diff() and array_intersect().

3.3. Establishing Pathway Relationship Between Two Chaperones in Two-Component Functional Modules Using Coexpression and KEGG Pathway Information

Two chaperones in two-component functional modules might either act on target protein along a single pathway or multiple pathways or both (Fig. 1). Genes coding for proteins that function in the same pathway are likely coexpressed (13). Therefore, it is rational to estimate pathway relationships between a pair of chaperones in a functional module based on their gene coexpression data. To this end, gene coexpression results are combined with KEGG pathway information (14). It should be emphasized that this analysis is restricted to two-component chaperone modules.

3.3.1. Measuring Gene Coexpression

The degree of gene coexpression is measured with Pearson correlation coefficient (PCC). PCC is calculated using the formula:

$$\mathrm{PCC} = \frac{\Sigma XY - (\Sigma X \Sigma Y / N)}{\sqrt{\left(\Sigma X^2 - (\Sigma X)^2 / N\right)\left(\Sigma Y^2 - (\Sigma Y)^2 / N\right)}}. \quad (6)$$

Many software packages such Microsoft Excel and R, implement the calculation. Use of R is potentially more efficient because the calculation can be parallelized on a computer cluster or multicore desktops. This is particularly useful if multiple gene expression datasets are to be analyzed. The R function for calculating PCC is cor(X, Y), where the parameters X and Y are vectors representing the expression values of two genes. N is the size of the vectors.

3.3.2. KEGG Pathway Relationship of Proteins

KEGG (Kyoto Encyclopedia of Genes and Genomes) (14) pathways are downloadable from http://www.genome.jp/kegg/. The pathway list table maps relevant genes to a pathway(s). The pathway relationships between each pair of genes are calculated. These relationships include what we term single pathway (two proteins involved in one pathway), multiple pathway (two proteins involved in two pathways), and both single and multiple pathways. A pseudocode for this purpose is as follows:

```
Read the pathway list into an array in which the protein is the key and
a 2nd dimensional array containing the pathways it is involved in is the
value
For each protein1
    For each protein2 after protein1 in the array
        Find intersect of the pathway arrays of protein1 and protein2
        If no overlap between the two pathway arrays
            The pathway relationship is single pathway
        Else
            If there is only one pathway in the two list
                The pathway relationship is multiple pathways
            Else
                The pathway relationship is both
            Endif
        Endif
    Endfor
Endfor
```

3.3.3. Combining Gene Coexpression Results with KEGG Pathway Relationship

The coexpression results (PCCs) are binned for all pairs of genes that are involved in the KEGG pathways. PCCs range from −1 to 1, and the bins start from −1 and are incremented 0.1 at a time. For each bin, the number of gene pairs functioning in single pathway, multiple pathways, or both are counted. The probability for each pathway relationship is calculated as the ratio of each number to the sum of three numbers in a bin. Using the expression data from Cho et al. (15) and Gasch et al. (16), it is found that the probability of single pathway relationship increases with PCC, the probability of multiple pathways relationship decreases with increase in PCC, and the probability of involvement in both single and multiple pathways does not show as dramatic a change as for other relationships, as expected. This association between gene coexpression and protein pathway relationships provides a reference for inferring the latter based on the former.

3.3.4. Determining Probability of Pathway Relationship Between Two Chaperones in a Two-Component Functional Module

With the association of pathway relationship and coexpression correlation as a reference, we are able to assign the probability of pathway relationship to chaperone functional modules based on coexpression strength of two chaperones in the module. This involves the following two steps.

Step 1: Assign the probability for each pathway relationship based on gene coexpression data

As previously described, gene expression data are binned based on coexpression strength, and for each bin the probability for each pathway relationship is calculated. Given a pair of chaperones in a functional module together with their coexpression coefficient from an expression dataset, the probabilities of pathway relationships are transferred to this pair of chaperones. If multiple gene expression datasets are available, then the coexpression data is integrated as described in step 2 below.

Step 2: Integration of the information from multiple expression datasets

A data integration method (Hon Nian Chua, National University of Singapore, personal communication) is used for combining the probabilities of pathway relationships from multiple expression datasets. This involves the following equation:

$$P = 1 - \prod_{k \in Du,v} \left(1 - P(k)\right), \quad (7)$$

where Du,v is the set of expression data that contains both chaperones, $P(k)$ is the probability that the two chaperones have a particular pathway relationship determined using KEGG pathways as training dataset, and P is the integrated probability that the two chaperones have a particular pathway relationship. Three probabilities are then derived for single, multiple, or single and multiple (both) folding pathways. The final probabilities are calculated by normalization, that is, each probability is divided by the total of the three. The results of such an analysis for our chaperone proteomic data are shown in Fig. 2.

In conclusion, the protocol discussed here describes (1) how to filter large-scale interaction data, (2) how to identify chaperone functional modules from this large-scale interaction data, (3) how to compare modules obtained from the large-scale interaction data with modules derived from a consensus chaperone network, and (4) how to determine the probability of the two chaperones in a two-component chaperone module act along single, multiple, or both single and multiple folding pathways (Figs. 1 and 2). Our approach provides important insights into the organization of chaperone networks inside the cell and provides first hints into how cellular protein folding is regulated by molecular chaperones. Finally, the protocol we describe above can be used for any other biological system for which large-scale interaction data might be available.

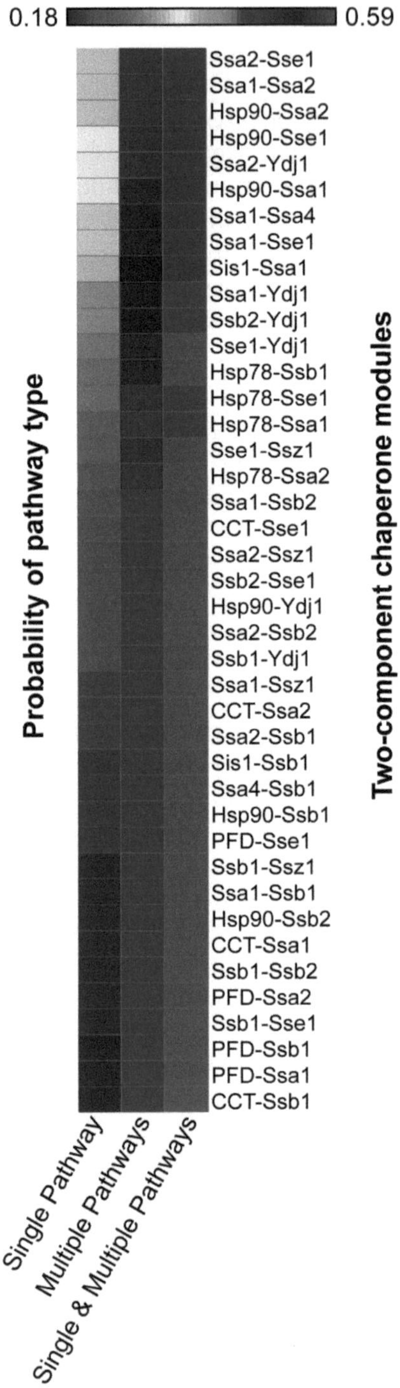

Fig. 2. Pathway relationship of two-component chaperone modules obtained from our yeast chaperone interaction proteomic data (4).

4. Note

1. Cytoscape accepts a variety of data format, from simple sif format, which does not contain node and edge features, such as shape and color, to gml format which does contain node and edge features. A brief but complete gml file is shown below:

```
graph
[
    label ""
    directed    1
    node
    [
        id    0
        label "Hsp70"
        graphics
    [
        x     1916.0
        y     4409.0
        w     20.0
        h     20.0
        type  "ellipse"
        width 1.00000
        fill  "#E1E1E1"
        outline      "#000000"
    ]
]
```

```
node
[
        id    1
        label "Hsp90"
        graphics
        [
                x      2021.0
                y      4333.0
                w      20.0
                h      20.0
                type  "ellipse"
                width 1.00000
                fill  "#FF0000"
                outline      "#000000"
        ]
]
edge
        [
                source       0
                target       1
                label "pp"
                graphics
                [
                        width 2
                        type  "line"
                        fill  "#0000E1"
                ]
        ]
]
```

This gml script describes an edge between two nodes. The node features, namely, label, position, size, shape, and color, and the edge features, namely, width and color, are defined. To create a complex network file with many nodes and edges, one can write a script in a language of his/her choice. The script reads the data file and writes a gml file. For example, if the interactors shared between chaperones are to be visualized, chaperones are depicted as nodes with different colors indicating the chaperone groups, and the edges represent shared interactors colored to indicate the number of shared interactors. The created gml file can then be imported into CytoScape.

Acknowledgments

This work was supported by grants from the Canadian Institutes of Health Research (MOP-81256) to W.A.H., and Genome Canada through the Ontario Institute of Genomics to Z.Z.

References

1. Doyle SM, Wickner S. Hsp104 and ClpB: protein disaggregating machines. Trends Biochem Sci 2009;34:40–8.
2. Hartl FU, Hayer-Hartl M. Converging concepts of protein folding in vitro and in vivo. Nat Struct Mol Biol 2009;16:574–81.
3. Sghaier H, Le Ai TH, Horiike T, Shinozawa T. Molecular chaperones: proposal of a systematic computer-oriented nomenclature and construction of a centralized database. In Silico Biol 2004;4:311–22.
4. Gong Y, Kakihara Y, Krogan N, et al. An atlas of chaperone-protein interactions in *Saccharomyces cerevisiae*: implications to protein folding pathways in the cell. Mol Syst Biol 2009;5:275.
5. Shaner L, Wegele H, Buchner J, Morano KA. The yeast Hsp110 Sse1 functionally interacts with the Hsp70 chaperones Ssa and Ssb. J Biol Chem 2005;280:41262–9.
6. Matsumoto R, Rakwal R, Agrawal GK, et al. Search for novel stress-responsive protein components using a yeast mutant lacking two cytosolic Hsp70 genes, SSA1 and SSA2. Mol Cells 2006;21:381–8.
7. Cline MS, Smoot M, Cerami E, et al. Integration of biological networks and gene expression data using Cytoscape. Nat Protoc 2007; 2: 2366–82.
8. Breitkreutz BJ, Stark C, Reguly T, et al. The BioGRID Interaction Database: 2008 update. Nucleic Acids Res 2008;36:D637-40.
9. Mewes HW, Dietmann S, Frishman D, et al. MIPS: analysis and annotation of genome information in 2007. Nucleic Acids Res 2008; 36:D196-201.
10. Aoki KF, Kanehisa M. Using the KEGG database resource. Curr Protoc Bioinformatics 2005;Chapter 1:Unit 1 12.
11. Young JC, Agashe VR, Siegers K, Hartl FU. Pathways of chaperone-mediated protein folding in the cytosol. Nat Rev Mol Cell Biol 2004; 5:781–91.
12. Knuth DE. The Art of Computer Programming 3ed. Boston: Addison-Wesley; 1997.
13. Adler P, Peterson H, Agius P, Reimand J, Vilo J. Ranking genes by their co-expression to subsets of pathway members. Ann N Y Acad Sci 2009;1158:1–13.
14. Kanehisa M, Araki M, Goto S, et al. KEGG for linking genomes to life and the environment. Nucleic Acids Res 2008;36:D480-4.
15. Cho RJ, Campbell MJ, Winzeler EA, et al. A genome-wide transcriptional analysis of the mitotic cell cycle. Mol Cell 1998;2:65–73.
16. Gasch AP, Spellman PT, Kao CM, et al. Genomic expression programs in the response of yeast cells to environmental changes. Mol Biol Cell 2000;11:4241–57.

Chapter 16

Hsp70: Anti-apoptotic and Tumorigenic Protein

Anne-Laure Rérole, Gaëtan Jego, and Carmen Garrido

Abstract

Heat shock protein 70 (Hsp70) is a powerful chaperone whose expression is induced in response to a wide variety of physiological and environmental insults, including anticancer chemotherapy, thus allowing the cell to survive to lethal conditions. Hsp70 cytoprotective properties may be explained by its anti-apoptotic function. Indeed, this protein can inhibit key effectors of the apoptotic machinery at the pre- and postmitochondrial level. In cancer cells, the expression of Hsp70 is abnormally high, and Hsp70 may participate in oncogenesis and in resistance to chemotherapy. In rodent models, Hsp70 overexpression increases tumor growth and metastatic potential. Depletion or inhibition of Hsp70 frequently reduces the size of the tumors and even can cause their complete involution. But Hsp70 can also be found in the extra-cellular medium. Its role is then immunogenic and the term chaperokine to define the extracellular chaperones has been advanced. Hsp70 tumorigenic functions as well as the strategies that are being developed in cancer therapy in order to inhibit Hsp70 are commented in this chapter.

Key words: Hsp70, Apoptosis, Cancer, Anticancer therapy, Immunomodulator

1. Introduction

The heat shock response, first discovered in 1962 in Drosophila after a heat shock, is a universal mechanism induced by almost all cells and tissues after many different stresses. It is associated with an increase in the expression of a family of proteins called heat shock proteins (Hsps) (1, 2). Most Hsps have strong cytoprotective effects and behave as molecular chaperones for other cellular proteins (3, 4). Acute or chronic stress results in protein misfolding, protein aggregation, or disruption of regulatory complexes (5). The action of chaperones, through their properties in protein homeostasis, is thought to restore balance. Mammalian Hsps have been classified into five families according to their molecular size: Hsp100, Hsp90, Hsp70, Hsp60, and small Hsps (15–30 kDa)

Stuart K. Calderwood and Thomas L. Prince (eds.), *Molecular Chaperones: Methods and Protocols*,
Methods in Molecular Biology, vol. 787, DOI 10.1007/978-1-61779-295-3_16, © Springer Science+Business Media, LLC 2011

that includes Hsp27. Each family of Hsps is composed of members that are expressed either constitutively or regulated inductively, and that are targeted to different subcellular compartments. High molecular weight Hsps are ATP-dependent chaperones while small Hsps act in an ATP-independent fashion. Chaperones are instrumental for signaling and protein traffic, even in the absence of stress. However, the need of Hsps increases after proteotoxic damage. Hsp70 is one of the most strongly and universally induced chaperones and has strong cytoprotective properties. Overexpressed Hsp70 prevents cell death triggered by various stimuli, including hyperthermia, oxidative stress, or the addition of cytotoxic drugs (6–9). Several among these stimuli induce Hsp70 overexpression themselves (10, 11), providing an example of how pro-apoptotic stimuli, delivered below a threshold level, can elicit protective responses.

Hsps have also been shown to play a role in cell differentiation. For instance, we and other groups have demonstrated that Hsp70 played a key role in erythroid differentiation. During red blood cells formation, Hsp70 accumulates within the nucleus of the erythroblast (12) and directly associates with GATA-1 protecting this transcription factor required for erythropoiesis from its inactivation. As a result, erythroblasts continue their differentiation process instead of dying by apoptosis (13).

Hsp70 basal levels are unusually high in malignant cells, and Hsp70 has been accused of participating in oncogenesis and/or in chemotherapy resistance, presumably due to its capacity to disable cell death (14). Hsp70 can act at multiple control points of the apoptotic pathways to ensure that stress-induced damage does not inappropriately trigger cell death. In this chapter, we describe the different pathways through which Hsp70 modulates cancer cell death and the emerging strategies that are being developed for cancer therapy. The immunogenic function of extracellular Hsp70 is also discussed.

2. Cytoprotective Functions of Hsp70

2.1. Hsp70 Is a Molecular Chaperone

Human cells contain several Hsp70 family members, including stress-inducible Hsp70, constitutively expressed Hsc70, mitochondrial Hsp75, and GRP78, which is localized in the endoplasmic reticulum (15–17). Under normal conditions, Hsp70 proteins function as ATP-dependent molecular chaperones that assist the folding of newly synthesized polypeptides, the assembly of multiprotein complexes and the transport of proteins across cellular membranes (18–20). Hsp70 contains two distinct functional regions (Fig. 1): a peptide-binding domain (PBD) and the amino-terminal ATPase domain (ABD). The PBD, that includes a

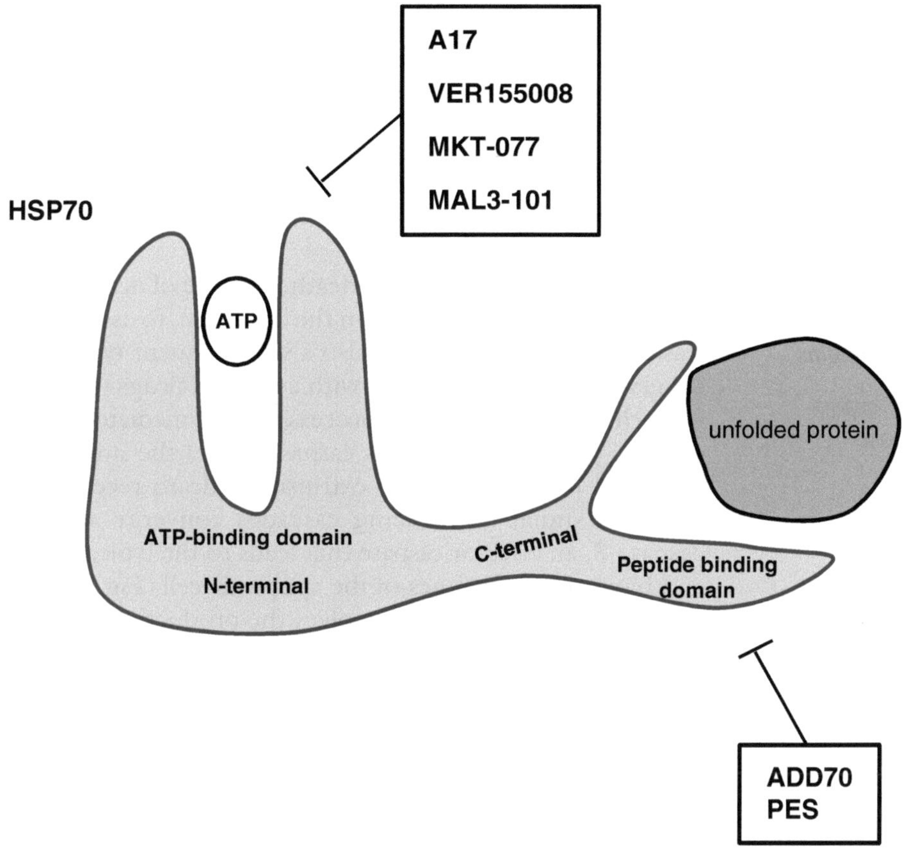

Fig. 1. Structure of Hsp70 with sites of action of drugs discussed in the review.

carboxyl-terminal chaperone EEVD motif, is responsible for substrate binding and refolding. The ABD, in turn, facilitates the release of the client protein after ATP hydrolysis. A conserved proline in the ATPase domain is essential to alternate Hsp70 conformations in response to ATP binding and hydrolysis (21). Distinct co-chaperones bind to Hsp70 and regulate its chaperone function. They can be classified in three groups. (a) The J-domain co-chaperones, like Hsp40, are a relatively large group that binds to the Hsp70 ABD and stimulate the low ATPase activity of this chaperone (22, 23). (b) The nucleotide exchange factor co-chaperones catalyze the release of ADP which is required for the completion of Hsp70 ATPase cycle. Members of this group are Bag-1, Hsp110, or HspBP1. (c) The TPR domain co-chaperones (Hop, CHIP) bind to the C-terminal EEVD motif presented in both Hsp70 and Hsp90 (24). They are essential for combinational assembly of Hsp70 and Hsp90 complexes. CHIP, that has an ubiquitin ligase activity, has been implied in the ubiquitination of at least some Hsp client proteins (Fig. 1) (24).

Under stressful conditions, elevated Hsp70 levels allow cells to cope with increased concentrations of unfolded or denatured proteins (7). Some of the important house-keeping functions attributed to Hsp70 include: (1) import of proteins into cellular compartments; (2) folding of proteins in the cytosol, endoplasmic reticulum, and mitochondria; (3) degradation of unstable proteins; (4) dissolution of protein complexes; (5) control of regulatory proteins; (6) refolding of misfolded proteins, and (7) translocation of precursor proteins into mitochondria (25).

2.2. Hsp70 Interferes with the Action of Key Apoptotic Proteins

2.2.1. Cell Signaling and Apoptosis

Apoptosis, or programed cell death, is a type of death essential during embryogenesis and, later in the organism, to assure cell homeostasis (26, 27). Apoptosis is also a very frequent type of cell death observed after the treatment with cytotoxic drugs (28). Two pathways characterize apoptotic processes, both mediated by a family of cysteine proteases known as caspases: (29) the intrinsic or mitochondrial pathway and the extrinsic or death receptors pathway. The two signal transducing cascades converge at the level of caspase-3, an effector caspase that leads to the typical morphologic and biochemical changes of the apoptotic cell (Fig. 2) (30).

The intrinsic pathway involves the production or activation of pro-apoptotic molecules upon intracellular stress signals. These

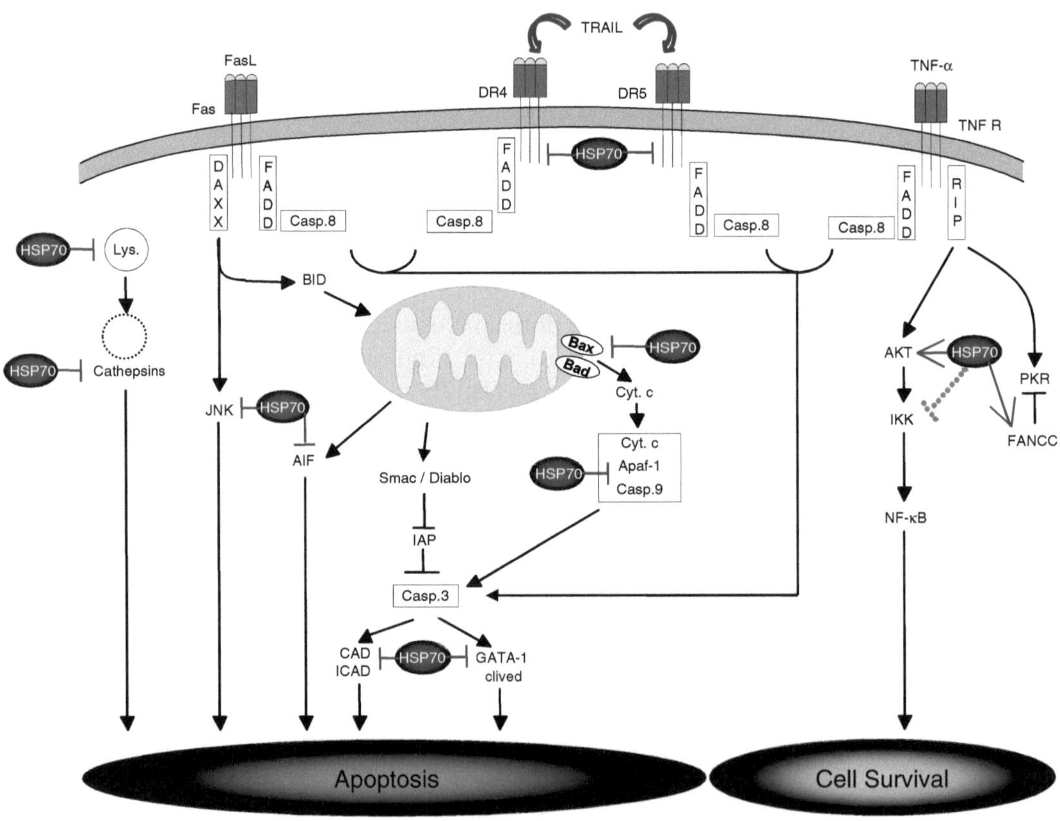

Fig. 2. Targets of Hsp70 in the apoptotic and cell survival pathways. Controversial effect is shown by a *dotted line*.

molecules converge on the mitochondria to trigger the release of mitochondrial apoptogenic molecules under control of the B-cell lymphocytic leukemia proto-oncogene (Bcl-2) family of proteins (31). Bcl-2 proteins include anti-apoptotic members, such as Bcl-2 and Bcl-xL, multidomain pro-apoptotic members, mainly Bax and Bak (32, 33) and a series of BH3 domain-only pro-apoptotic proteins, such as Bid (34), that function upstream of Bax and Bak (Fig. 2) (35). One of the released mitochondrial molecule is cytochrome *c*, which interacts with cytosolic Apoptotic protease activation factor-1 (Apaf-1) to form the apoptosome that recruit and activate procaspase-9 and finally activate caspase-3 (36, 37). Two other mitochondrial proteins, Smac/Diablo and Htra2/Omi, activate apoptosis by neutralizing the inhibitory activity of the inhibitory apoptotic proteins (IAPs) that associate with and inhibit some of the activated caspases (38–41).

The extrinsic pathway is triggered through plasma membrane proteins of the tumor necrosis factor (TNF) receptor family known as death receptors and leads to the direct activation of the receptor-proximal caspase-8 or caspase-10 in the death-inducing signaling complex (42). Caspase-8 either directly activates the downstream cascade of caspases or cleaves Bid into an active truncated form named tBidthatconnectstheextrinsictotheintrinsicapoptoticpathwaysthrough mitochondria permeabilization (Fig. 2) (43, 44).

Hsps have been shown to block apoptosis by interfering with caspase activation. Overexpression of Hsp70 inhibits apoptosis and prevents caspase activation in many different cellular models upon a variety of cellular stresses, including accumulation of misfolded proteins, ROS or DNA damage (45–47). On the contrary, depletion of Hsp70, either by antisense constructions or siRNA strategies, increases the cells' sensitivity to apoptotic stimuli (48, 49). In some cellular contexts, Hsp70 depletion is sufficient to trigger apoptosis through caspase-3 activation, in the absence of any additional stressful stimulus (49, 50). Therefore, Hsps are involved, either directly or indirectly, in the modulation of caspase activities. More precisely, Hsps can block both the intrinsic and the extrinsic apoptotic pathways through the interaction with key proteins at three levels: (1) upstream the mitochondria, thereby modulating signaling pathways; (2) at the mitochondrial level, thus controlling the release of apoptogenic molecules; (3) and at the post-mitochondrial level, at the latest point in the apoptotic process ever discussed for a protective protein.

2.2.2. Pathways of Apoptosis

Hsp70 and Mitochondrial-Dependent Apoptosis

Hsp70 is an evolutionary conserved protein whose expression enhances the ability of cells to survive to a panel of lethal conditions. Gene ablation studies demonstrate that Hsp70 plays an important role in apoptosis. Cells lacking *hsp70.1* and *hsp70.3*, the two genes that code for inducible Hsp70, are highly sensitive to apoptosis induced by a wide range of lethal stimuli (51). Ablation of the testis-specific isoform of Hsp70 (*hsp70.2*) results in germ cell apoptosis (52). Elevated Hsp70 levels block the apoptotic pathway at different levels, both upstream and downstream of the death-associated mitochondrial events.

Hsp70 interacts with several targets in upstream signaling pathways. Growth factors, such as nerve growth factor or platelet-derived growth factor, induce cell survival by activating the phosphatidylinositol 3-kinase pathway (PI3-K). Activated PI3-K phosphorylates inositol lipids in the plasma membrane that attract the serine/threonine kinase Akt/PKB, a protein that generates a survival signal in response to growth factor stimulation. Akt targets multiple proteins of the apoptotic machinery, including Bad and caspase-9 (53–56). Hsp70 binds and stabilizes Akt/PKB (Fig. 2) (57). Interestingly, the endothelial-specific HspA12B, a distant member of Hsp70 family that is required for zebrafish vasculature development, is involved in endothelial cell migration and tube formation through sustaining Akt activity (58). Thus, Hsp70 family members could play a role both in the control of cell survival and differentiation. Hsp70 also binds to nonphosphorylated protein kinase C (PKC) via the kinase's unphosphorylated carboxyl-terminus, priming the kinase for rephosphorylation and stabilizing the protein (57).

Hsp70 can inhibit stress-activated kinases, such as apoptosis signal regulating kinase 1 (Ask1), e.g., Hsp70 downregulation facilitates H_2O_2-induced Ask1 activation and subsequent apoptosis in NIH3T3 cells (9). Hsp70 also binds to c-Jun N-terminal Kinase (JNK), and prevents its activation independently of ATP (46, 59, 60). For example, MEF *hsp70.1*$^{-/-}$ resist to JNK-mediated apoptosis induced by hyperosmolarity, whereas AEG3482, a promising novel compound, inhibits JNK activity through increased the expression of Hsp70 (61, 62). Hsp70 also negatively interferes with p38 kinase activity (63).

Hsp70 has also been shown to affect some transcription factors involved in the expression of the Bcl-2 family. Bcl-2 and Bax are transcriptional targets of the tumor suppressor protein p53: the transcription of Bcl-2 is repressed by p53, whereas that of Bax is induced. As a consequence, p53 expression induces apoptosis in response to DNA damage. Many tumor cells have a mutated p53 and either Hsp70 or Hsc70 could form stable complexes with this mutant protein. Hsp70 could also mask the nuclear localization sequence of p53, thus preventing its nuclear import (64, 65).

The role of Hsp70 in regulating nuclear factor-κB (NF-κB) function is more controversial. It has been postulated that cytosolic Hsp70 could inhibit NF-κB while plasma membrane-associated Hsp70 could activate this transcription factor (66, 67). However, the stimuli that induce accumulation of cytosolic and membrane-bound Hsp70 are the same (68). In endothelial cells, elevation of Hsp70 to a significant level favors TNF-α-mediated apoptosis via inhibition of the NF-κB survival pathway (69). Hsp70 could block NF-κB activation through inhibition of inhibitor-κBα kinase (IKK) activation and subsequent degradation of I-κBα (70, 71). As a consequence, Hsp70 might promote the elimination of cells with damaged DNA. Inhibitors of growth (ING) proteins, which are tumor suppressors whose expression is downregulated in a variety of human cancers, transduce stress signals after DNA damage and bind histones, thus regulating chromatin structure and p53 activity. These proteins induce Hsp70 that in turn promotes TNF-α receptor-mediated apoptosis by binding I-κBα kinase (IKK) and impairing NF-κB survival signaling (Fig. 2) (72).

At the mitochondrial level, Hsp70 coupled to Hsp40 blocks Bax translocation, thus preventing mitochondrial outer membrane permeabilization and inhibiting the release of cytochrome c and other mitochondrial apoptogenic molecules, such as apoptosis inducing factor (AIF) (73). This Hsp70 function depends on both its chaperone and its ATP hydrolytic domains (Fig. 2) (74).

Hsp70 acts also at the postmitochondrial level. It has been found that Hsp70 inhibited apoptosis downstream of the release of cytochrome c and upstream of the activation of caspase-3 (75). Indeed, Hsp70 has been demonstrated to directly bind to Apaf-1, thereby preventing the recruitment of procaspase-9 to the apoptosome (76). The ATPase domain of Hsp70 was described to be necessary for this interaction (77). In TNF-induced apoptosis, Hsp70 does not preclude the activation of caspase-3 but prevents downstream morphological changes that are characteristic of dying cells like activation of phospholipase A2 and changes in nuclear morphology (78). One protein target of caspase-3 is the caspase-activated DNase (CAD), responsible of chromosomal DNA digestion during the final phases of apoptosis. The enzymatic activity and proper folding of CAD has been reported to be regulated by Hsp70, its co-chaperone Hsp40 and ICAD, and by the inhibitor of CAD. ICAD recognizes an intermediate folding state conferred by Hsp70–Hsp40 (79). It has also been reported in TCR-stimulated T-cells that Hsp70 binds CAD and enhances its activity (80). Poly [ADP-ribose] polymerase (PARP), a nuclear protein involved in DNA repair, is an early target of caspase-3 and its cleavage that leads to its inactivation is necessary to prevent necrosis and inflammation during apoptosis (81, 82). It has recently been shown that Hsp70 interacts simultaneously with PARP1 and XRCC1, another DNA repair protein, in the nucleus of cells undergoing single-strand breaks after heat treatment (83). These results suggest that Hsp70 could play a role in DNA integrity through its association with PARP1 to create the repair protein complex. Another final target of caspase-3 is the transcription factor GATA-1. We have demonstrated in human primary erythroid precursors that Hsp70 can protect GATA-1 from caspase-3 cleavage. As a consequence, erythroid cells do not die by apoptosis but instead differentiate (Fig. 2) (13).

Hsp70 and the Extrinsic Death Receptor Pathway

Hsp70 inhibits TNF-α-induced cell death and this protective effect is lost in Bid homozygous-deleted MEF cells; Hsp70 can block the cleavage of Bid by activated caspase-8 (84). Exposure of hematopoietic cells to TNF-α induces the activity of the pro-apoptotic double-stranded RNA-dependent protein kinase (PKR). An inhibitor of PKR is the Fanconi ANemia Complementation group C gene product (FANCC). Hsp70 interacts with the FANCC protein via its ATPase domain and, together with Hsp40, inhibits TNF-induced apoptosis through the ternary complex Hsp70, FANCC, and PKR (85, 86).

In human chronic myeloid leukemia cells, Hsp70 can mediate Bcr-Abl induced resistance to TNFα-related apoptosis-inducing ligand (TRAIL)-induced apoptosis by preventing the formation of death-inducing signaling complexes involving the death receptors DR4 and DR5 (Fig. 2) (87). The role of Hsp70 in Fas-mediated apoptosis is more controversial with opposite effects that depend on the cell context (88, 89).

Hsp70 and Alternatives, Caspase-Independent, Apoptosis-Like Pathways

Upon activation of the intrinsic pathway, not only cytochrome c is released from the mitochondria. AIF and endonuclease G (EndoG), two mitochondria intermembrane proteins, translocate to the nucleus and trigger caspase-independent nuclear changes (Fig. 2) (90). Hsp70 has been shown to prevent cell death in conditions in which caspase activation does not occur, due to the addition of exogenous caspase inhibitors (91), or in cells in which Apaf-1 or caspase-9 were genetically inactivated (92), indicating that the cytochrome c/Apaf-1/caspase was not the sole pathway of the anti-apoptotic action of Hsp70. Indeed, Hsp70 directly binds to AIF and inhibits AIF-induced chromatin condensation. Hsp70 neutralizes the apoptogenic effects of AIF in cell-free systems, in intact cells microinjected with recombinant Hsp70 and/or AIF protein, as well as in cells transiently transfected with AIF cDNA. Of note, endogenous levels of Hsp70 seem to be sufficiently high to control AIF-mediated apoptosis since downregulation of Hsp70 by an antisense construct sensitized the cells to serum withdrawal and AIF (92). This protective effect of Hsp70 might be physiologically relevant as AIF sequestration by Hsp70 reduces neonatal hypoxic/ischemic brain injury (93). In addition, Hsp70 inhibits erythroblast apoptotis by blocking AIF nuclear import (94). Hsp70 associates also with EndoG to prevent DNA fragmentation, but this association could involve AIF as a molecular bridge (95).

Lysosomes also function as integrators of cell death signals in many different cell death scenarios (96). Lysosomal proteases, of which the most studied are the cathepsins, translocate from the lysosomal lumen to the cytosol in response to a wide variety of apoptotic stimuli, such as TNF-α, Fas, microtubule stabilizing agents, staurosporine, p53 activation, oxidative stress, growth factor deprivation, and lysosomotropic agents (Fig. 2) (97–104).

Upon release to the cytosol, cathepsins can trigger mitochondrial outer membrane permeabilization (97, 101, 102, 104, 105). Cathepsins release can also ensue an apoptotic morphology that is independent of the mitochondria release of cytochrome c or AIF (98, 106). Hsp70 is found in the endolysosomal membranes of many tumors and stressed cells, where it inhibits the release of lysosomal cathepsins into the cytosol (107, 108). Lysosomes positive for Hsp70 display an increased size and resistance against chemical and physical membrane destabilization (107).

It has also been shown that Hsp70 can protect against cell death induced by ultraviolet (UV A and UV B) exposure in murine fibroblasts. This protection is mediated through the inhibition of IL-6 release induced by UV-light (109).

In Conclusion, Hsp70 Can Be Considered as the Quintessential Inhibitor of Apoptosis

Hsp70 can block practically all cell death pathways: classical mitochondrial apoptosis and non-canonical apoptosis. In the classical mitochondrial apoptosis it acts: at a premitochondrial stage by inhibiting stress inducing signaling; at the mitochondrial stage, by preventing mitochondrial membrane permeabilization through the blockage of Bax translocation; and finally, at the postmitochondrial level by interacting with AIF and Apaf-1 or by protecting essential nuclear proteins from caspase-3 cleavage (Fig. 2).

Does the effect of Hsp70 in the apoptotic process result from the chaperoning of the different apoptotic proteins that interact with Hsp70? As indicated in Fig. 1, Hsp70 has two domains. The domain necessary for the chaperone function is the carboxyl-terminal domain. All the reports that studied the domain of interaction with Hsp70 indicate that the apoptotic partners of Hsp70 bind to the PBD of the protein. However, for some interaction/protective properties, the ATPase function of Hsp70 is needed but not for others. For instance, while the ATPase function seems necessary for Apaf-1 and AIF (74, 76), it is dispensable for JNK or GATA-1 binding/protection. In this way, in erythroblasts, in which Hsp70 blocks apoptosis by protecting GATA-1 from caspase-3 cleavage, an Hsp70 mutant that only contains the PBD is as efficient as wild type Hsp70 to assure their protection (13). Based on these data, we can hypothesize that Hsp70 might chaperone proteins in an ATP-dependent as well as ATP-independent manner.

What drives the interaction of Hsp70 with a given partner? It may be determined by its cellular localization. For example, the presence of Hsp70 in the lysosomal membrane is necessary for inhibiting the lysosomal death pathway. The migration of Hsp70 in the nucleus is essential to determine whether an erythroblast is going to die by apoptosis or to differentiate into a red blood cell (13). In tumor cells, Hsp70 expression at the plasma membrane facilitates their clearance by the immune system. Membranary Hsp70 facilitates granzyme B penetration in a perforin-independent way and ensures cytotoxic T-lymphocytes killing through FasL/Fas (68, 110).

3. Hsp70 and Cancer

3.1. Hsp70: Tumorigenicity and Cancer Cell Resistance

In normal cells, under nonstressed conditions, the expression of Hsp70 is usually very low. In contrast, high Hsp70 expression is a property of cancer cells and essential for the survival of many cancers. Hsp70 expression further increases after chemotherapy and, as a consequence of Hsp70 strong cytoprotective properties, its cellular content has been correlated with the cells' resistance to death stimuli, including anticancer therapy. Clinical studies suggested Hsp70 as a bad prognosis factor. Increased expression of Hsp70 has been reported in high-grade malignant tumors, such as endometrial cancer, osteosarcoma, and renal cell tumors (111). Hsp70, along with PSA, are good tumor markers to identify patients with early stage prostate cancer (112). Hsp70 is very abundant in chronic myeloid leukemia characterized by Bcr-Abl expression (113). This overexpression of Hsp70 is associated with resistance to imatinib, an effective chemotherapeutic agent used by its ability to block Bcr-Abl tyrosine kinase activity (114). A relationship between Hsp70 and gastric cancer has also been reported through the effect of Helicobacter Pylori in Hsp70 expression (115). Hsp70.2, a member of the Hsp70 family that is normally expressed only in spermatogenesis, is present at high levels in breast cancer and inhibits the onset of senescence (114). High expression of Hsp70 in breast, endometrial, gastric cancer, or acute leukemia has been associated with metastasis, poor prognosis, and resistance to chemotherapy or radiation therapy (45, 116, 117). More recently, nuclear accumulation of Hsp70 has been defined as an objective marker for the presence of epithelial dysplasia and the presence of antibodies against Hsp70 in the serum has been described as a diagnostic marker in patients with hepatocellular carcinoma (118, 119).

Studies in Bcr-Abl human leukemia cells show that Hsp70 is a promising therapeutic target for reversing cancer cell drug resistance probably by its ability to inhibit apoptosis both upstream and downstream of the mitochondrial signaling (87, 113). In experimental in vivo models, Hsp70 overexpression has been shown to increase the tumorigenicity of cancer cells (colon cancer, melanoma and pancreatic adenocarcinoma) while Hsp70 downregulation strongly decreases tumorigenicity (49, 50, 120). In this way, Hsp70 antisense constructs have chemosensitizing properties and may even kill cancer cells (in the context of adenoviral infection) in the absence of additional stimuli (50). The cytotoxic effect of Hsp70 down-modulation is particularly strong in transformed cells yet undetectable in normal, nontransformed cell lines or primary cells (51). The fact that depletion of Hsp70 selectively sensitizes or kills cancer cells has been rationalized by assuming that tumor cells, as compared to their normal counterparts, exhibit a constitutively stressed phenotype with an enhanced dependency on the cytoprotective function of Hsp70.

3.2. The Inhibition of Hsp70 in Cancer Therapy

Neutralizing Hsp70 is therefore an attractive strategy for anticancer therapy. However, in contrast to another Hsp, Hsp90, for which many specific inhibitors are available and some are currently being tested in clinical trials, very few have been described for Hsp70. To block Hsp70, until quite recently, authors used inhibitors of the heat shock response that act at the level of the heat shock transcription factor 1 (HSF1) and thereby block the transcription of all stress-inducible Hsps. These inhibitors include the flavonoid quercetin, the diterpene triperoxide, triptolide (48, 121–125), and the benzo(a)pyrene, described to inhibit Hsp70 expression in endothelial cells (126). Although it is true that these products by affecting HSF1 function block Hsp70, one of the Hsps involved in the stress response, they remain rather unspecific. Besides, in most cancer cells Hsp70 is constitutively and abnormally abundant. Another nonspecific inhibitor of Hsp70 is resveratrol, described to inhibit Hsp70 expression in K562 cells via downregulation of Akt kinase activity and upregulation of ERK1/2 kinase activity (127).

3.2.1. Specific Inhibitors of Hsp70

Targeting the PBD

Our group has demonstrated that rationally designed decoy targets of Hsp70 derived from AIF can sensitize cancer cells to apoptosis induction by neutralizing Hsp70 function. These AIF-derived peptides all carry the AIF region from aa 150 to aa 228, previously defined as required for Hsp70 binding in its PBD and lack AIF pro-apoptotic function (128). These inhibitors, called ADD70 (for AIF derived decoy for Hsp70), bind to and inhibit Hsp70. In vivo experiments, in a syngeneic rat colon cancer cell model and in a mice model of melanoma (B16F10), showed that ADD70 decreased the size of the tumors in rats and induced an important growth delay of the mice tumors. In addition, ADD70 sensitizes both the rat colon cancer cells and mouse melanoma cells to the chemotherapeutic agent cisplatin. This ADD70 antitumorigenic effect was only observed in syngeneic animals but not in immunodeficient animals. We have demonstrated that ADD70 antitumorigenic effects are associated with an increase in tumor-infiltrating cytotoxic $CD8^+$ T-cells (129).

Concerning chemical molecules specific for Hsp70, recently Leu et al. described a small-molecule inhibitor that interacts with C-terminal of Hsp70 called 2-phenylethynesulfonamide (PES). PES interaction with Hsp70 chaperone activity leads to the aggregation of misfolded proteins, and destabilization of lysosome membranes thus inducting an autophagic cell death (130). However, other reports indicate that PES induces a clear apoptotic, caspase-dependent, cell death (131). The ability of PES to induce an apoptotic or autophagic cell death could be cell-type dependent.

Targeting the Amino-Terminal ATPase Domain

By analogy with the Hsp90 inhibitors, compounds that displace ATP from Hsp70 are expected to be powerful tools. Williamson et al. designed and synthesized adenosine-derived molecules based on the X-ray crystal structure of a commercially available ATP analog. Although further studies are necessary to determine their specificity and potency, some in vitro results are encouraging (132).

VER-155008 is an adenosine-derived compound that fits into the ATPase domain of Hsp70 (and Hsc70), thereby inhibiting the chaperone activity of these proteins. The product, also sensitizes colon HCT116 carcinoma cells to die, but has not been yet tested in vivo models (133).

A high-throughput screening for ATP turnover mediated by human Hsp70 allowed the identification of azure C, methylene blue, and myricetin as inhibitors of Hsp70. However, their specificity for inducible Hsp70 family has not been analyzed (134).

The ABD is also the site of action of MKT-077, a cationic rhodacyanine dye with selectivity toward cancer cells. MKT-077 localizes in the mitochondria where it interacts with mitochondrial Hsp70. This product has been explored on phase I clinical trial as an antitumor agent. Although the product does not specifically bind to Hsp70 (i.e., it also binds to actin), the drug-like nature of this compound deserves further investigation (135).

Screening for Hsp70 ATPase activity and a subset of the NCI drug collection allowed the identification of NSC 630668, a dihydropyrimidine. More interesting is the second generation compound MAL3-101, which inhibits Hsp70 ATPase activity and blocks proliferation of SK-BR-3 cancer cells (136). MAL2-11B was shown to inhibit the activity of a viral J-domain protein, T antigen, suggesting it may be a new class of polyomavirus inhibitors (137). The molecular mechanism of action of these molecules requires further study.

Our group has recently selected combinatorial peptide aptamers from two peptide aptamer libraries, consisting of an *Escherichia coli* thioredoxin scaffold displaying variable peptide loops of 13 or 8 aminoacids. The screening was based on the ability of the aptamers to interact with Hsp70 ATP-binding domain in a yeast two hybrid assay. One of them (A17) increases the sensitivity to apoptosis induction by anticancer drugs and, in vivo, has a strong antitumor effect. A17 inhibits Hsp70 chaperone activity, yet does not affect that of Hsc70 or Hsp90 (unpublished results).

Targeting Hsp70 Co-chaperones

An interesting way of inhibiting only certain functions of Hsp70 is to find molecules that specifically disrupt the interaction of Hsp70 with a particular co-chaperone. Although this strategy in drug discovery has not yet been exploited, using an alphaScreen high-throughput approach, pyrimidotriazinediones were found to interfere with Hop/Hsp70 interaction. These compounds were toxic for WST-1 cells in vitro (138). More recently, it has been reported that drugs targeting Hip impair Hsp70 chaperone proteostasis and stimulate neurodegeneration (139).

3.2.2. Perspectives; Hsp70 Combined with Hsp90 Inhibitors

The inhibition of Hsp70 is also considered an interesting strategy in combination with other drugs since most of them, including Hsp90 inhibitors, induce Hsp70 that can circumvent the drugs' tumor cell killing effect (51, 129). Many Hsp inhibitors, such as 17-AAG, are now in advanced clinical trials. An increase in Hsp70 expression in the treated patients has been reported.

Hsp70 accumulation can reduce cell death induced by Hsp90 inhibitors and therefore decrease their antitumor efficacy. This Hsp70 induction has been reported to be related to the calcium mobilization that results from Hsp90 inhibitors treatment (140) and to TGF-beta signaling (141). Depletion of Hsp70 by siRNA strongly increase the cells' sensitivity to 17-AAG (142). Simultaneous silencing of both Hsp70 and Hsc70 induce proteasomal-dependent degradation of Hsp90 client proteins (143). Confirming these results, it has recently been described that Hsp70 inhibitor VER-155008 potentiates the effect of 17-AAG inducing apoptosis in HCT116 colon carcinoma cells (133). We obtained a similar increase in 17-AAG anticancer activity in colon cancer cells when using ADD70 to block Hsp70 (129). Future will tell whether Hsp70 inhibitors could be directly injected in the tumor. This would open new perspectives in their use as local chemosensitization agents in Hsp70-expressing human tumors.

Hsp70 inhibitors, similarly to Hsp90 inhibitors, could also be combined with HDAC inhibitors which have been shown to increase cell surface expression of Hsp70 on hematopoietic cancer cells. Although it is too early to know clearly the consequences of such extracellular expression (we have recently shown that increased cell membrane Hsp70 activates suppressive myeloid immune cells) (144), it will be interesting to combine both inhibitors to study their effect in tumor growth in a syngeneic experimental model.

4. Anticancer Therapeutical Approaches Based on Extracellular Hsp70

4.1. Extracellular Hsp70 has an Immunological Function

Apart from its protective roles in the cytosol Hsp70 has been found to play a role in the stimulation of the immune system when located in the extracellular space or on the plasma membrane. The immunogenic function of extracellular Hsps is mainly described through their role chaperoning antigenic peptides. The word "chaperokine" has been evoked (145).

In humans, Hsp70 presence in the serum is associated with stress or pathological conditions, including cancer. Tumor cells were identified as a natural source for extracellular located Hsp70. Previous studies identify a role of Hsp70 expressed at the cancer cell surface during tumor growth and metastatic spread (146). An active release of Hsc70 from tumor cells was observed following treatment with interferon-γ (147). In vitro, members of the Hsp70 family have been detected in the medium of antigen presenting cells (APCs). In different cancer cell lines (neuroblastoma, lung and colon adenocarcinoma), a high expression on the cell surface of different Hsps, including Hsp70, have been described (148).

Cytosolic Hsps do not contain leader peptides enabling membrane localization. One possibility is that cytosolic Hsps are transported to the plasma membrane in concert with other proteins possessing transmembrane domains. Another possibility for membrane anchorage might be a direct interaction of Hsps with lipid components. Hsp70 has been shown to associate with phosphatidylserine (PS) in PC12 tumor cells (149). More recently, Hsp70 has been described to co-localize with PS in the outer membrane leaflet of hypoxic tumor cells (150). Thus, it has been hypothesized that in stressed tumor cells translocation of Hsp70 to the outer membrane leaflet would be allowed by a mechanism associated with PS flipping. In what concern Hsp extracellular localization, today there are two mechanisms reported by which Hsps may be released form cells: (1) a passive release mechanism that includes: necrotic cell death, severe blunt trauma, surgery and following infection with lytic viruses; (2) an active release mechanism which involves the nonclassical protein release pathway in which Hsps are released from highly immunologically potent exosomes and as free Hsp (151, 152). De Maio's group has demonstrated in vitro, in HepG2 cells exposed to a heat shock, that Hsp70 was release to the extracellular environment associated in a membrane-associated form. Extracellular membranes containing Hsp70 strongly activated macrophages, much more than free recombinant Hsp70, as indicated by their ability to induce TNF-α production (153).

Hsp70 appeared to augment the immunogenicity of tumors. In numerous studies, Hsp70 has been shown to activate innate and adaptive immune reactions (154, 155). Hsp70 chaperones antigenic peptides and channels them, in a receptor-mediated manner, into the major histocompatibility complex (MHC) class I presentation pathway of professional APCs, which then prime peptide-specific cytotoxic T-lymphocyte (CTL) (75). Therefore, Hsp70 derived from tumors can be used as tumor-specific vaccines (156). Hsp70 also elicits the release of pro-inflammatory cytokines from innate immune cells and augments the expression of co-stimulatory molecules (157, 158). Furthermore, Hsp70 has been shown to activate natural killer (NK) cell lytic machinery against tumors expressing Hsp70 at the cell surface (159). These features have led to view Hsp70 as an endogenous adjuvant and immunological danger signal (160).

Members of the Hsp70 family together with those of the Hsp90 family were identified as key regulators of the host's immune system. In vivo, in animal models, Hsp-peptide complexes isolated from tumors developed an antitumor response while those isolated from normal tissues did not induce an anticancer immune response (161). Therefore, the tumor-specific peptides chaperoned by Hsps were at the basis of the immunogenicity. Following cross-presentation of Hsp-chaperoned peptides on MHC class I molecules, a T-cell response is initiated (162–172). Exogenously delivered antigenic peptides complexed to Hsps, like Hsp70, are able to enter endogenous Ag-processing pathway and prime $CD8^+$ T-cells for antigen production (173). Hsps have been found to be important players in the process of cross-presentation of tumor-derived, antigenic peptides that are uptaken by the APCs (162, 174, 175). Hsp-chaperoned antigenic peptides derived from tumor cells are internalized in the APCs by endocytosis via "Hsp receptors" (i.e., CD91, CD40, TLR2/4 ± CD14, CD35, LOX-1, SR-A). Following uptake and processing, the peptides are presented to the MHC class I molecules and thus induce a $CD8^+$ T-cell response against cancer. Recombinant Hsp70 enhances tumor antigen cross-presentation by means of complex formation and augmenting antigen uptake. Innate immune activation by recombinant Hsp70 was not observed (176).

Hsp70-peptide complexes have also been described to enhance MHC class II restricted peptide presentation and $CD4^+$ T-cells activation (177). Even in the absence of immunogenic antigens, Hsps can provide danger signals for the host's immune system (178). In this way, it is believed that Hsp70-peptide complexes can suppress tumor growth via at least two pathways: one antigen-dependent and the other antigen-independent. The c-terminal domain of Hsp70 seems to be crucial in eliciting antigen-independent responses, including NK cell stimulation against tumor challenges (68). This can explain why, in experimental models, immunization with multiple Hsp70-fusion peptide complexes resulted in a better antitumor effect (179). Hsp70 has been suggested to have cytokine-inducing effects. In APCs cells, it induces the release of pro-inflammatory cytokines like IL-1β, IL-6, or TNF-α (68). In established melanomas, Hsp70 has been shown to act as a potent immune adjuvant inducing the production of TNF-α (180). However, this role of Hsps inducing cytokines production has been reported to be, at least in some cases, consequence of LPS contaminations in the Hsps preparations (181, 182).

4.2. Immuno-Therapeutical Approaches Based on Hsp70

Most Hsps-based immunotherapeutical approaches against cancer exploit their carrier function for immunogenic peptides (183, 184). In experimental models of metastatic prostate cancer, Hsp70 is used as a potent adjuvant to induce cancer auto-immunity (185). It has been shown that immunization with leukemic cells A20-derived Hsp70 induced the production of leukemia A20-specific antibodies which, in turn, induced complement-dependent cytotoxicity against A20 cells and enable eradication of leukemia cells in mice (186). Therapeutic vaccination with dendritic cells pulsed with tumor-derived Hsp70 induces immunity against B16 melanoma, which is increased in combination with COX-2 inhibitors (187). $CD8^+$ and $CD4^+$ T-cell epitopes, fused to the carboxyl-terminus of Hsc70, can enhance tumor immunity beyond the effect of the $CD8^+$ epitope alone and eradicates tumors (188). Fusion of HPV16 E7 to C-terminus half of Hsp70 can induce potent antigen-specific CTL activity in experimental models (189). Moreover, a recombinant N domain of calreticulin/E7/C-terminal half of Hsp70, through synergistic effects, can both generate potent antitumor immunity and antiangiogenesis (189).

However, although some encouraging clinical results are being obtained, we are far from the conclusive results obtained in tumor rodent models treated Hsps-based vaccines (161). An improved formulation of chaperone protein-based tumor vaccine may be the extraction of Hsp70-peptide complexes from dendritic tumor fusion cells. It has been shown that a vaccine consisting of Hsp70 extracted from fusions of dendritic and tumor cells reverses immunotolerance of cancer cells more efficiently than its counterpart vaccine extracted from tumor cells alone (190). A recent pilot study in cancer patients (renal carcinoma, breast cancer, and astrocytoma) has recently demonstrated that an autologous antitumor vaccine using hydroxyapatite particles (to attract APCs) with at least three Hsps (Hsp70, Hsp27, and GRP96) could be safely administrated (191).

Also exploiting the immunogenic properties of extracellular Hsp70, Multhoff's group has developed a different approach. NK cells, important effectors of the innate immune system, play a role in the antitumor immunity generated by chaperone-rich cells lysate vaccination (192). Hsp70 is a triggering factor for NK cells (193, 194). Mapping of the Hsp70 sequence revealed that the 14-mer peptide (TKD), derived from Hsp70 C-terminal PBD (aa 450–463), has similar immunostimulatory capacity on NK cells like full length Hsp70 (195, 196). Incubation of NK cells with cytokines plus soluble Hsp70 protein or TKD peptide enhances the cell surface density of activating NK cell receptors, including CD94 (193). Concomitantly, the cytolytic and migratory capacity of resting NK cells was found to be initiated (197). Membrane-bound Hsp70 serves as a tumor-selective target structure, since Hsp70 is frequently presented on the plasma membrane of tumors and metastases but not in normal tissues. Incubation of peripheral blood lymphocytes with TKD peptide plus IL-2 initiates the cytolytic and migratory capacity of NK cells toward Hsp70 membrane-positive tumor cells in vitro and in a xenograft tumor mouse model (198). A phase I clinical trial has been performed in patients with therapy-refractory, metastasized colorectal and nonsmall lung cell carcinoma with very encouraging results (199). The advantage of the approach is the excellent safety profile and the unlimited availability of the synthetic Hsp70 peptide which stimulates NK cells, as compared to the complexes Hsp-peptides/vaccines isolated from patients, very limited in amount.

5. Concluding Remarks

Hsp70 has different functions depending on its location, intracellular versus extracellular. Intracellular Hsp70 have a protective function and its subcellular location may determine its interaction with different partners. Hsp70 is one of the most cytoprotective proteins ever described since it can block almost all main cell death pathways as described in this chapter. Hsp70 cytoprotective function is essential for cancer cell survival. Hsp70 may have oncogene-like functions and likewise mediate "non-oncogene addiction" of stressed tumor cells that must adapt to hostile microenvironment thereby becoming dependent for their survival on Hsp70. Hsp70-targeting drugs have therefore emerged as potential anticancer agents. The search for inhibitors of Hsp70 has dramatically increased over the last 2 years. Although many preclinical data are available, their clinical evaluation has not started yet, unlike the inhibitors of another Hsp, Hsp90, that some are already in phase II/III clinical trials. However, the clinical results obtained so far are not the expected ones. Those disappointing results may be associated to the strong induction of Hsp70 in the tumors that they provoke which can circumvent Hsp90 chemosensitizing effect. Therefore, a good solution could be to combine Hsp90 and Hsp70 inhibitors.

Extracellular Hsp70 is believed to be immunogenic. Experimentally, Hsp70 has a well-demonstrated effect as an adjuvant and, together with other chaperones, it may be a very useful tool for vaccine productions and treatment of metastatic cancers. However, a high extracellular level of Hsp70 has been found in cancer patients and has been associated with bad prognosis, in contrast to what should be expected considering its potential immune anticancer effect. More paradoxical is the observation made by our group that Hsp70 inhibition or depletion provokes tumor regression in syngeneic animals by inducing an immune antitumor response (129). This paradox can be explained by our recent results demonstrating that cancer cells secrete exosomes that abundantly express Hsp70 on the cell surface; this external Hsp70 activates myeloid suppressive cells explaining why Hsp70 inhibition in cancer models can result in induction of an antitumor immune response (144).

In conclusion, the dual function of Hsp70 (protective, immunogenic) strongly increases the interest for this molecule in cancer therapy.

Acknowledgments

This work was supported by grants from the "Ligue Nationale Contre le Cancer" and the committees form the "Nièvre" and "Saône et Loire." CG's group has the label de "La Ligue Contre le Cancer." The work was also supported by grants from the "Institut National du Cancer" and "Conseil Regional de Bourgogne."

References

1. Ritossa F. A new puffing pattern induced by temperature shock and DNP and in Drosophila. Experientia 1962;18:571–3.
2. Tissieres A, Mitchell HK, Tracy UM. Protein synthesis in salivary glands of Drosophila melanogaster: Relation to chromosome puffs. Journal of molecular biology 1974;85(3): 389–98.
3. Li GC, Werb Z. Correlation between synthesis of heat shock proteins and development of thermotolerance in Chinese hamster fibroblasts. Proceedings of the National Academy of Sciences of the United States of America 1982;79(10):3218–22.
4. Gething MJ, Sambrook J. Protein folding in the cell. Nature 1992;355(6355):33–45.
5. Feige U, Morimoto RI, Yahara I, Polla BS. Stress-inducible cellular responses. 1996: 492p.
6. Parcellier A, Gurbuxani S, Schmitt E, Solary E, Garrido C. Heat shock proteins, cellular chaperones that modulate mitochondrial cell death pathways. Biochemical and biophysical research communications 2003;304(3):505–12.
7. Nollen EA, Brunsting JF, Roelofsen H, Weber LA, Kampinga HH. In vivo chaperone activity of heat shock protein 70 and thermotolerance. Mol Cell Biol 1999;19(3):2069–79.
8. Polla BS, Kantengwa S, Francois D, et al. Mitochondria are selective targets for the protective effects of heat shock against oxidative injury. Proc Natl Acad Sci USA 1996;93(13): 6458–63.
9. Park HS, Cho SG, Kim CK, et al. Heat shock protein hsp72 is a negative regulator of apoptosis signal-regulating kinase 1. Mol Cell Biol 2002;22(22):7721–30.
10. Garrido C, Gurbuxani S, Ravagnan L, Kroemer G. Heat shock proteins: endogenous modulators of apoptotic cell death. Biochem Biophys Res Commun 2001;286(3):433–42.
11. Qian SB, McDonough H, Boellmann F, Cyr DM, Patterson C. CHIP-mediated stress recovery by sequential ubiquitination of substrates and Hsp70. Nature 2006;440(7083): 551–5.
12. Zermati Y, Garrido C, Amsellem S, et al. Caspase activation is required for terminal erythroid differentiation. J Exp Med 2001; 193(2):247–54.
13. Ribeil JA, Zermati Y, Vandekerckhove J, et al. Hsp70 regulates erythropoiesis by preventing caspase-3-mediated cleavage of GATA-1. Nature 2007;445(7123):102–5.
14. Jaattela M. Escaping cell death: survival proteins in cancer. Exp Cell Res 1999;248(1): 30–43.
15. Jaattela M. Heat shock proteins as cellular lifeguards. Ann Med 1999;31(4):261–71.
16. Tavaria M, Gabriele T, Kola I, Anderson RL. A hitchhiker's guide to the human Hsp70 family. Cell Stress Chaperones 1996;1(1):23–8.
17. Daugaard M, Rohde M, Jaattela M. The heat shock protein 70 family: Highly homologous proteins with overlapping and distinct functions. FEBS Lett 2007;581(19):3702–10.
18. Beckmann RP, Mizzen LE, Welch WJ. Interaction of Hsp 70 with newly synthesized proteins: implications for protein folding and assembly. Science 1990;248(4957):850–4.
19. Murakami H, Pain D, Blobel G. 70-kD heat shock-related protein is one of at least two distinct cytosolic factors stimulating protein import into mitochondria. J Cell Biol 1988;107(6 Pt 1):2051–7.
20. Shi Y, Thomas JO. The transport of proteins into the nucleus requires the 70-kilodalton heat shock protein or its cytosolic cognate. Mol Cell Biol 1992;12(5):2186–92.
21. Vogel M, Bukau B, Mayer MP. Allosteric regulation of Hsp70 chaperones by a proline switch. Mol Cell 2006;21(3):359–67.

22. Hennessy F, Nicoll WS, Zimmermann R, Cheetham ME, Blatch GL. Not all J domains are created equal: implications for the specificity of Hsp40-Hsp70 interactions. Protein Sci 2005;14(7):1697–709.
23. Young JC, Agashe VR, Siegers K, Hartl FU. Pathways of chaperone-mediated protein folding in the cytosol. Nat Rev Mol Cell Biol 2004; 5(10):781–91.
24. Bukau B, Weissman J, Horwich A. Molecular chaperones and protein quality control. Cell 2006;125(3):443–51.
25. Voos W. A new connection: chaperones meet a mitochondrial receptor. Mol Cell 2003;11(1):1–3.
26. Kerr JF, Wyllie AH, Currie AR. Apoptosis: a basic biological phenomenon with wide-ranging implications in tissue kinetics. Br J Cancer 1972;26(4):239–57.
27. Baehrecke EH. How death shapes life during development. Nat Rev Mol Cell Biol 2002;3(10):779–87.
28. Solary E, Droin N, Bettaieb A, Corcos L, Dimanche-Boitrel MT, Garrido C. Positive and negative regulation of apoptotic pathways by cytotoxic agents in hematological malignancies. Leukemia 2000;14(10):1833–49.
29. Nicholson DW, Thornberry NA. Caspases: killer proteases. Trends Biochem Sci 1997;22(8):299–306.
30. Earnshaw WC, Martins LM, Kaufmann SH. Mammalian caspases: structure, activation, substrates, and functions during apoptosis. Annu Rev Biochem 1999;68:383–424.
31. Kaufmann S. The Intrinsic Pathway of Apoptosis. Apoptosis, Senescence, and Cancer 2007:3–30.
32. Wei MC, Zong WX, Cheng EH, et al. Proapoptotic BAX and BAK: a requisite gateway to mitochondrial dysfunction and death. Science 2001;292(5517):727–30.
33. Zong WX, Lindsten T, Ross AJ, MacGregor GR, Thompson CB. BH3-only proteins that bind pro-survival Bcl-2 family members fail to induce apoptosis in the absence of Bax and Bak. Genes Dev 2001;15(12):1481–6.
34. Breckenridge DG, Xue D. Regulation of mitochondrial membrane permeabilization by BCL-2 family proteins and caspases. Curr Opin Cell Biol 2004;16(6):647–52.
35. Cheng EH, Wei MC, Weiler S, et al. BCL-2, BCL-X(L) sequester BH3 domain-only molecules preventing BAX- and BAK-mediated mitochondrial apoptosis. Mol Cell 2001;8(3): 705–11.
36. Li P, Nijhawan D, Budihardjo I, et al. Cytochrome c and dATP-dependent formation of Apaf-1/caspase-9 complex initiates an apoptotic protease cascade. Cell 1997;91(4): 479–89.
37. Zhivotovsky B, Orrenius S, Brustugun OT, Doskeland SO. Injected cytochrome c induces apoptosis. Nature 1998;391(6666):449–50.
38. Du C, Fang M, Li Y, Li L, Wang X. Smac, a mitochondrial protein that promotes cytochrome c-dependent caspase activation by eliminating IAP inhibition. Cell 2000;102(1): 33–42.
39. Verhagen AM, Silke J, Ekert PG, et al. HtrA2 promotes cell death through its serine protease activity and its ability to antagonize inhibitor of apoptosis proteins. The Journal of biological chemistry 2002;277(1):445–54.
40. Verhagen AM, Vaux DL. Cell death regulation by the mammalian IAP antagonist Diablo/ Smac. Apoptosis 2002;7(2):163–6.
41. Creagh EM, Murphy BM, Duriez PJ, Duckett CS, Martin SJ. Smac/Diablo antagonizes ubiquitin ligase activity of inhibitor of apoptosis proteins. The Journal of biological chemistry 2004;279(26):26906–14.
42. Schulze-Osthoff K, Ferrari D, Los M, Wesselborg S, Peter ME. Apoptosis signaling by death receptors. Eur J Biochem 1998; 254(3):439–59.
43. Luo X, Budihardjo I, Zou H, Slaughter C, Wang X. Bid, a Bcl2 interacting protein, mediates cytochrome c release from mitochondria in response to activation of cell surface death receptors. Cell 1998;94(4):481–90.
44. Li H, Zhu H, Xu CJ, Yuan J. Cleavage of BID by caspase 8 mediates the mitochondrial damage in the Fas pathway of apoptosis. Cell 1998;94(4):491–501.
45. Garrido C, Brunet M, Didelot C, Zermati Y, Schmitt E, Kroemer G. Heat shock proteins 27 and 70: anti-apoptotic proteins with tumorigenic properties. Cell cycle (Georgetown, Tex 2006;5(22):2592–601.
46. Mosser DD, Caron AW, Bourget L, et al. The chaperone function of hsp70 is required for protection against stress-induced apoptosis. Mol Cell Biol 2000;20(19):7146–59.
47. Mosser DD, Morimoto RI. Molecular chaperones and the stress of oncogenesis. Oncogene 2004;23(16):2907–18.
48. Aghdassi A, Phillips P, Dudeja V, et al. Heat shock protein 70 increases tumorigenicity and inhibits apoptosis in pancreatic adenocarcinoma. Cancer Res 2007;67(2):616–25.
49. Gurbuxani S, Bruey JM, Fromentin A, et al. Selective depletion of inducible HSP70 enhances immunogenicity of rat colon cancer cells. Oncogene 2001;20(51):7478–85.

50. Nylandsted J, Rohde M, Brand K, Bastholm L, Elling F, Jaattela M. Selective depletion of heat shock protein 70 (Hsp70) activates a tumor-specific death program that is independent of caspases and bypasses Bcl-2. Proc Natl Acad Sci USA 2000;97(14):7871–6.
51. Schmitt E, Parcellier A, Gurbuxani S, et al. Chemosensitization by a non-apoptogenic heat shock protein 70-binding apoptosis-inducing factor mutant. Cancer Res 2003; 63(23):8233–40.
52. Dix DJ, Allen JW, Collins BW, et al. Targeted gene disruption of Hsp70-2 results in failed meiosis, germ cell apoptosis, and male infertility. Proc Natl Acad Sci USA 1996;93(8): 3264–8.
53. Biggs WH, 3rd, Meisenhelder J, Hunter T, Cavenee WK, Arden KC. Protein kinase B/ Akt-mediated phosphorylation promotes nuclear exclusion of the winged helix transcription factor FKHR1. Proc Natl Acad Sci USA 1999;96(13):7421–6.
54. Ozes ON, Mayo LD, Gustin JA, Pfeffer SR, Pfeffer LM, Donner DB. NF-kappaB activation by tumour necrosis factor requires the Akt serine-threonine kinase. Nature 1999; 401(6748):82–5.
55. Datta SR, Dudek H, Tao X, et al. Akt phosphorylation of BAD couples survival signals to the cell-intrinsic death machinery. Cell 1997;91(2):231–41.
56. Cardone MH, Roy N, Stennicke HR, et al. Regulation of cell death protease caspase-9 by phosphorylation. Science 1998;282(5392):1318–21.
57. Gao T, Newton AC. The turn motif is a phosphorylation switch that regulates the binding of Hsp70 to protein kinase C. J Biol Chem 2002;277(35):31585–92.
58. Hu G, Tang J, Zhang B, et al. A novel endothelial-specific heat shock protein HspA12B is required in both zebrafish development and endothelial functions in vitro. J Cell Sci 2006;119(Pt 19):4117–26.
59. Meriin AB, Yaglom JA, Gabai VL, et al. Protein-damaging stresses activate c-Jun N-terminal kinase via inhibition of its dephosphorylation: a novel pathway controlled by HSP72. Mol Cell Biol 1999;19(4):2547–55.
60. Park HS, Lee JS, Huh SH, Seo JS, Choi EJ. Hsp72 functions as a natural inhibitory protein of c-Jun N-terminal kinase. Embo J 2001;20(3):446–56.
61. Lee JS, Lee JJ, Seo JS. HSP70 deficiency results in activation of c-Jun N-terminal Kinase, extracellular signal-regulated kinase, and caspase-3 in hyperosmolarity-induced apoptosis. J Biol Chem 2005;280(8):6634–41.
62. Salehi AH, Morris SJ, Ho WC, et al. AEG3482 is an antiapoptotic compound that inhibits Jun kinase activity and cell death through induced expression of heat shock protein 70. Chem Biol 2006;13(2):213–23.
63. Gabai VL, Yaglom JA, Volloch V, et al. Hsp72-mediated suppression of c-Jun N-terminal kinase is implicated in development of tolerance to caspase-independent cell death. Mol Cell Biol 2000;20(18):6826–36.
64. Akakura S, Yoshida M, Yoneda Y, Horinouchi S. A role for Hsc70 in regulating nucleocytoplasmic transport of a temperature-sensitive p53 (p53Val-135). J Biol Chem 2001;276(18): 14649–57.
65. Zylicz M, King FW, Wawrzynow A. Hsp70 interactions with the p53 tumour suppressor protein. Embo J 2001;20(17):4634–8.
66. Asea A, Rehli M, Kabingu E, et al. Novel signal transduction pathway utilized by extracellular HSP70: role of toll-like receptor (TLR) 2 and TLR4. J Biol Chem 2002;277(17): 15028–34.
67. Mijatovic T, Mathieu V, Gaussin JF, et al. Cardenolide-induced lysosomal membrane permeabilization demonstrates therapeutic benefits in experimental human non-small cell lung cancers. Neoplasia 2006;8(5):402–12.
68. Schmitt E, Gehrmann M, Brunet M, Multhoff G, Garrido C. Intracellular and extracellular functions of heat shock proteins: repercussions in cancer therapy. J Leukoc Biol 2007;81(1): 15–27.
69. Ran R, Lu A, Zhang L, et al. Hsp70 promotes TNF-mediated apoptosis by binding IKK gamma and impairing NF-kappa B survival signaling. Genes Dev 2004;18(12):1466–81.
70. Shanley TP, Ryan MA, Eaves-Pyles T, Wong HR. Heat shock inhibits phosphorylation of I-kappaBalpha. Shock 2000;14(4):447–50.
71. Yoo CG, Lee S, Lee CT, Kim YW, Han SK, Shim YS. Anti-inflammatory effect of heat shock protein induction is related to stabilization of I kappa B alpha through preventing I kappa B kinase activation in respiratory epithelial cells. J Immunol 2000;164(10):5416–23.
72. Feng X, Bonni S, Riabowol K. HSP70 induction by ING proteins sensitizes cells to tumor necrosis factor alpha receptor-mediated apoptosis. Mol Cell Biol 2006;26(24):9244–55.
73. Stankiewicz AR, Lachapelle G, Foo CP, Radicioni SM, Mosser DD. Hsp70 inhibits heat-induced apoptosis upstream of mitochondria by preventing Bax translocation. J Biol Chem 2005;280(46):38729–39.
74. Ruchalski K, Mao H, Li Z, et al. Distinct hsp70 domains mediate apoptosis-inducing factor

release and nuclear accumulation. J Biol Chem 2006;281(12):7873–80.
75. Castellino F, Boucher PE, Eichelberg K, et al. Receptor-mediated uptake of antigen/heat shock protein complexes results in major histocompatibility complex class I antigen presentation via two distinct processing pathways. J Exp Med 2000;191(11):1957–64.
76. Beere HM, Wolf BB, Cain K, et al. Heat-shock protein 70 inhibits apoptosis by preventing recruitment of procaspase-9 to the Apaf-1 apoptosome. Nat Cell Biol 2000;2(8): 469–75.
77. Pandey P, Saleh A, Nakazawa A, et al. Negative regulation of cytochrome c-mediated oligomerization of Apaf-1 and activation of procaspase-9 by heat shock protein 90. The EMBO journal 2000;19(16):4310–22.
78. Jaattela M, Wissing D, Kokholm K, Kallunki T, Egeblad M. Hsp70 exerts its anti-apoptotic function downstream of caspase-3-like proteases. Embo J 1998;17(21):6124–34.
79. Sakahira H, Nagata S. Co-translational folding of caspase-activated DNase with Hsp70, Hsp40, and inhibitor of caspase-activated DNase. J Biol Chem 2002;277(5):3364–70.
80. Liu QL, Kishi H, Ohtsuka K, Muraguchi A. Heat shock protein 70 binds caspase-activated DNase and enhances its activity in TCR-stimulated T cells. Blood 2003;102(5): 1788–96.
81. Duriez PJ, Shah GM. Cleavage of poly(ADP-ribose) polymerase: a sensitive parameter to study cell death. Biochem Cell Biol 1997;75(4):337–49.
82. Decker P, Muller S. Modulating poly (ADP-ribose) polymerase activity: potential for the prevention and therapy of pathogenic situations involving DNA damage and oxidative stress. Curr Pharm Biotechnol 2002;3(3): 275–83.
83. Kotoglou P, Kalaitzakis A, Vezyraki P, et al. Hsp70 translocates to the nuclei and nucleoli, binds to XRCC1 and PARP-1, and protects HeLa cells from single-strand DNA breaks. Cell Stress Chaperones 2009;14(4):391–406.
84. Gabai VL, Mabuchi K, Mosser DD, Sherman MY. Hsp72 and stress kinase c-jun N-terminal kinase regulate the bid-dependent pathway in tumor necrosis factor-induced apoptosis. Mol Cell Biol 2002;22(10):3415–24.
85. Pang Q, Keeble W, Christianson TA, Faulkner GR, Bagby GC. FANCC interacts with Hsp70 to protect hematopoietic cells from IFN-gamma/TNF-alpha-mediated cytotoxicity. Embo J 2001;20(16):4478–89.
86. Pang Q, Christianson TA, Keeble W, Koretsky T, Bagby GC. The anti-apoptotic function of Hsp70 in the interferon-inducible double-stranded RNA-dependent protein kinase-mediated death signaling pathway requires the Fanconi anemia protein, FANCC. J Biol Chem 2002;277(51):49638–43.
87. Guo F, Sigua C, Bali P, et al. Mechanistic role of heat shock protein 70 in Bcr-Abl-mediated resistance to apoptosis in human acute leukemia cells. Blood 2005;105(3):1246–55.
88. Schett G, Steiner CW, Groger M, et al. Activation of Fas inhibits heat-induced activation of HSF1 and up-regulation of hsp70. Faseb J 1999;13(8):833–42.
89. Liossis SN, Ding XZ, Kiang JG, Tsokos GC. Overexpression of the heat shock protein 70 enhances the TCR/CD3- and Fas/Apo-1/CD95-mediated apoptotic cell death in Jurkat T cells. J Immunol 1997;158(12):5668–75.
90. Cande C, Vahsen N, Garrido C, Kroemer G. Apoptosis-inducing factor (AIF): caspase-independent after all. Cell Death Differ 2004;11(6):591–5.
91. Creagh EM, Carmody RJ, Cotter TG. Heat shock protein 70 inhibits caspase-dependent and -independent apoptosis in Jurkat T cells. Exp Cell Res 2000;257(1):58–66.
92. Ravagnan L, Gurbuxani S, Susin SA, et al. Heat-shock protein 70 antagonizes apoptosis-inducing factor. Nat Cell Biol 2001;3(9):839–43.
93. Matsumori Y, Hong SM, Aoyama K, et al. Hsp70 overexpression sequesters AIF and reduces neonatal hypoxic/ischemic brain injury. J Cereb Blood Flow Metab 2005;25(7):899–910.
94. Lui JC, Kong SK. Heat shock protein 70 inhibits the nuclear import of apoptosis-inducing factor to avoid DNA fragmentation in TF-1 cells during erythropoiesis. FEBS Lett 2007;581(1):109–17.
95. Kalinowska M, Garncarz W, Pietrowska M, Garrard WT, Widlak P. Regulation of the human apoptotic DNase/RNase endonuclease G: involvement of Hsp70 and ATP. Apoptosis 2005;10(4):821–30.
96. Jaattela M, Tschopp J. Caspase-independent cell death in T lymphocytes. Nat Immunol 2003;4(5):416–23.
97. Guicciardi ME, Deussing J, Miyoshi H, et al. Cathepsin B contributes to TNF-alpha-mediated hepatocyte apoptosis by promoting mitochondrial release of cytochrome c. J Clin Invest 2000;106(9):1127–37.
98. Foghsgaard L, Wissing D, Mauch D, et al. Cathepsin B acts as a dominant execution protease in tumor cell apoptosis induced by tumor necrosis factor. J Cell Biol 2001;153(5): 999–1010.

99. Brunk UT, Svensson I. Oxidative stress, growth factor starvation and Fas activation may all cause apoptosis through lysosomal leak. Redox Rep 1999;4(1–2):3–11.
100. Broker LE, Huisman C, Span SW, Rodriguez JA, Kruyt FA, Giaccone G. Cathepsin B mediates caspase-independent cell death induced by microtubule stabilizing agents in non-small cell lung cancer cells. Cancer Res 2004; 64(1):27–30.
101. Bidere N, Lorenzo HK, Carmona S, et al. Cathepsin D triggers Bax activation, resulting in selective apoptosis-inducing factor (AIF) relocation in T lymphocytes entering the early commitment phase to apoptosis. J Biol Chem 2003;278(33):31401–11.
102. Yuan XM, Li W, Dalen H, et al. Lysosomal destabilization in p53-induced apoptosis. Proc Natl Acad Sci USA 2002;99(9): 6286–91.
103. Brunk UT, Dalen H, Roberg K, Hellquist HB. Photo-oxidative disruption of lysosomal membranes causes apoptosis of cultured human fibroblasts. Free Radic Biol Med 1997;23(4):616–26.
104. Boya P, Gonzalez-Polo RA, Poncet D, et al. Mitochondrial membrane permeabilization is a critical step of lysosome-initiated apoptosis induced by hydroxychloroquine. Oncogene 2003;22(25):3927–36.
105. Roberg K, Kagedal K, Ollinger K. Microinjection of cathepsin d induces caspase-dependent apoptosis in fibroblasts. Am J Pathol 2002;161(1):89–96.
106. Vancompernolle K, Van Herreweghe F, Pynaert G, et al. Atractyloside-induced release of cathepsin B, a protease with caspase-processing activity. FEBS Lett 1998;438(3): 150–8.
107. Nylandsted J, Gyrd-Hansen M, Danielewicz A, et al. Heat shock protein 70 promotes cell survival by inhibiting lysosomal membrane permeabilization. J Exp Med 2004;200(4): 425–35.
108. Bivik C, Rosdahl I, Ollinger K. Hsp70 protects against UVB induced apoptosis by preventing release of cathepsins and cytochrome c in human melanocytes. Carcinogenesis 2007;28(3):537–44.
109. Simon MM, Reikerstorfer A, Schwarz A, et al. Heat shock protein 70 overexpression affects the response to ultraviolet light in murine fibroblasts. Evidence for increased cell viability and suppression of cytokine release. J Clin Invest 1995;95(3):926–33.
110. Sashchenko LP, Dukhanina EA, Shatalov YV, et al. Cytotoxic T lymphocytes carrying a pattern recognition protein Tag7 can detect evasive, HLA-negative but Hsp70-exposing tumor cells, thereby ensuring FasL/Fas-mediated contact killing. Blood 2007;110(6):1997–2004.
111. Ciocca DR, Calderwood SK. Heat shock proteins in cancer: diagnostic, prognostic, predictive, and treatment implications. Cell Stress Chaperones 2005;10(2):86–103.
112. Abe M, Manola JB, Oh WK, et al. Plasma levels of heat shock protein 70 in patients with prostate cancer: a potential biomarker for prostate cancer. Clin Prostate Cancer 2004; 3(1):49–53.
113. Ray S, Lu Y, Kaufmann SH, et al. Genomic mechanisms of p210BCR-ABL signaling: induction of heat shock protein 70 through the GATA response element confers resistance to paclitaxel-induced apoptosis. J Biol Chem 2004;279(34):35604–15.
114. Pocaly M, Lagarde V, Etienne G, et al. Overexpression of the heat-shock protein 70 is associated to imatinib resistance in chronic myeloid leukemia. Leukemia 2007;21(1): 93–101.
115. Targosz A, Pierzchalski P, Krawiec A, et al. Helicobacter pylori inhibits expression of heat shock protein 70 (HSP70) in human epithelial cell line. Importance of Cag A protein. J Physiol Pharmacol 2006;57(2):265–78.
116. Vargas-Roig LM, Gago FE, Tello O, Aznar JC, Ciocca DR. Heat shock protein expression and drug resistance in breast cancer patients treated with induction chemotherapy. Int J Cancer 1998;79(5):468–75.
117. Brondani Da Rocha A, Regner A, Grivicich I, et al. Radioresistance is associated to increased Hsp70 content in human glioblastoma cell lines. Int J Oncol 2004;25(3):777–85.
118. Seoane JM, Varela-Centelles PI, Ramirez JR, Cameselle-Teijeiro J, Romero MA, Aguirre JM. Heat shock proteins (HSP70 and HSP27) as markers of epithelial dysplasia in oral leukoplakia. Am J Dermatopathol 2006;28(5): 417–22.
119. Takashima M, Kuramitsu Y, Yokoyama Y, et al. Proteomic analysis of autoantibodies in patients with hepatocellular carcinoma. Proteomics 2006;6(13):3894–900.
120. Jaattela M. Over-expression of hsp70 confers tumorigenicity to mouse fibrosarcoma cells. Int J Cancer 1995;60(5):689–93.
121. Westerheide SD, Kawahara TL, Orton K, Morimoto RI. Triptolide, an inhibitor of the human heat shock response that enhances stress-induced cell death. J Biol Chem 2006;281(14):9616–22.

122. Phillips PA, Dudeja V, McCarroll JA, et al. Triptolide induces pancreatic cancer cell death via inhibition of heat shock protein 70. Cancer Res 2007;67(19):9407–16.
123. Antonoff MB, Chugh R, Borja-Cacho D, et al. Triptolide therapy for neuroblastoma decreases cell viability in vitro and inhibits tumor growth in vivo. Surgery 2009;146(2): 282–90.
124. Bae JH, Kim JY, Kim MJ, et al. Quercetin enhances susceptibility to NK cell-mediated lysis of tumor cells through induction of NKG2D ligands and suppression of HSP70. J Immunother 2010;33(4):391–401.
125. Li M, Wang J, Jing J, et al. Synergistic promotion of breast cancer cells death by targeting molecular chaperone GRP78 and heat shock protein 70. J Cell Mol Med 2009;13(11–12): 4540–50.
126. Gong Z, Yang J, Yang M, et al. Benzo(a) pyrene inhibits expression of inducible heat shock protein 70 in vascular endothelial cells. Toxicol Lett 2006;166(3):229–36.
127. Banerjee Mustafi S, Chakraborty PK, Raha S. Modulation of Akt and ERK1/2 pathways by resveratrol in chronic myelogenous leukemia (CML) cells results in the downregulation of Hsp70. PLoS One 2010;5(1):e8719.
128. Gurbuxani S, Schmitt E, Cande C, et al. Heat shock protein 70 binding inhibits the nuclear import of apoptosis-inducing factor. Oncogene 2003;22(43):6669–78.
129. Schmitt E, Maingret L, Puig PE, et al. Heat shock protein 70 neutralization exerts potent antitumor effects in animal models of colon cancer and melanoma. Cancer Res 2006;66(8):4191–7.
130. Leu JI, Pimkina J, Frank A, Murphy ME, George DL. A small molecule inhibitor of inducible heat shock protein 70. Mol Cell 2009;36(1):15–27.
131. Steele AJ, Prentice AG, Hoffbrand AV, et al. 2-Phenylacetylenesulfonamide (PAS) induces p53-independent apoptotic killing of B-chronic lymphocytic leukemia (CLL) cells. Blood 2009;114(6):1217–25.
132. Williamson DS, Borgognoni J, Clay A, et al. Novel adenosine-derived inhibitors of 70 kDa heat shock protein, discovered through structure-based design. Journal of medicinal chemistry 2009;52(6):1510–3.
133. Massey AJ, Williamson DS, Browne H, et al. A novel, small molecule inhibitor of Hsc70/Hsp70 potentiates Hsp90 inhibitor induced apoptosis in HCT116 colon carcinoma cells. Cancer Chemother Pharmacol 2009.
134. Jinwal UK, Miyata Y, Koren J, 3 rd, et al. Chemical manipulation of hsp70 ATPase activity regulates tau stability. J Neurosci 2009;29(39):12079–88.
135. Deocaris CC, Widodo N, Shrestha BG, et al. Mortalin sensitizes human cancer cells to MKT-077-induced senescence. Cancer letters 2007;252(2):259–69.
136. Evans CG, Wisen S, Gestwicki JE. Heat shock proteins 70 and 90 inhibit early stages of amyloid beta-(1–42) aggregation in vitro. The Journal of biological chemistry 2006;281(44): 33182–91.
137. Wright CM, Seguin SP, Fewell SW, et al. Inhibition of Simian Virus 40 replication by targeting the molecular chaperone function and ATPase activity of T antigen. Virus research 2009;141(1):71–80.
138. Yi F, Regan L. A novel class of small molecule inhibitors of Hsp90. ACS chemical biology 2008;3(10):645–54.
139. Roodveldt C, Bertoncini CW, Andersson A, et al. Chaperone proteostasis in Parkinson's disease: stabilization of the Hsp70/alpha-synuclein complex by Hip. The EMBO journal 2009;28(23):3758–70.
140. Chang YS, Lee LC, Sun FC, Chao CC, Fu HW, Lai YK. Involvement of calcium in the differential induction of heat shock protein 70 by heat shock protein 90 inhibitors, geldanamycin and radicicol, in human non-small cell lung cancer H460 cells. J Cell Biochem 2006;97(1):156–65.
141. Yun CH, Yoon SY, Nguyen TT, et al. Geldanamycin inhibits TGF-beta signaling through induction of Hsp70. Archives of biochemistry and biophysics 2010;495(1):8–13.
142. Powers MV, Clarke PA, Workman P. Death by chaperone: HSP90, HSP70 or both? Cell cycle (Georgetown, Tex 2009;8(4):518–26.
143. Powers MV, Clarke PA, Workman P. Dual targeting of HSC70 and HSP72 inhibits HSP90 function and induces tumor-specific apoptosis. Cancer Cell 2008;14(3):250–62.
144. Chalmin F, Ladoire S, Mignot G, et al. Membrane-associated Hsp72 from tumor-derived exosomes mediates STAT3-dependent immunosuppressive function of mouse and human myeloid-derived suppressor cells. The Journal of clinical investigation 2010;120(2): 457–71.
145. Asea A, Kabingu E, Stevenson MA, Calderwood SK. HSP70 peptidembearing and peptide-negative preparations act as chaperokines. Cell Stress Chaperones 2000;5(5): 425–31.
146. Bausero MA, Page DT, Osinaga E, Asea A. Surface expression of Hsp25 and Hsp72 differentially regulates tumor growth and metastasis. Tumour Biol 2004;25(5–6):243–51.

147. Bausero MA, Gastpar R, Multhoff G, Asea A. Alternative mechanism by which IFN-gamma enhances tumor recognition: active release of heat shock protein 72. J Immunol 2005; 175(5):2900–12.
148. Shin BK, Wang H, Yim AM, et al. Global profiling of the cell surface proteome of cancer cells uncovers an abundance of proteins with chaperone function. J Biol Chem 2003;278(9): 7607–16.
149. Arispe N, Doh M, Simakova O, Kurganov B, De Maio A. Hsc70 and Hsp70 interact with phosphatidylserine on the surface of PC12 cells resulting in a decrease of viability. Faseb J 2004;18(14):1636–45.
150. Schilling D, Gehrmann M, Steinem C, et al. Binding of heat shock protein 70 to extracellular phosphatidylserine promotes killing of normoxic and hypoxic tumor cells. Faseb J 2009;23(8):2467–77.
151. Asea A. Mechanisms of HSP72 release. J Biosci 2007;32(3):579–84.
152. Mambula SS, Calderwood SK. Heat shock protein 70 is secreted from tumor cells by a nonclassical pathway involving lysosomal endosomes. J Immunol 2006;177(11): 7849–57.
153. Vega VL, Rodriguez-Silva M, Frey T, et al. Hsp70 translocates into the plasma membrane after stress and is released into the extracellular environment in a membrane-associated form that activates macrophages. J Immunol 2008;180(6):4299–307.
154. Srivastava P. Roles of heat-shock proteins in innate and adaptive immunity. Nat Rev Immunol 2002;2(3):185–94.
155. Nicchitta CV. Re-evaluating the role of heat-shock protein-peptide interactions in tumour immunity. Nat Rev Immunol 2003;3(5): 427–32.
156. Udono H, Srivastava PK. Heat shock protein 70-associated peptides elicit specific cancer immunity. J Exp Med 1993;178(4):1391–6.
157. Asea A, Kraeft SK, Kurt-Jones EA, et al. HSP70 stimulates cytokine production through a CD14-dependant pathway, demonstrating its dual role as a chaperone and cytokine. Nat Med 2000;6(4):435–42.
158. Kuppner MC, Gastpar R, Gelwer S, et al. The role of heat shock protein (hsp70) in dendritic cell maturation: hsp70 induces the maturation of immature dendritic cells but reduces DC differentiation from monocyte precursors. Eur J Immunol 2001;31(5):1602–9.
159. Multhoff G, Botzler C, Jennen L, Schmidt J, Ellwart J, Issels R. Heat shock protein 72 on tumor cells: a recognition structure for natural killer cells. J Immunol 1997;158(9): 4341–50.
160. Todryk SM, Melcher AA, Dalgleish AG, Vile RG. Heat shock proteins refine the danger theory. Immunology 2000;99(3):334–7.
161. Castelli C, Rivoltini L, Rini F, et al. Heat shock proteins: biological functions and clinical application as personalized vaccines for human cancer. Cancer Immunol Immunother 2004;53(3):227–33.
162. Arnold-Schild D, Hanau D, Spehner D, et al. Cutting edge: receptor-mediated endocytosis of heat shock proteins by professional antigen-presenting cells. J Immunol 1999;162(7): 3757–60.
163. Basu S, Binder RJ, Ramalingam T, Srivastava PK. CD91 is a common receptor for heat shock proteins gp96, hsp90, hsp70, and calreticulin. Immunity 2001;14(3):303–13.
164. Binder RJ, Han DK, Srivastava PK. CD91: a receptor for heat shock protein gp96. Nat Immunol 2000;1(2):151–5.
165. Binder RJ, Vatner R, Srivastava P. The heat-shock protein receptors: some answers and more questions. Tissue Antigens 2004;64(4):442–51.
166. Sondermann H, Becker T, Mayhew M, Wieland F, Hartl FU. Characterization of a receptor for heat shock protein 70 on macrophages and monocytes. Biol Chem 2000;381(12):1165–74.
167. Binder RJ, Blachere NE, Srivastava PK. Heat shock protein-chaperoned peptides but not free peptides introduced into the cytosol are presented efficiently by major histocompatibility complex I molecules. J Biol Chem 2001;276(20):17163–71.
168. Doody AD, Kovalchin JT, Mihalyo MA, Hagymasi AT, Drake CG, Adler AJ. Glycoprotein 96 can chaperone both MHC class I- and class II-restricted epitopes for in vivo presentation, but selectively primes CD8+ T cell effector function. J Immunol 2004;172(10):6087–92.
169. Schild H, Arnold-Schild D, Lammert E, Rammensee HG. Stress proteins and immunity mediated by cytotoxic T lymphocytes. Curr Opin Immunol 1999;11(1):109–13.
170. Singh-Jasuja H, Toes RE, Spee P, et al. Cross-presentation of glycoprotein 96-associated antigens on major histocompatibility complex class I molecules requires receptor-mediated endocytosis. J Exp Med 2000;191(11): 1965–74.
171. Srivastava PK, Menoret A, Basu S, Binder RJ, McQuade KL. Heat shock proteins come of age: primitive functions acquire new roles

in an adaptive world. Immunity 1998;8(6): 657–65.
172. Wells AD, Malkovsky M. Heat shock proteins, tumor immunogenicity and antigen presentation: an integrated view. Immunol Today 2000;21(3):129–32.
173. Flechtner JB, Cohane KP, Mehta S, et al. High-affinity interactions between peptides and heat shock protein 70 augment CD8+ T lymphocyte immune responses. J Immunol 2006;177(2):1017–27.
174. Habich C, Baumgart K, Kolb H, Burkart V. The receptor for heat shock protein 60 on macrophages is saturable, specific, and distinct from receptors for other heat shock proteins. J Immunol 2002;168(2):569–76.
175. Vabulas RM, Ahmad-Nejad P, da Costa C, et al. Endocytosed HSP60s use toll-like receptor 2 (TLR2) and TLR4 to activate the toll/interleukin-1 receptor signaling pathway in innate immune cells. J Biol Chem 2001; 276(33):31332–9.
176. Bendz H, Ruhland SC, Pandya MJ, et al. Human heat shock protein 70 enhances tumor antigen presentation through complex formation and intracellular antigen delivery without innate immune signaling. J Biol Chem 2007;282(43):31688–702.
177. Haug M, Schepp CP, Kalbacher H, Dannecker GE, Holzer U. 70-kDa heat shock proteins: specific interactions with HLA-DR molecules and their peptide fragments. Eur J Immunol 2007;37(4):1053–63.
178. Matzinger P. The danger model: a renewed sense of self. Science 2002;296(5566): 301–5.
179. Zhang H, Huang W. Fusion proteins of Hsp70 with tumor-associated antigen acting as a potent tumor vaccine and the C-terminal peptide-binding domain of Hsp70 being essential in inducing antigen-independent anti-tumor response in vivo. Cell Stress Chaperones 2006;11(3):216–26.
180. Sanchez-Perez L, Kottke T, Daniels GA, et al. Killing of normal melanocytes, combined with heat shock protein 70 and CD40L expression, cures large established melanomas. J Immunol 2006;177(6):4168–77.
181. Gao B, Tsan MF. Recombinant human heat shock protein 60 does not induce the release of tumor necrosis factor alpha from murine macrophages. J Biol Chem 2003;278(25): 22523–9.
182. Reed RC, Berwin B, Baker JP, Nicchitta CV. GRP94/gp96 elicits ERK activation in murine macrophages. A role for endotoxin contamination in NF-kappa B activation and nitric oxide production. J Biol Chem 2003; 278(34):31853–60.
183. Suto R, Srivastava PK. A mechanism for the specific immunogenicity of heat shock protein-chaperoned peptides. Science 1995; 269(5230):1585–8.
184. Janetzki S, Palla D, Rosenhauer V, Lochs H, Lewis JJ, Srivastava PK. Immunization of cancer patients with autologous cancer-derived heat shock protein gp96 preparations: a pilot study. Int J Cancer 2000;88(2): 232–8.
185. Kottke T, Sanchez-Perez L, Diaz RM, et al. Induction of hsp70-mediated Th17 autoimmunity can be exploited as immunotherapy for metastatic prostate cancer. Cancer Res 2007;67(24):11970–9.
186. Jimbo J, Sato K, Hosoki T, et al. Induction of leukemia-specific antibodies by immunotherapy with leukemia-cell-derived heat shock protein 70. Cancer Sci 2008;99(7): 1427–34.
187. Toomey D, Conroy H, Jarnicki AG, Higgins SC, Sutton C, Mills KH. Therapeutic vaccination with dendritic cells pulsed with tumor-derived Hsp70 and a COX-2 inhibitor induces protective immunity against B16 melanoma. Vaccine 2008;26(27–28):3540–9.
188. Mizukami S, Kajiwara C, Ishikawa H, Katayama I, Yui K, Udono H. Both CD4+ and CD8+ T cell epitopes fused to heat shock cognate protein 70 (hsc70) can function to eradicate tumors. Cancer Sci 2008;99(5): 1008–15.
189. Liu B, Ye D, Song X, et al. A novel therapeutic fusion protein vaccine by two different families of heat shock proteins linked with HPV16 E7 generates potent antitumor immunity and antiangiogenesis. Vaccine 2008;26(10):1387–96.
190. Enomoto Y, Bharti A, Khaleque AA, et al. Enhanced immunogenicity of heat shock protein 70 peptide complexes from dendritic cell-tumor fusion cells. J Immunol 2006;177(9): 5946–55.
191. Ciocca DR, Frayssinet P, Cuello-Carrion FD. A pilot study with a therapeutic vaccine based on hydroxyapatite ceramic particles and self-antigens in cancer patients. Cell Stress Chaperones 2007;12(1):33–43.
192. Zeng Y, Chen X, Larmonier N, et al. Natural killer cells play a key role in the antitumor immunity generated by chaperone-rich cell lysate vaccination. Int J Cancer 2006;119(11): 2624–31.
193. Gross C, Hansch D, Gastpar R, Multhoff G. Interaction of heat shock protein 70 peptide

with NK cells involves the NK receptor CD94. Biol Chem 2003;384(2):267–79.

194. Gross C, Schmidt-Wolf IG, Nagaraj S, et al. Heat shock protein 70-reactivity is associated with increased cell surface density of CD94/CD56 on primary natural killer cells. Cell Stress Chaperones 2003;8(4):348–60.

195. Botzler C, Li G, Issels RD, Multhoff G. Definition of extracellular localized epitopes of Hsp70 involved in an NK immune response. Cell Stress Chaperones 1998;3(1):6–11.

196. Multhoff G, Mizzen L, Winchester CC, et al. Heat shock protein 70 (Hsp70) stimulates proliferation and cytolytic activity of natural killer cells. Exp Hematol 1999;27(11):1627–36.

197. Gastpar R, Gehrmann M, Bausero MA, et al. Heat shock protein 70 surface-positive tumor exosomes stimulate migratory and cytolytic activity of natural killer cells. Cancer Res 2005;65(12):5238–47.

198. Stangl S, Wortmann A, Guertler U, Multhoff G. Control of metastasized pancreatic carcinomas in SCID/beige mice with human IL-2/TKD-activated NK cells. J Immunol 2006;176(10):6270–6.

199. Krause SW, Gastpar R, Andreesen R, et al. Treatment of colon and lung cancer patients with ex vivo heat shock protein 70-peptide-activated, autologous natural killer cells: a clinical phase i trial. Clin Cancer Res 2004;10(11):3699–707.

Chapter 17

Determination of Cell Survival or Death

Alexander E. Kabakov, Vladimir A. Kudryavtsev, and Vladimir L. Gabai

Abstract

Cell death (in particular, apoptosis and necrosis) is accompanied by appearance of certain hallmarks that are manifested as specific alterations in cellular membranes, cytoplasm, nucleus and mitochondria. Some of those hallmarks are easily detectable in situ and, therefore, they can be applied for the assessment of dying or dead cells. In turn, there are also signs of viable cells that include a set of features, such as normal functioning of their membranes and organelles, ability to proliferate, etc. The present chapter provides descriptions of several convenient methods for quantitative determination of dead (apoptotic and necrotic) cells and also methods for determination of survived and viable cells. Here, we describe in details the methods of annexin V/propidium iodide (PI) staining, TUNEL assay, Hoechst/PI staining, MTS tetrazolium assay, and colony formation assay, with the principles, advantages, and drawbacks of each technique.

Key words: Apoptosis, Necrosis, Annexin V, TUNEL, Colony formation assay, Tetrazolium assay

1. Introduction

It is generally accepted that heat shock proteins (HSPs) are "cell-death determinants" that influence molecular pathways resulting in cell survival or death (see for review refs. 1–3). Consequently, up- or downregulation in the HSP expression or activities may affect the outcome of pathophysiological insults or some cytotoxic treatments by either promoting or preventing cell death.

Researchers working in the field of cellular stress and HSPs often need to quantify the cell death/survival after stressful conditions. Furthermore, many pharmacologists and oncologists want to test the efficacy of tumor cell killing by chemotherapy or radiotherapy, and human tumors with the enhanced HSP expression are known to be more resistant to anticancer drugs and radiation exposure (4). The present chapter describes several popular assays allowing to quantitatively assess the cell death/survival in numerous biosamples.

Stuart K. Calderwood and Thomas L. Prince (eds.), *Molecular Chaperones: Methods and Protocols*, Methods in Molecular Biology, vol. 787, DOI 10.1007/978-1-61779-295-3_17, © Springer Science+Business Media, LLC 2011

This material may be especially useful as a set of practical recommendations for performing cell death/survival tests in terms of the HSP-related basic studies or medical applications.

Besides cell death/viability assays described in this chapter, there are several other assays worthy to mention. For instance, there are methods for determination/quantification of apoptosis based on activity of caspases or cleavage of their substrates (e.g., cleavage of poly-ADP-ribose polymerase). As to necrotic cell death determination, there are assays based on uptake/exclusion of some dyes (like trypan blue or eosin), or on release of cytosolic enzymes (like lactate dehydrogenase or creatine kinase) by dead cells. Many of these assays are now available as kits from various companies.

2. Materials

2.1. Method 1

1. Dulbecco's phosphate-buffered saline (DPBS, pH 7.4).
2. FITC-annexin V.
3. Binding buffer 10×, (0.1 M HEPES/NaOH, pH 7.4; 1.4 M NaCl; 25 mM $CaCl_2$).
4. Propidium iodide (PI) 10× (PI) (50 μg/ml).
5. Tubes for a fluorescence-activated cell sorter (FACS).
6. Flow cytometer.

2.2. Method 2

1. 4% paraformaldehyde.
2. 70% ice-cold ethanol.
3. TdT reaction buffer (5×); 1 M potassium or sodium cacodylate, 125 mM Tris–HCl, pH 6.6.
4. 1.25 mg/ml BSA (store at −20°C).
5. F-dUTP final concentration – 2 mM.
6. Terminal deoxynucleotidyl Transferase (TdT).
7. PI 10× (50 μg/ml).
8. 24-well plastic tissue culture plates.
9. Fluorescence microscope.

2.3. Method 3

1. Hoechst 33342 100× stock solution in PBS (100 μg/ml).
2. PI 100× stock solution in PBS (500 μg/ml).
3. Fluorescence microscope.

2.4. Method 4

1. MTS/PMS ready to use solution can be purchased as CellTiter 96 Aqueous One Solution Cell Proliferation Assay (Promega Corporation). This solution can be stored long-term for at −20°C, protected from light, and thawed immediately before use.

For frequent use, solution may be stored at 4°C, protected from light, for up to 6 weeks.

2. 96-well tissue culture plate.
3. Tissue culture incubator with a 5% CO_2 atmosphere.
4. 96-well plate reader capable of monitoring absorbance at 490 nm (Bio Rad).

2.5. Method 5

1. DPBS (pH 7.4), trypsin–EDTA (0.05%), crystal violet, ethanol (190 Proof).
2. Prepare 0.5% crystal violet solution in 70% ethanol (v/v). Use gloves during the preparation to avoid staining or hands. This solution is stable for at least a month at room temperature.

3. Methods

3.1. Determination of Fractions of Apoptotic and Necrotic Cells by Staining with Labeled Annexin V and PI

Annexin V conjugated to a fluorescent label (e.g., FITC or PE) or biotin is used to determine the fraction of cells within a population that are undergoing apoptosis. This method is based on the property of cells to lose membrane asymmetry at the early stage of apoptosis. In apoptotizing cells, the membrane phospholipid phosphatidylserine (PS) is translocated from the inner leaflet of the plasma membrane to the outer leaflet, thus exposing PS at the cell surface (5, 6). Annexin V is a 35–36 kDa Ca^{2+}-dependent phospholipids-binding protein that has a high affinity to PS and, consequently, this feature can be exploited for identifying apoptotic cells with exposed PS (6, 7).

Importantly, the loss of membrane integrity which accompanies necrotic cell death leads to total accessibility of membrane PS for binding to annexin V; therefore, staining with fluorochrome-labeled annexin V is usually used in conjunction with a vital dye. PI is a standard (red fluorescence) probe enabling to distinguish viable cells from the dead ones. Viable cells with the intact plasma membrane exclude PI, whereas the membranes of dead (necrotic) cells are permeable to PI and such "cell corpses" become stained by red. The PI intercalates into the major groove of double-stranded DNA (and double-stranded RNA) and produces a highly fluorescent adduct that can be excited at 488 nm with a broad emission peaked around 600 nm. The excitation of PI at 488 nm facilitates its use on the benchtop cytometers [PI can also be excited in the UV field (351–364 nm line from the argon laser) which should be considered when performing multicolor analysis on the multibeam cell sorters]. The red fluorescence of PI is well combined with the green fluorescence of FITC-annexin V in routine double-label tests by flow cytometric analysis for cell death determination. If some of the cells are stained with FITC-annexin V, while being negative for PI, they are undergoing apoptosis. In turn,

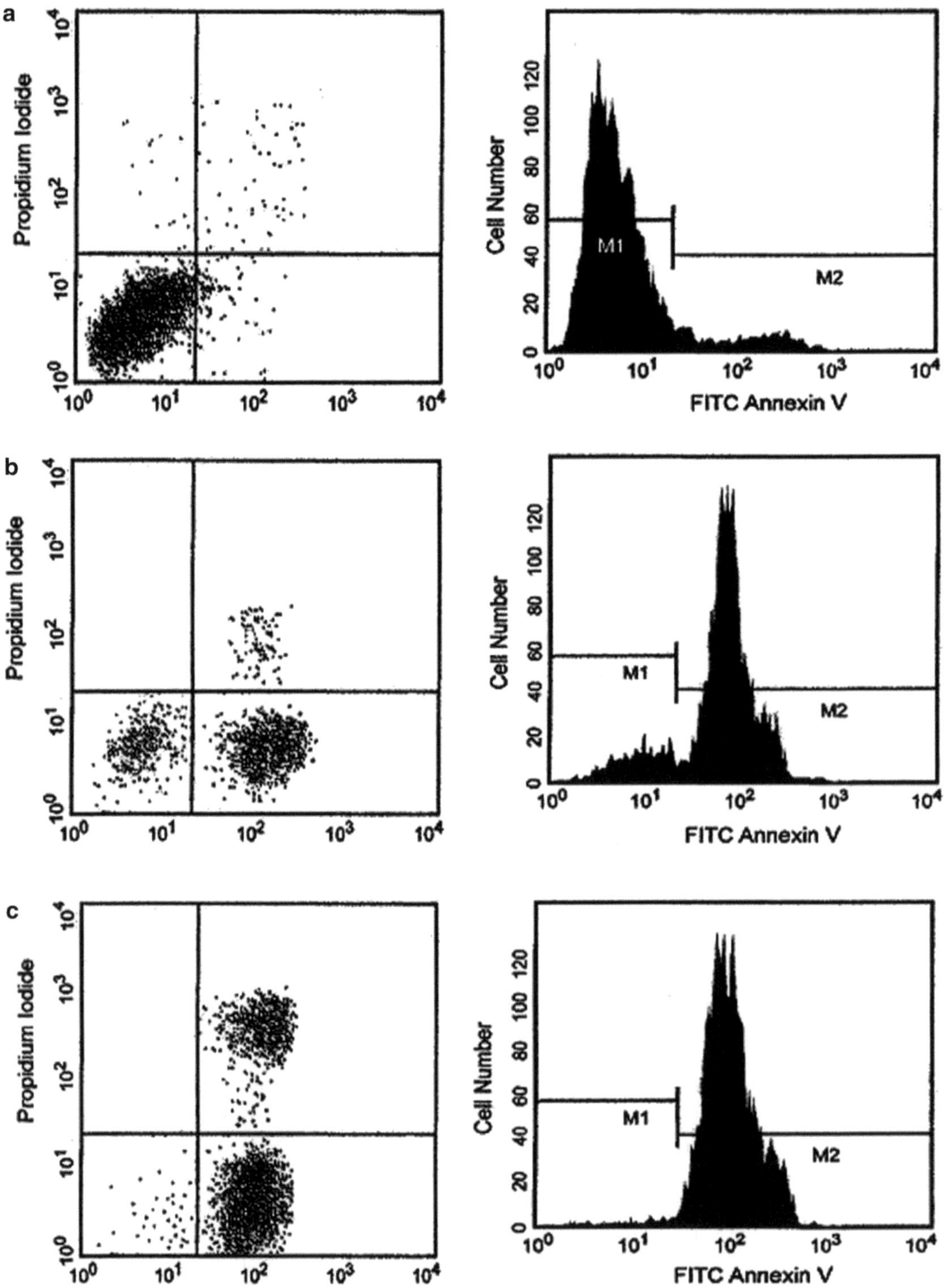
a
b
c
Propidium Iodide
Cell Number
FITC Annexin V
M1
M2
10^0
10^1
10^2
10^3
10^4
0
20
40
60
80
100
120

cells positive for both FITC-annexin V and PI are undergoing necrosis (virtually, they are already dead). Finally, cells negative for both PI and FITC-annexin V are alive (i.e., no measurable apoptosis or necrosis at the moment). An example is shown in Fig. 1.

Protocol:

1. Harvest cells, wash twice in PBS (4°C) and resuspend at a concentration of 1×10^6 cells/ml in the 1× binding buffer.
2. Aliquot cells (100 μl) into FACS tubes and add 5 μl FITC-Annexin V and/or 10 μl PI.
3. Mix gently and incubate for 15 min at room temperature in the dark.
4. Add 400 μl of the binding buffer to each tube and analyze immediately by flow cytometry on FACS.

Recommended samples for measurements and controls to set up compensation and quadrants:

1. Cells + FITC-annexin V
2. Cells + PI
3. Cells + annexin V + PI (see Note 1)

3.2. Visualization of Apoptotic Cells by TUNEL Assay

Terminal Transferase dUTP Nick End Labeling (TUNEL) assay is a method for in situ detection of DNA degradation in apoptotic cells because one of the hallmarks of late phases of apoptosis is the fragmentation of nuclear chromatin which leads to an appearance of a multitude of 3′-hydroxyl ends within broken DNA strands (10). This can be used to identify apoptotic cells by labeling the nuclear DNA breaks with biotin- or fluorescein-tagged deoxyuridine triphosphate nucleotides (F-dUTP) (11). The enzyme TdT catalyzes a template-independent addition of deoxyribonucleoside triphosphates to the 3′-hydroxyl ends of double- or single-stranded DNA. As the apoptotic pathway generates numerous DNA breaks with exposed 3′-hydroxyl ends, the TdT reaction serves to label

Fig. 1. Flow cytometric analyses for the cell death determination with FITC-annexin V and propidium iodide (PI) – demonstrating how this method was used to compare cancer cell death following the treatment with 10 nM taxol (acytostatic and apoptosis-inducing anticancer agent) or co-treatment with 10 nM taxol and 100 nM 17AAG (an inhibitor of the HSP90 chaperone activity). As 17AAG is able to sensitize malignant cells to chemotherapy (see for review refs. 8, 9), it was expected that this HSP90 inhibitor enhances the cytotoxic effect of taxol on leukemic HL-60 cells. In fact, while 100 nM 17AAG was low toxic for these cells (data not shown), the increased cytotoxicity in the case of co-treatment (10 + 100 nM 17AAG) as compared with the effect of 10 nM taxol alone is clearly seen in Fig. 1**b**, **c**. Leukemic HL-60 cells were either untreated (**a**) or incubated with 10 nM taxol (**b**), or with 10 nM taxol + 100 nM 17AAG (**c**). After a 20-h incubation with the drug(s), the cells were stained with FITC-annexin V/PI and analyzed by flow cytometry. About 99% of the untreated cells (**a**) were FITC-annexin V- and PI-negative, thus indicating that they are viable and not undergoing apoptosis or necrosis. A major part of the cells treated with taxol alone (**b**) was undergoing apoptosis (FITC-annexin V-positive) and the minor cell fraction has already died (PI-positive) while many cells remained still viable. In the case of co-treatment taxol + 17AAG (**c**), the enhancement of cytotoxicity is seen: almost no viable cells remained, and there is a considerably increased fraction of the already dead (PI-positive) cells (*left panels* show a distribution of the stained cell subpopulations while the *right panels* show the intensity of FITC fluorescence per cell).

these nuclear DNA break sites with F-dUTP. The labeled (apoptotic) cells are visualized on a fluorescent microscope as cells with heterogeneously stained (green) nuclei.

The TUNEL technique reveals only apoptotic cells while total cell numbers in each viewed preparation are also required to calculate a percentage of apoptosis. PI staining can be used for this purpose: after routine fixing and permeabilizing steps followed by the PI treatment, all cells are rendered red (PI-positive) and well visible in any microscope vision field taken for analysis.

Instead of (or in addition to) F-dUTP, the deoxythymidine analog 5-bromo-2′-deoxyuridine 5-triphosphate (BrdUTP) can also be used as a substrate in the TdT reaction to label the DNA break sites. Once incorporated into the nuclear DNA ends BrdU can be detected with an anti-BrdU antibody and standard immunofluorescence techniques. As compared with the use of F-dUTP alone, an approach with BrdUTP/anti-BrdU yields an enhancement of fluorescence shining, thanks to the additional (antibody conjugate-conferred) labeling.

Nonapoptotic cells do not incorporate much of the F-dUTP or BrdUTP labels owing to the absence of large amounts of the exposed 3′-hydroxyl ends in their nuclear DNA. If determination of necrotic cells is also required in the same cell preparation, prestaining with PI is possible before the fixing step. The TUNEL assay is performed on fixed biomaterial so that this method is equally applicable for both tissue sections and cell culture samples. However, the TUNEL procedure for the staining of tissue sections requires more controls and additional treatments (see ref. 11).

Protocol:

1. Grow cultures of adherent cells in 24-well plastic plates (or onto coverslips placed into such plates) and treat them by an apoptosis-inducing stimulus.
2. Wash adherent cells twice with PBS.
3. Fix cells by the addition of 5 ml of 4% paraformaldehyde and incubate on ice for 15 min.
4. Again wash cells twice in PBS.
5. Permeabilize cells by the addition of ice-cold ethanol (70%) for at least 4 h on ice (in the fixed state, cells may be stored for several weeks in 70% ethanol at −20°C).
6. Wash fixed cell preparations in PBS and add 100 μl of TdT reaction solution: 20 μl TdT reaction buffer, 2 mM F-dUTP, 20 U TdT, 19 μl $CoCl_2$ (10 mM), 67 μl H_2O.
7. Incubate cells for 40 min at 37°C with occasional shaking (or at room temperature overnight).
8. Wash with PBS.

9. Add 1 ml of PI-staining solution, incubate for at least 30 min at room temperature in the dark.
10. Analyze on a microscope for FITC/PI fluorescence with appropriate controls (see Note 2).

3.3. Detection of Apoptotic and Necrotic Cells by Staining with Hoechst 33342 and PI

Apoptotizing cells exhibit characteristic morphology of their nuclei that are due to the apoptotic pathway-induced alterations in chromatin and nuclear matrix (10). This attributive morphology of the apoptotic nuclei seems hard to be detected by light microscopy, but it is clearly seen on a fluorescent microscope after staining of cell samples with a chromatin-targeting fluorescent probe (e.g., Hoechst, acridine orange, etc.). In particular, Hoechst 33342 is a well-known fluorescent probe that brightly stains chromatin in cell nuclei; therefore, this blue fluorescence dye enables to reveal the apoptosis-related alterations in the chromatin state. Among the morphological signs of apoptosis, there are such typical ones as condensation (clumping) and/or fragmentation of chromatin within the nuclear compartment and shrinking or fragmentation of the whole nucleus. All these patterns are easily recognized on a fluorescent microscope as abnormalities in the nucleus size or shape, heterogeneous staining of the nucleus with an appearance of extremely bright zones and/or unusual dark zones, formation of a brightly stained rim adjacent to the nuclear envelope, splitting the nucleus body, etc. In contrast, the most of nonapoptotic cells have normal morphology of their nuclei which are homogeneously stained with Hoechst 33342 (see for example Fig. 2).

It seems very convenient to use double-label staining with Hoechst 33342 and PI because blue fluorescence of the former and red fluorescence of the latter are nicely combined and enable to detect both the fraction of apoptotic cells and the fraction of necrotic (PI-positive) cells within the same cell preparation. Overall, this is a relatively simple method, especially as compared with the annexin V staining or the TUNEL assay (see above). An example is shown in Fig. 2.

Protocol:

1. Add Hoechst 33342 (final concentration – 1 μg/ml) and PI (final concentration – 5 μg/ml) to cells and incubate for 30 min in the dark at 37°C.
2. View and photograph (as digitized color images) the stained cell preparations on a fluorescent microscope displaying Hoechst 33342/PI fluorescence.
3. Upon analyzing each stained preparation, count all cells with apoptotic nuclei and all PI-positive (necrotic) cells within 10–20 occasionally chosen microscope fields of vision.

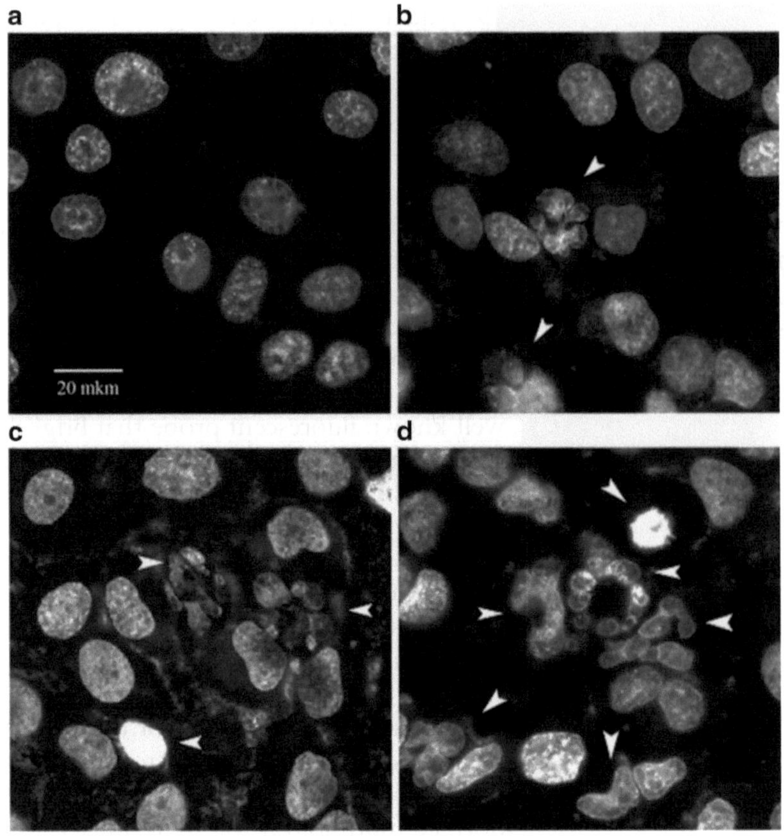

Fig. 2. Visualization of apoptotic nuclei in culture cell preparations by staining with Hoechst 33342. The figure illustrates how the staining with Hoechst 33342 was employed on an in vitro model of tumor cell thermosensitization. HeLa cells derive from a cervical tumor and are known to be thermoresistant (12). To overcome their thermoresistance, we used an approach with joint inhibition of both the chaperone activity of constitutively expressed HSP90 and the de novo induction of HSPs that occurs in the treated cells. The HSP90 chaperone activity and the de novo HSP induction were inhibited by simultaneous treating the cells with 200 nM 17AAG (8, 9) and 40 nM triptolide (13), respectively. An evidence for either inhibitory action came from results of immunoblotting (not shown). The treated cells and control samples were stained with Hoechst 33342 (1 μg/ml) and photographed on a fluorescent microscope. As compared with untreated control (**a**) or effects of hyperthermia (43°C for 60 min) alone (**b**), or effect of the combined hyperthermia + triptolide (**c**), the cell sample exposed to all the three treatments (hyperthermia + 17AAG + triptolide) displays the greater number of apoptotic nuclei (**d**) thus confirming the fact of thermosensitization. Adherent HeLa cells were untreated (**a**) or subjected to hyperthermia (43°C for 60 min) (**b**), or double treatment: 200 nM 17AAG + hyperthermia (**c**), or triple treatment: 200 nM 17AAG + 40 nM triptolide + hyperthermia (**d**). 24 h after the treatments, the nonfixed cells were incubated with Hoechst 33342 (1 μg/ml) for 30 min and then analyzed on a fluorescent microscope. *Arrows* denote typical apoptotic nuclei with fragmented chromatin or abnormal shape, or abnormal brightness. One can see the thermosensitizing effect of the triple treatment (**d**) when a conjunction of hyperthermia with simultaneous inhibition of the HSP90 chaperone activity (by 17AAG) and HSP induction (by triptolide) enhances apoptosis (see Note 3).

3.4. Assay of Cell Viability by MTS Tetrazolium Compound

Background. There are several widely used assays based on the ability of live, but not dead cells to reduce tetrazolium compounds which leads to the formation of colored formazan products that can be measured by spectrophotometer. Here, we describe assay based on MTS [3-(4,5-dimethylthiazol-2-yl)-5-(3-carboxymethoxyphenyl)-2-(4-sulfophenyl)-2 H-tetrazolium, inner salt], which in the presence of electron coupling reagent phenazine methosulfate (PMS) is bioreduced by cells into formazan product. Main advantage of this assay comparing to other similar assays (e.g., widely used MTT assay) is that MTS/PMS colored formazan product is soluble in a standard tissue culture medium and can be measured directly from 96-well assay plate without additional processing (i.e., solubilization of insoluble formazan product as in MTT assay). This advantage is especially important when it is necessary to perform fast screen for toxicity of large number of compounds. Tetrazolium compounds, including MTS, are believed to be reduced mainly by mitochondria electron transport chain; therefore, these assays actually measure metabolic activity of cells, not death per se. Because of this, tetrazolium-based assay can be used in long-term experiments to assess cell proliferation along with viability (see Note 4 below). However, when used in short-term experiments (less than 24-h incubation with compounds), these assays can be employed for the determination of cell viability since dead cells cannot reduce tetrazolium compounds.

Protocol:

Below is the protocol for the measurement of cell viability of attached cells in 96-well plates, but the assay can be used for suspension cultures and in larger or smaller size of plates.

1. Seed cells in 96-well plate to achieve approximately 50% confluency before the addition of tested compounds. Leave some wells empty (without cells, only with medium) as a blank control.
2. Incubate cells with compounds of interest for certain time (up to 24 h, see Notes below).
3. At the end of incubation, add 1/10 volume of MTS/PMS One Solution Reagent (e.g., 20–200 μL of medium) to all wells, including blank control.
4. Incubate plates at 37°C in humidified, 5% CO_2 atmosphere for 1–4 h to allow color development. The time of incubation strongly depend on cell line used and should be adjusted experimentally.
5. Record the absorbance at 490 nm using 96-well plate reader.
6. Calculate cell viability using absorbance in control (untreated) cells as 100% and absorbance in blank wells without cells as 0%.

3.5. Clonogenic Assay

It now appears that apoptosis is the major form of cell demise mainly in lymphoid cells, whereas in epithelial cells and fibroblasts clinically relevant doses of genotoxic drugs or radiation do not cause significant apoptosis, necrosis, or authophagic cell death. Instead, they may cause either growth arrest leading to DNA repair and cell survival, or, if DNA repair is unsuccessful, it leads to mitotic catastrophe or premature senescence (see refs. 14–16 for review). Numerous in vitro studies clearly demonstrated that mitotic catastrophe and/or senescence rather than apoptosis are the main cause of elimination of epithelial tumor cells of different origin (16). As a combined measure of different modes of cell death the most reliable method is apparently clonogenic assay, since it measures ability of a cell to divide and form a colony. Therefore, this assay is basically independent on the way how cells are killed, e.g., by apoptosis, necrosis, autophagic cell death, mitotic catastrophe, or some unknown at present time mechanism, and senescent cells cannot form colonies as well. Invented more than 5 decades ago, this assay is still a gold standard for the evaluation of cell sensitivity to radiation and various drugs, and there is a good correlation between clonogenic ability of cells in vitro and tumor response in vivo. Of note, however, that only fraction of cells (usually around 10–40%) of untreated tumor population can form colonies, i.e., divide indefinitely, but obviously this dividing cell subpopulation is most important for tumor growth and its sensitivity to drugs and radiation. This fraction of dividing cells capable of forming colonies may be related to stem cells capacity. An example is shown in Fig. 3.

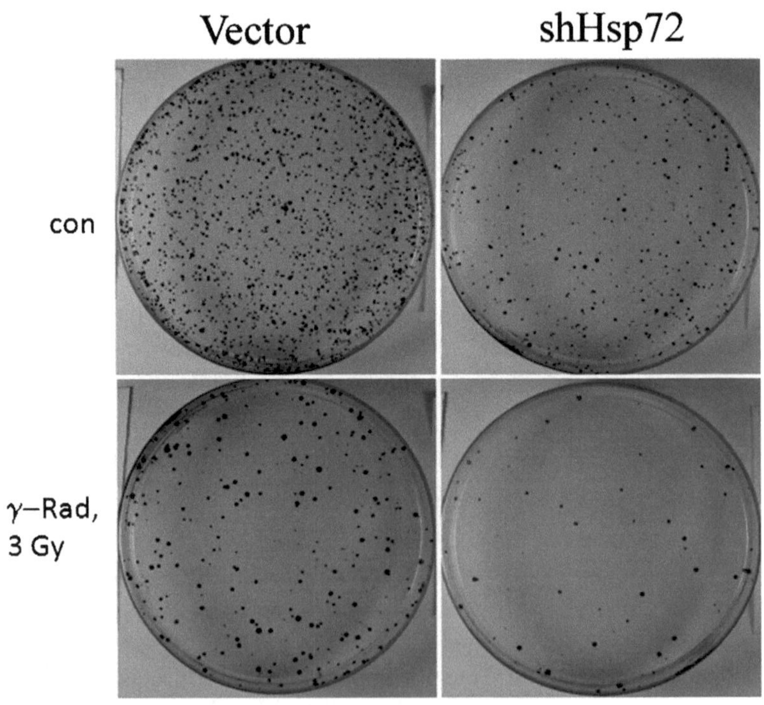

Fig. 3. HCT116 cells were infected with shHsp72 retrovirus to downregulate Hsp72 (see refs. 17, 18 for details), then irradiated with 3 Gy of gamma radiation, and colony formation was assessed 8 days later.

Protocol:
Although the clonogenic assay may be adapted for suspension culture, here we describe its most common application for attached cells.

1. Wash control and treated cells in DPBS, add trypsin–EDTA (0.2–0.25 ml per 35-mm plate), and incubate at 37°C for several minutes to completely detach cells.
2. Add growth medium to plates and resuspend cells using pipeting to have final volume of 1 ml.
3. Count cell number using hemocytometer, or some other cell counting device.

   Plate different dilutions of cells on 60-mm plates. For most cells lines, it should be between 200 and 1,000 cells plated in control (without treatments). If you expect that your treatment significantly reduce number of surviving cells, you should plate larger number of treated cells (starting from 10-fold up to 1,000-fold). The main idea is that number of forming colonies should not be too high (it will be difficult to count), or too low (it will increase error). Optimal number of colonies per 60 mm plate should be 20–100 depending on cell line.
4. Incubate plates with cells under their optimal conditions of growth for 6–14 days checking size of colonies in control (untreated cells) starting from day 6. When colonies will reach size of more than 50 cells (i.e., cells divide more than six times, 26 = 64).

   Remove medium and stain formed colonies with crystal iolet solution for 5 min.

   Remove staining solution with aspirator and wash several times with DPBS to remove background staining.
5. Count colonies in control and treated cells manually or making pictures and using special software programs. Only colonies containing more than 50 cells should be counted.

   Calculate cell survival by taking into account number of seeded cells and initial clonogenic ability of control cells. For example, if in control 200 cells give 40 colonies (colony-forming units, CFU) and after treatment 400 cells give 10 CFU, cell survival (or survival fraction) after treatment will be 10/400: 40/200 = 0.125, or 12.5% (see Note 5).

4. Notes

1. The described method of FITC-annexin V staining followed by flow cytometric analysis is intended mainly for nonadherent cell types (e.g., leukocytes/lymphocytes, suspension

cell cultures, etc.). This is absolutely inapplicable for tissue sections, and this is not a routine procedure for adherent cell types (e.g., HeLa, NIH 3T3, etc.) because specific membrane damage may occur during EDTA/trypsin treatments, cell detachment, and harvesting. Nonetheless, techniques that adapt the annexin V staining for flow cytometry on adherent cell types have also been reported (19, 20). Another point of concern is that Ca^{2+} ions in an incubation medium promote cell-to-cell sticking and aggregation thus complicating performance of flow cytometric analysis. Herein, periodical vortexing and/or gentle resuspending by pipettes of each cell sample can minimize the problem of cell aggregation.

2. The exact incubation times for fixation, permeabilization, and TdT reaction for the TUNEL assay may vary in dependence on cell type; in some cases, it may be necessary to empirically optimize all these parameters for certain types of cells and applications. Also, remains of apoptotic cells are sometimes phagocytosed by neighboring (viable) cells (21) which, being alive prior to the fixation/staining procedure, afterward seem TUNEL-positive through TdT/dUTP targeting to the phagocytosed apoptotic nucleus fragments – such "falsely apoptotic" cells can, however, be recognized on the presence of their own full-size and TUNEL-negative nuclei.

3. Sometimes, mitotic cells stained with Hoechst 33342 may resemble the apoptotic cells so that some experience is needed to well distinguish one mechanism from another. Also, a big variety of patterns of the apoptotic nuclei complicates the use of standard image-analyzing software programs while eye vision-based analysis of many samples is rather hard physically and has elements of subjectivity. Finally, cells with the high pumping-drug-out activity (e.g., cell cultures overexpressing *MDR1* gene) are too poorly stained with Hoechst 33342 through rapid exclusion of the dye.

4. The MTS assay, although convenient cannot discriminate between apoptotic, necrotic, or other forms of cell death. It works as cytotoxicity assay only for short-term incubations with compounds of interest. If used for longer time periods (more than 24 h), cytostatic, not cytotoxic effects of compounds will lead to decreased formazan formation just because of decreased cell number. For instance, if cells multiply once every 24 h, after 72-h incubation cell number in control will increase three times, so three times lower reading in treated cells would not mean that 2/3 of cell population of treated cells are dead – they may just stop dividing.

   Optimal absorbance in control should be in the range of 0.3–1.0 OD, therefore cell number and incubation time should be adjusted accordingly.

Since the MTS assay measures metabolic activity of cells, some compounds (e.g., mitochondrial inhibitors) may interfere with the assay (decreasing or increasing formazan formation) without affecting actual cell viability.

As with all colorimetric assays, it may not be suitable for assessing toxicity of some colored compounds which absorb in the same wavelength (around 490 nm) as formazan. If the absorption of the compounds at this wavelength is not very high, such interference can be partially avoided by including additional wells containing only compounds + MTS/PMS (without cells) and using them as blank control.

5. Although the clonogenic assay is the most sensitive and reliable method for assessing cell death/survival, it has some limitation. By definition, it cannot be used for nondividing (differentiated) cells (e.g., lymphocytes, hepatocytes, cardiomyocytes, etc.) since these cells cannot form colonies at all. Some proliferating cell lines also may have poor colony-forming ability (less than 5–10%) which also makes these cell lines less convenient for this assay. It is also labor and time consuming since it takes at least 1–2 weeks to obtain results. Clonogenic ability of cell lines may differ significantly, so it may be necessary to adjust concentration of seeded cells to achieve statistically reliable results. Some cells are poorly attached and need fixation (e.g., by 4% paraformaldehyde for 10 min) before staining. Precise counting of cells before plating is critical for the assay. Especially, clamping of cells before counting should be avoided by intense pipetting and vortexing.

References

1. Xanthoudakis, S., and Nicholson, D.W. (2000) Heat-shock proteins as death determinants. *Nature Cell Biol* **2**, E163-E165.
2. Beer, H.M. (2005) Death versus survival: functional interaction between the apoptotic and stress-inducible heat shock protein pathways. *J Clin Invest* **115**, 2633–9.
3. Arya, R., Mallik, M., and Lakhotia, S.C. (2007) Heat shock genes – integrating cell survival and death. *J Biosci* **32**, 595–610.
4. Ciocca, D.R., and Calderwood, S.K. (2005) Heat shock proteins in cancer: diagnostic, prognostic, predictive, and treatment implications. *Cell Stress Chaperones* **10**, 86–103.
5. Raynal, P., and Pollard, H.B. (1994) Annexins: the problem of assessing the biological role for a gene family of multifunctional calcium- and phospholipids-binding proteins. *Biochim Biophys Acta* **1197**, 63–93.
6. Martin, S.J., Reutelingsperger, C.P., McGahon, A.J., Rader, J.A., van Schie, R.C., Laface, D.M., and Green, D.R. (1995) Early redistribution of plasma membrane phosphatidylserine is a general feature of apoptosis regardless of the initiating stimulus: inhibition by overexpression of bcl-2 and Abl. *J Exp Med* **182**, 1545–1556.
7. Vermes, I., Haanen, C., Steffens-Nakken, H., and Reutelingsperger, C. (1995) A novel assay for apoptosis. Flow cytometric detection of phosphatidylserine expression on early apoptotic cells using fluorescein labelled Annexin V. *J Immunol Meth* **184**, 39–51.
8. Hwang, M., Moretti, L., and Lu, B. (2009) Hsp90 inhibitors: multi-target antitumor effects and novel combinatorial therapeutic approaches in cancer therapy. *Curr Med Chem* **16**, 3081–3092.

9. Kabakov, A.E. (2009) Geldanamycin derivatives as promising anticancer drugs: therapy via Hsp90 inhibition. In: *Anticancer Drugs: Design, Delivery and Pharmacology* (Spencer, P., and Holt, W., eds.), Nova Science Publishers, New York, pp. 87–113.
10. Samejima, K., and Earnshaw, W.C. (2005) Trashing the genome: The role of nucleases during apoptosis. *Nat Rev Mol Cell Biol* **6**, 677–688.
11. Gavrieli, Y., Sherman, Y., and Ben-Sasson, S.A. (1992) Identification of programmed cell death in situ via specific labeling of nuclear DNA fragmentation. *J Cell Biol* **119**, 493–501.
12. Rossi, A., Ciafre, S., Balsamo, M., Pierimarchi, P., and Santoro, M.G. (2006) Targeting the heat shock factor 1 by RNA interference: A potent tool to enhance hyperthermochemotherapy efficacy in cervical cancer. *Cancer Res* **66**, 7678–7685.
13. Westerheide, S.D., Kawahara, T.L.A., Orton, K., and Morimoto, R.I. (2007) Triptolide, an inhibitor of the human heat shock response that enhances stress-induced cell death. *J Biol Chem* **281**, 9616–9622.
14. Roninson, I.B., Broude, E.V., and Chang, B.-D. (2001) If not apoptosis, then what? Treatment-induced senescence and mitotic catastrophe in tumor cells. *Drug Resistance Updates* **4**, 303–313.
15. Roninson, I.B. (2003) Tumor Cell Senescence in Cancer Treatment. *Cancer Res* **63**, 2705–2715.
16. Schmitt, C.A. (2007) Cellular senescence and cancer treatment. *Biochim Biophys Acta (BBA) - Reviews on Cancer* **1775**, 5–20.
17. Yaglom, J.A., Gabai, V.L. and Sherman, M.Y. (2007) High Levels of Heat Shock Protein Hsp72 in Cancer Cells Suppress Default Senescence Pathways. *Cancer Res* **67**, 2373–2381.
18. Gabai, V.L., Sherman, M.Y., and Yaglom, JA. (2010). Hsp72 depletion suppresses gamma-H2AX activation by genotoxic stresses via p53/p21 signaling. *Oncogene* **29**, 1952–1962.
19. Casciola-Rosen, L., Rosen, A., Petri, M, and Schlissel, M. (1996) Surface blebs on apoptotic cells are sites of enhanced procoagulant activity: Implications for coagulation events and antigenic spread in systemic lupus erythematosus. *Proc Natl Acad Sci USA* **93**, 1624–9.
20. van Engeland, M., Ramaekers, F.C., Schutte, B., and Reutelingsperger, C.P. (1996) A novel assay to measure loss of plasma membrane asymmetry during apoptosis of adherent cells in culture. *Cytometry* **24**, 131–9.
21. Lauber, K., Blumenthal, S.G., Waibel, M., and Wesselborg, S. (2004) Clearance of apoptotic cells: getting rid of the corpses. *Mol Cell* **14**, 277–287.

Chapter 18

Immunohistochemistry of Human Hsp60 in Health and Disease: From Autoimmunity to Cancer

Francesco Cappello, Everly Conway de Macario, Giovanni Zummo, and Alberto J.L. Macario

Abstract

Hsp60 (also called Cpn60) is a chaperonin with essential functions for cell physiology and survival. Additionally, its involvement in the pathogenesis of a number of diseases (e.g., some autoimmune disorders and cancer) is becoming evident with new research. For example, the distribution and levels of Hsp60 in cells and tissues have been found altered in many pathologic conditions, and the significance of these alterations is being investigated in a number of laboratories. The aim of this ongoing research is to determine the meaning of these Hsp60 alterations with regard to pathogenetic mechanisms, diagnosis, classification of lesions, and assessing of prognosis and response to treatment.

Hsp60 occurs in the mitochondria, i.e., its typical residence according to classic knowledge, and also in other locales, such as the cytosol, the cell membrane, the intercellular space, and biological fluids (e.g., blood and cerebrospinal fluid). Detection and quantitative determinations in all these locations are becoming essential components of laboratory pathology in clinics and research. Consequently, immunohistochemistry targeting Hsp60 is also becoming essential for pathologists and researchers interested in disorders involving this chaperonin.

In this chapter, we briefly summarize some recent discoveries on the participation of Hsp60 in the pathogenesis of human diseases and describe in detail how to perform immunohistochemical reactions for detecting the chaperonin, determining its location, and measuring its levels of expression.

Key words: Hsp60, Chaperonin Hsp60, Hsp60 immunohistochemistry, Hsp60 immunostaining, Hsp60 in tissues, Hsp60 locations, Hsp60 in cancer, Hsp60 and autoimmunity, Molecular mimicry, Hsp60 antibodies

1. Introduction

Hsp60 is a chaperonin conserved in evolution, typically described as a mitochondrial molecule (named Cpn60) related to bacterial GroEL that works together with Hsp10 (evolutionarily related to

Stuart K. Calderwood and Thomas L. Prince (eds.), *Molecular Chaperones: Methods and Protocols*,
Methods in Molecular Biology, vol. 787, DOI 10.1007/978-1-61779-295-3_18, © Springer Science+Business Media, LLC 2011

bacterial GroES) in assisting the correct folding of other mitochondrial proteins (1, 2). However, we now know that in mammalian cells up to 40% of cellular Hsp60 occurs in extramitochondrial sites (e.g., cytosol and cell membrane) in normal and pathological tissues (3, 4).

A number of in vivo studies have revealed that the levels of cytosolic Hsp60 gradually increase or decrease during the carcinogenetic steps that go from normal tissue through dysplasia to carcinoma in various organs (4). These data stimulated the use of Hsp60 as a biomarker for assessing diagnosis and prognosis of preneoplastic and neoplastic lesions (4).

In addition, the presence of Hsp60 on the cell membrane (mHsp60) has been noted in normal (5), stressed (6), and tumor (7, 8) cells and was thought to be associated with membrane transport and signaling (9). An increase in mHsp60 levels (which may be accompanied by Hsp60 release into circulation) is considered a danger signal for the immune system since it can lead to activation of innate and/or adaptive immune responses (10).

Interestingly, mHsp60 occurs on the cell surface of certain tumors (8) associated with p21ras protein (11) and/or alpha-3-beta-1 integrin, the latter involved in the adhesion of metastatic breast cancer cells to lymph nodes and bone tissue (12). It has also been shown that mHsp60 plays a role in the metastatization of pancreatic carcinoma (7) and that it occurs on the surface of oral tumor cells, participating in the mechanism of the tumor cell lysis induced by gamma delta T lymphocytes (13). In vitro experiments have demonstrated that photodynamic therapy induces an increase of mHsp60 which may be taken to indicate that Hsp60 could be used as target antigen in anticancer immunotherapy (14–16).

Other than with carcinogenesis, Hsp60 has been associated with several diseases (17), particularly those with autoimmune components, like atherosclerosis (18), systemic autoimmune vasculitis (19), degenerative joint diseases (20) autoimmune glomerulonephritis (21), juvenile dermatomyositis (22), and both plaque and guttate psoriasis (23). In some of these diseases, molecular mimicry of human and microbial Hsp60 has been postulated as the pathogenic mechanism (24). For instance, Hsp60 from eukaryotes shares around 50% identical amino acids with prokaryotic (e.g., bacterial pathogens) counterparts (24, 25). This high similarity in primary structure implies common antigenic sites, which would react with cross-reactive antibodies (24, 26). Exogenous Hsp60 from a microbe elicits an immune response in humans that although directed primarily against the microbial molecule also recognizes the endogenous chaperonin (27). Thus, Hsp60 provides a link between infection and development of autoimmune diseases (24), as postulated for arthritis (28), multiple sclerosis (29), and diabetes (30).

The findings summarized above and the identification of the chaperonopathies as an important group of newly recognized disorders (17, 31) have stimulated interest in Hsp60 in physicians, including oncologists, immunologists, and pathologists, who study this molecule, to learn about its presence and degree of expression in human tissues and to correlate their findings with disease onset, progression, and response to treatment.

In this chapter, we describe methods to perform immunohistochemistry for Hsp60 in normal and pathological human tissues to detect and localize the chaperonin and to assess its levels.

Although molecular techniques (e.g., PCR) are nowadays gaining in importance not only in research, but also in diagnostics (e.g., in surgical pathology), immunohistochemistry still represents the "gold standard," the reference technique in many laboratories for protein identification in situ. This is because of the potential of immunohistochemistry to identify, localize, and assess the quantity of a protein while at the same time providing information on the cell type (normal, tumoral, epithelial, stromal, etc.), in which the protein (e.g., Hsp60) is expressed, and the characteristics (e.g., normal or pathologic) of the structures in the vicinity. Thus, immunohistochemistry provides a range of information not only on the protein itself, but also on its surroundings. The methods we describe are applicable to a large variety of tissues and proteins, such as Hsp60 and also other Hsp chaperones by just substituting the pertinent-specific antibodies.

Essentially, the methodology we describe is based on the principle that if Hsp60 is present in a histological section fixed on to a glass slide it will react with anti-Hsp60 antibody (i.e., the primary antibody). To make visible this reaction, a biotinylated secondary antibody directed against the primary antibody serves as bridge for a signal-emitting compound, such as the enzyme peroxidase. The latter is linked to streptavidin, which has high affinity for the biotin in the secondary antibody and, therefore, it binds to the complex primary/biotinylated secondary antibody when the complex is bound to the histological section because the ligand for the primary antibody, i.e., Hsp60, is present in the tissue. The following step consists of revealing and assessing the signal. If the sign-emitting molecule is an enzyme, such as peroxidase, its presence will be revealed by adding an enzyme substrate that, upon enzymatic action, yields a compound that produces a colored precipitate (the visible signal) in the tissue, exactly on the spot in which Hsp60 is located. In a simplistic way, it can be said that if there is no signal, the conclusion is that Hsp60 is not present in the tissue tested. On the contrary, if there is signal, the conclusion is that the bigger the signal the more Hsp60 occurs in the tissue. However, controls must be run with every test to rule out false positive and false negative results, as it is explained later.

2. Materials

1. Ten staining jars.
2. One staining humid tray.
3. Hydrophobic marker pen (PAP pen).
4. Xylene (stored at room temperature).
5. Alcohol, both absolute and 96°, and distilled water (all stored at room temperature).
6. Aqueous 3% H_2O_2 (stored at 4°C).
7. Dulbecco's phosphate-buffered saline (PBS) (stored at room temperature).
8. 10% bovine serum albumin (BSA) or ready-to-use protein blocking solution, serum-free (both stored at 4°C).
9. Primary antibody for Hsp60 (stored as indicated by the supplier) to be diluted in the diluent provided with the antibodies or a similar commercially available alternative (both stored at 4°C). Make a fresh dilution, following the directions in the antibody instructions datasheet; this dilution is stable at 4°C for variable periods of times (few hours to few days), depending on the antibody.
10. Ready-to-use biotinylated secondary antibodies diluted in PBS (stored at 4°C).
11. Ready-to-use streptavidin–peroxidase complex diluted in PBS (stored at 4°C).
12. Ready-to-use chromogen (i.e., aminoethilcarbazole or diaminobenzidine). These reagents must be used with caution since they are highly toxic and light sensitive, and must always be manipulated wearing protective gloves. Store at 4°C in the dark.
13. Sodium hypochlorite.
14. Ready-to-use aqueous hematoxylin (stored at 4°C).
15. Ready-to-use aqueous mounting medium (stored at room temperature).

3. Methods

Immunohistochemical procedure to demonstrate the presence of Hsp60 on formalin-fixed paraffin-embedded histological sections on glass slides. All steps are carried out at room temperature. Steps 1–4, 8, 11, 13, 15, 18, 19, and 20 are to be done in appropriate staining jars (see Notes 1–3). Steps 6, 7, 9, 10, 12, 14, and 16 have to be performed in a humid staining tray (put a sheet of wet laboratory paper on the bottom of the tray) protected with a cover to transform the tray into a closed chamber; "slide" in what follows means glass slide with a histological thin section onto it.

1. Remove wax from the slides ("dewaxing" procedure) by immersing them in xylene for 10 min, twice (see Note 4).
2. Clear with absolute alcohol for 5 min, once (see Note 5).
3. Dehydrate slides passing them through a series of alcohols, from 96° through 80°/70°/50° to 30°C, one time, 5 min for each alcohol.
4. Place slides in distilled water for 5 min to remove excess alcohol.
5. Blot away (with absorbent paper) excess water from the slides: first allow water to drain and then gently dry the slide with appropriate absorbent paper. Encircle the histological section on the slide with the PAP pen to make it visible throughout the steps that follow (see Note 6).
6. Place slides horizontally in the humid tray on the wet paper at the bottom or, better still, place the slides on glass rods placed parallel to each other like rails separated by the distance necessary to provide support at each end of the slides (see Note 7).
7. Apply on top of the section a drop of aqueous 3% H_2O_2 for 10 min (to block the tissue's endogenous peroxidase) (see Note 8).
8. Rinse slides with fresh PBS for 5 min. Drain excess PBS and dry the slides as above (step 5) (see Note 9).
9. Apply 10% BSA (or protein-blocking serum-free solution) for 10 min.
10. Do not rinse the slides with PBS but, after blotting excess of blocking solution, apply directly diluted Hsp60 primary antibody on to the section. Simultaneously, apply normal serum to the negative control section in the negative control slide. Likewise, in the positive control slide, apply a primary antibody known to give a positive reaction in the tissue under examination (for example, antibody anti-cytokeratins give a positive reaction in colon cancer specimens). Keep all the reagents on the slides for 1 h in the humid tray now closed with a cover, so it becomes a humid chamber (evaporation must be avoided at all costs!) (see Notes 8, 10, and 11).
11. Rinse slides twice with PBS, each for 5 min. Drain PBS and dry slides (see step 5) (see Note 9).
12. Apply biotinylated secondary antibody to all sections for 15 min (in the humid chamber) (see Note 8).
13. Rinse slides twice with PBS, each for 5 min. Drain PBS and dry the slides (see step 5) (see Note 9).

14. Apply streptavidin–peroxidase complex to all sections for 15 min (in the humid chamber) (see Note 8).
15. Rinse slides twice with PBS, each for 5 min. Drain PBS and dry slides (as above) (see Note 9).
16. Apply chromogen for 9 min (or as indicated in the instructions datasheet provided by the supplier) in the humid chamber (see Notes 8, 12, and 13).
17. Rinse off chromogen from the slides into the staining tray and add into the tray 5–10 drops of sodium hypochlorite to inactivate the chromogen (see Note 13).
18. Rinse slides twice with fresh water, each for 5 min. Drain water and dry slides (see above).
19. Apply hematoxylin for 1–2 min for counterstaining.
20. Rinse slides twice with fresh water, each for 5 min. Drain water and dry slides (see above).
21. Mount coverslips with aqueous mounting medium (see Note 14).
22. Examine the slides microscopically: positive stain (presence of Hsp60) appears brown or red (depending on the chromogen) (see Note 15).
23. A common light microscope equipped with 4×, 10×, 20×, and 40× objectives and 10× eyepieces is suitable to the examination of the stained slides. The magnifications obtained by combining these objectives with 10× eyepieces are 40-, 100-, 200-, and 400-fold, respectively, each providing complementary information in a way that the final result, after examining the section at all these magnifications, is a complete set of immunohistochemical information. The lowest magnification provides a panoramic view of the section and the general characteristics of the reaction (e.g., highly localized or generalized and, if localized, in only one or few spots as opposed to many spots). As the magnification increases, more details of the cells positive and negative for Hsp60 become evident. It also becomes possible to tell whether the positive reaction (indicating the presence of Hsp60) is located, for example, in the epithelium but not in the connective tissue or, when there is a tumor, it is possible to determine where the positivity occurs, namely, in the tumor or in the surrounding normal tissue. At the highest magnification, it is possible to ascertain whether the positivity is in the cell nucleus, in the cytoplasm, or in the membrane (see Note 15).

An example of immunostaining of Hsp60 in the bronchial mucosa is given in Fig. 1.

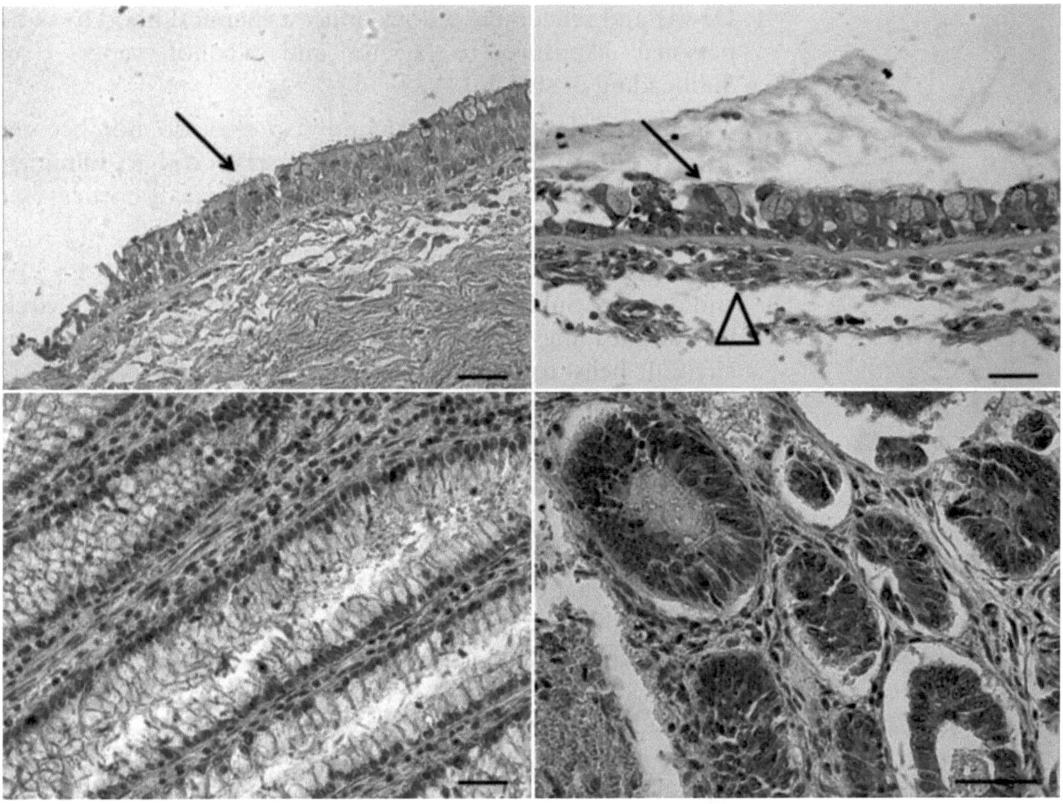

Fig. 1. Panel of immunohistochemical results after immunostaining for Hsp60. *Upper left*: Normal bronchial mucosa shows Hsp60 positivity (*black spots*) localized only to the cytoplasm of epithelial cells (*arrow*). *Upper right*: A chronically inflamed mucosa (the specimen is from a patient with chronic obstructive pulmonary disease) shows a higher (as compared with normal, noninflamed tissue) positivity (*dark gray*) for Hsp60 in cytoplasm of both epithelial (*arrow*) and lamina propria (*arrow head*) cells of bronchial mucosa. *Lower left*: Normal colonic mucosa cells are negative for Hsp60. *Lower right*: Colon adenocarcinoma cells show intense cytoplasmic positivity (*dark gray*) for Hsp60. Scale bar: 50 μm.

4. Notes

1. Perform the same immunostaining in two serial sections to compare them. If immunopositivity differs in the two series, technical problems are likely and the procedure must be repeated with new slides.
2. Before the start, mark with a pencil the side and part of the slide in which the histological section lies to be sure throughout the procedure of the location of the section. Also, enumerate the slides to be tested using Arabian numbers.
3. It is better to prepare the antibody dilution as first step before to start dewaxing the slides. Keep the diluted antibody at 4°C, ready for use.

4. Dewax and rehydrate sections under a chemical hood to avoid personal exposure to xylene and alcohol vapors (see Subheading 3, step 1).
5. Make alcohol dilutions frequently, so they do not become altered due to evaporation and hydration and to minimize these problems, keep the alcohols in sealed bottles (see Subheading 3, step 2).
6. After rinsing the slides with PBS, carefully drain excess PBS and dry (see Subheading 3, step 5) outside the PAP pen circle (see Subheading 3, step 5), but keep wet the section inside the circle. It helps to use a small square of adsorbent paper (2 cm each side) to dry slides.
7. Verify using a laboratory level that the staining tray is perfectly horizontal so that the reagent solutions do not cover the section unevenly: the reagent solutions must cover the section completely and evenly (see Subheading 3, step 6).
8. Make sure that each reagent (i.e., aqueous 3% H_2O_2, primary antibody dilution, biotinylated secondary antibody, streptavidin, and chromogen) added on the slide covers the histological section completely; the section must never be dry (see Subheading 3, steps 7, 10, 12, 14, and 16).
9. PBS solution should be made fresh the same day of use or at most 1 or 2 days before. Check the pH (it must be 7.2–7.4) (see Subheading 3, steps 8, 11, 13, and 15).
10. Many antibodies need antigen retrieval to improve the demonstration of antigens. It consists of a pretreatment with an antigen retrieval solution that breaks the protein cross-links formed by formalin fixation and thereby uncovers hidden antigenic sites. Nevertheless, some antibodies for Hsp60 do not need antigen retrieval before staining; we strongly suggest buying this kind of antibodies, since antigen retrieval may alter integrity of the section and, thus, alter the quality of the results (see Subheading 3, step 10).
11. Calculate a final volume of 20–100 μl of diluted antibody solution for each section, depending on its size (see Subheading 3, step 10).
12. Some chromogens are not sold "ready-to-use," but they consist of two components that must be mixed a few minutes before use. The mixing procedure is quite simple and it can be done during the step (15 min) of streptavidin incubation (see Subheading 3, step 16).
13. Make sure to wear good-quality gloves without holes of any size when handling chromogen and sodium hypochlorite (see Subheading 3, steps 16 and 17).

14. For mounting the coverslip, put it on the bench, put a little drop of aqueous mounting medium over it, and put the slide with the immunostaining over the drop. Wait a few seconds until the drop spreads toward the borders of the slide, thus covering all the section. Blot away excess of liquid with a small square of adsorbent paper (see Subheading 3, step 21).
15. Microscopic examination (see Subheading 3, steps 22 and 23) to analyze the reaction on the histological section must be preceded by an evaluation at various magnifications of the reaction quality. Here, the pathologist's experience plays a deciding role. One must learn how to detect false positives, for example. If the negative control slide is positive, it is very likely that the reaction in the experimental slides produced false positives. Clues in this regard are given by looking at cells that are known beforehand that they do not contain the antigen one is looking for and, if these cells are positive, the conclusions are that the reaction did not go well and produced false positives and that the test must be repeated. On the contrary, if the positive control slide does not show the expected positivity in terms of intensity and distribution, it is highly likely that the reaction in the experimental slides will show a false negativity. In the experimental slides, if cells that should have been positive are not, the conclusion is that the test produced false negatives which is often confirmed by a general lack of reaction in all areas. The test should be repeated. Obviously, the preceding guidelines are not always applicable in their entirety, but at least some of them always are, and these should be the tenets for a complete evaluation of slide and reaction quality and, when this is satisfactory, the examination of the slide and the recording of results can be done with confidence.

References

1. Levy-Rimler G, Bell RE, Ben-Tal N, Azem A (2002) Type I chaperonins: not all are created equal. FEBS Lett 529: 1–5.
2. Hansen JJ, Bross P, Westergaard M, Nielsen MN, Eiberg H, Borglum AD, Mogensen J, Kristiansen K, Bolund L, Gregersen N (2003) Genomic structure of the human mitochondrial chaperonin genes: HSP60 and HSP10 are localised head to head on chromosome 2 separated by a bidirectional promoter. Hum Genet 112: 71–77.
3. Soltys BJ, Gupta RS (1999) Mitochondrial proteins at unexpected cellular locations: export of proteins from mitochondria from an evolutionary perspective. Int Rev Cytol 194: 133–196.
4. Cappello F, Conway de Macario E, Marasà L, Zummo G, Macario AJL. (2008) Hsp60 expression, new locations, functions and perspectives for cancer diagnosis and therapy. Cancer Biol Ther 7: 801–809.
5. Soltys BJ, Gupta RS (1997) Cell surface localization of the 60 kDa heat shock chaperonin protein (hsp60) in mammalian cells. Cell Biol Int 21: 315–320.
6. Lin L, Kim SC, Wang Y, Gupta S, Davis B, et al. (2007) HSP60 in heart failure: abnormal distribution and role in cardiac myocyte apoptosis. Am J Physiol Heart Circ Physiol 293: H2238–H2247.
7. Piselli P, Vendetti S, Vismara D, Cicconi R, Poccia F, et al. (2000) Different expression of CD44, ICAM-1, and HSP60 on primary tumor and metastases of a human pancreatic carcinoma growing in scid mice. Anticancer Res 20: 825–831.

8. Shin BK, Wang H, Yim AM, Le Naour F, Brichory F, et al. (2003) Global profiling of the cell surface proteome of cancer cells uncovers an abundance of proteins with chaperone function. J Biol Chem 278: 7607–7616.
9. Dziewanowska K, Carson AR, Patti JM, Deobold CF, Bayles KW, et al. (2000) Staphylococcal fibronectin binding protein interacts with heat shock protein 60 and integrins: role in internalization by epithelial cells. Infect Immun 68: 6321–6328.
10. Osterloh A, Meier-Stiegen F, Veit A, Fleischer B, von Bonin A, et al. (2004). Lipopolysaccharide-free heat shock protein 60 activates T cells. J Biol Chem 279: 47906–47911.
11. Ikawa S, Weinberg RA (1992) An interaction between p21ras and heat shock protein hsp60, a chaperonin. Proc Natl Acad Sci USA 89: 2012–2016.
12. Barazi HO, Zhou L, Templeton NS, Krutzsch HC, Roberts DD (2002) Identification of heat shock protein 60 as a molecular mediator of alpha 3 beta 1 integrin activation. Cancer Res 62: 1541–1548.
13. Laad AD, Thomas ML, Fakih AR, Chiplunkar SV (1999) Human gamma delta T cells recognize heat shock protein-60 on oral tumor cells. Int J Cancer 80: 709–714.
14. Hanlon JG, Adams K, Rainbow AJ, Gupta RS, Singh G (2001) Induction of Hsp60 by Photofrin-mediated photodynamic therapy. J Photochem Photobiol B 64: 55–61.
15. Jalili A, Makowski M, Switaj T, Nowis D, Wilczynski GM et al. (2004) Effective photoimmunotherapy of murine colon carcinoma induced by the combination of photodynamic therapy and dendritic cells. Clin Cancer Res 10: 4498–4508.
16. Korbelik M, Sun J, Cecic I (2005) Photodynamic therapy-induced cell surface expression and release of heat shock proteins: relevance for tumor response. Cancer Res 65: 1018–1026.
17. Macario AJL, Conway de Macario E (2004) The pathology of anti-stress mechanisms: A new frontier. Stress 7: 243–249.
18. Xu Q, Schett G, Perschinka H, Mayr M, Egger G, et al. (2000) Serum soluble heat shock protein 60 is elevated in subjects with atherosclerosis in a general population. Circulation 102: 14–20.
19. Alard JE, Dueymes M, Youinou P, Jamin C. (2008) HSP60 and anti-HSP60 antibodies in vasculitis: they are two of a kind. Clin Rev Allergy Immunol 35: 66–71.
20. Krenn V, Vollmers HP, von Landenberg P, Schmausser B, Rupp M (1996) Immortalized B-lymphocytes from rheumatoid synovial tissue show specificity for bacterial HSP60. Virchows Arch 427: 511–518.
21. Slot MC, Theunissen R, van Paassen P, Damoiseaux JG, Cohen Tervaert JW (2006) Evaluation of antibodies against human HSP60 in patients with MPO-ANCA associated glomerulonephritis: a cohort study. J Autoimmune Dis 5: 3–4.
22. Elst EF, Klein M, de Jager W, Kamphuis S, Wedderburn LR, et al. (2008) Hsp60 in inflamed muscle tissue is the target of regulatory autoreactive T cells in patients with juvenile dermatomyositis. Arthritis Rheum 58: 547–555.
23. Seung NR, Park EJ, Kim CW, Kim KH, Kim KJ, et al. (2007) Comparison of expression of heat-shock protein 60, Toll-like receptors 2 and 4, and T-cell receptor gammadelta in plaque and guttate psoriasis. J Cutan Pathol 34: 903–911.
24. Cappello F, Conway de Macario E, Di Felice V, Zummo G, Macario AJL (2009) *Chlamydia trachomatis* infection and anti-Hsp60 immunity: the two sides of the coin. PLoS Pathogens, 5:e1000552.
25. Karlin S, Brocchieri L (2000) Heat shock protein 60 sequence comparisons: duplications, lateral transfer, and mitochondrial evolution. Proc Natl Acad Sci USA 97: 11348–11353.
26. Zügel U, Kaufmann SH (1999) Role of heat shock proteins in protection from and pathogenesis of infectious diseases. Clin Microbiol Rev 12: 19–39.
27. Pockley AG (2003) Heat shock proteins as regulators of the immune response. Lancet 362: 469–476.
28. Res PC, Schaar CG, Breedveld FC, van Eden W, van Embden JD, et al. (1988) Synovial fluid T cell reactivity against 65kDa heat shock protein of mycobacteria in early chronic arthritis. Lancet 2: 478–480.
29. Georgopoulos C, Mc Farland H (1993) Heat shock proteins in multiple sclerosis and other autoimmune diseases. Immunol Today 14: 373–375.
30. Child D, Smith C, Williams C (1993) Heat shock protein and the double insult theory for the development of insulin-dependent diabetes. J Royal Soc Med (Eng) 86: 217–219.
31. Macario AJL, Conway de Macario E (2005) Sick chaperones, cellular stress and disease. New Eng J Med 353: 1489–1501.

Chapter 19

Preparation of a Heat-Shock Protein 70-Based Vaccine from DC–Tumor Fusion Cells

Desheng Weng, Stuart K. Calderwood, and Jianlin Gong

Abstract

We have developed an enhanced molecular chaperone-based vaccine through rapid isolation of heat-shock protein 70 peptide complexes (Hsp70.PC) after the fusion of tumor and dendritic cells (DCs) (Hsp70.PC-F). In this approach, the tumor antigens are introduced into the antigen-processing machinery of dendritic cells through the cell fusion process and, thus, we can obtain antigenic tumor peptides or their intermediates that have been processed by dendritic cells. Our results show that Hsp70.PC-F has increased immunogenicity compared to preparations from tumor cells alone and, therefore, constitutes an improved formulation of chaperone protein-based tumor vaccine.

Key words: Heat-shock protein 70, Dendritic cells, Cell fusion, Extraction of Hsp70 peptide complexes, Tumor vaccine

1. Introduction

The heat-shock protein 70 (Hsp70) family is intrinsic to cellular life, permitting proteins to perform essential enzymatic, signaling, and structural functions within the tightly crowded milieu of the cell and working to avert the catastrophe of protein aggregation during stress (1, 2). There are at least 12 members of the human Hsp70 family, including proteins expressed in the cytoplasm, endoplasmic reticulum, and mitochondria (1, 3, 4). For molecular chaperone function, Hsp70 family members are equipped with two major functional domains, including a carboxy-terminal region that binds peptides and denatured proteins and an N-terminal ATPase domain that controls the opening and closing of the peptide-binding domain (3). These two domains play important

Stuart K. Calderwood and Thomas L. Prince (eds.), *Molecular Chaperones: Methods and Protocols*, Methods in Molecular Biology, vol. 787, DOI 10.1007/978-1-61779-295-3_19, © Springer Science+Business Media, LLC 2011

roles in the functions of Hsp70 in tumor immunity, mediating the acquisition of cellular antigens and their delivery to immune effector cells (5, 6). Hsp70 expression becomes dysregulated in many types of cancer leading to elevated Hsp70 levels under nonstress conditions that protect emerging cancer cells from the apoptosis that accompanies many steps in transformation, but also creates an opportunity for vaccine design (4, 7–9).

The molecular chaperone-based tumor vaccine has been pioneered by Pramod Srivastava who has prepared autologous vaccines in mice and in human patients with the direct aim of targeting the unique antigens that characterize each individual neoplasm (10–14). In this approach, Hsp70 peptide complexes (Hsp70.PC) are isolated from the patients' tumors by affinity chromatography using ATP agarose and formulations of Hsp70 applied in a multidose regimen. The aim is for Hsp70.PC to facilitate antigen cross-presentation to the patients' T cells through host APC and for the unique mixture of peptides from the individual tumor to induce antitumor immunity. Despite immunologic and clinical responses obtained in a subset of patients with malignant tumors in the early phase I and/or II trials with molecular chaperone GP96.PC (vitespen) purified from patient-derived tumors (13–16), the randomized phase III trials, however, showed mixed results (17, 18).

We have attempted to produce an enhanced molecular chaperone-based vaccine through rapid isolation of Hsp70.PC from fusions of tumor and dendritic cells (DCs) (Hsp70.PC-F). In our animal studies, Hsp70.PC-F vaccines show superior immunological properties, such as enhanced induction of CTL against tumor cells and stimulation of DC maturation over counterparts from tumor cells (19). More importantly, immunization of mice with Hsp70.PC-F resulted in a T cell-mediated immune response, including significant increase of CD8 T cells and induction of effector and memory T cells able to break T cell unresponsiveness to a nonmutated tumor antigen and provide protection of mice against challenge with tumor cells. By contrast, the immune response to vaccination with Hsp70.PC derived from tumor cells alone is muted against such nonmutated tumor antigen. Hsp70.PC-F complexes differed from those derived from tumor cells in a number of key manners, most notably, enhanced association with immunologic peptides. In addition, the molecular chaperone Hsp90 was found to be associated with Hsp70.PC-F as indicated by co-immunoprecipitation, suggesting the ability to carry an increased repertoire of antigenic peptides by the two chaperones. These experiments indicate that Hsp70.PC derived from DC–tumor fusion cells have increased their immunogenicity and, therefore, constitute an improved formulation of chaperone protein-based tumor vaccine.

The rationale for the extraction of Hsp70.PC from DC–tumor fusion cells is based on the observation that DCs are the most potent antigen-presenting cells (20, 21). The fusion of DC and tumor cells through chemical (22–39), physical (25, 40–49), or biological (50, 51) means creates a heterokaryon which combines DC-derived costimulatory molecules, efficient antigen-processing and -presentation machinery, and an abundance of tumor-derived antigens. The DC and tumor cells become one hybrid cell sharing a unified cytoplasm. The integration of cytoplasm from DC and tumor cells renders the tumor antigens endogenous to the DC heterokaryon and, therefore, facilitates the entry of tumor antigens into the DC endogenous pathway of antigen-processing and -presentation machinery (28, 52, 53). It is likely that the antigen-processing machinery from DC can sort or select the immunogenic peptides to be processed and presented and work much more efficiently than that from tumor cells, thus increasing the quality and quantity of the Hsp-associated complexes.

2. Materials

2.1. Isolation of Tumor Cells from Patient-Derived Solid Sample or Malignant Fluid

1. DNase (0.1 mg/ml, Sigma-Aldrich, Saint Louis, MO).
2. Collagenase (1 mg/ml, Worthington Biochemical Corporation, Lakewood, NJ).
3. Ca^{2+}/Mg^{2+}-free Hanks balanced salt solution (HBSS medium, Mediatech Inc., Manassas, VA).
4. A sterile 50-μm nylon mesher (Sigma-Aldrich, Saint Louis, MO).
5. Heat-inactivated human AB serum (Sigma-Aldrich, Saint Louis, MO).
6. RPMI 1640 medium (Mediatech, Manassas, VA).
7. L-glutamine (2 mM, Mediatech, Manassas, VA).
8. Penicillin and Streptomycin (100 U/ml and 100 μg/ml) (Mediatech, Manassas, VA).

2.2. Generation of DC from Human Peripheral Blood Monocytes

1. Ficoll density gradient centrifugation (Ficoll-Paque™ plus, GE Healthcare Bio-Sciences AB, Sweden).
2. Granulocyte–macrophage colony-stimulating factor (hGM-CSF, 1,000 U/ml) (Genzyme, Framingham, MA).
3. Interleukin-4 (hIL-4, 500 U/ml) (R&D Systems, Minnea polis, MN).

2.3. Preparation of DC–Tumor Fusions

Polyethylene glycol (PEG, 50% MW1450) (Sigma-Aldrich, Saint Louis, MO).

2.4. Preparation of Hsp70.PC Extraction from DC–Tumor Fusions

1. Tris–HCl (pH7.4, 50 mM) (Boston Bioproduct, Ashland, MA).
2. NaCl (50 mM) (Sigma-Aldrich, Saint Louis, MO).
3. Nonidet P-40 (NP40, 1%, Sigma-Aldrich, Saint Louis, MO).
4. Protease inhibitor cocktail tablets, Complete Mini (Roche, Mannheim, Germany).
5. Sodium Orthovanadate ($NaVO_4$, 1 mM) (Boston Bioproduct, Ashland, MA).
6. Antibody against human Hsp70 (5C1A12, Developed by ProMab Biotechnologies, Inc., Albany, CA).
7. Dye Reagent Concentrate for protein assay (Bio-Rad, Hercules, CA).
8. Protein A Sepharose (GE Healthcare, Waukesha, WI).
9. Protein G Sepharose (GE Healthcare, Waukesha, WI).

2.5. Measurement of Levels of Endotoxin

Limulus amebocyte lysate (LAL kit, Cambrex Bio Science Inc., Walkersville, MD).

3. Methods

Fusion between DC and tumor cells can be achieved through chemical, physical, or biological means. In our laboratory, we use PEG to fuse DC and tumor cells. We have used the following protocol to prepare Hsp70.PC extracts from DC–tumor fusion cells.

3.1. Generation of DC from Human Peripheral Blood Monocytes

DC can be generated from human peripheral blood monocytes (PBMC) derived from patients or from healthy donors. We usually use Ficoll to separate PBMC and culture these cells in medium containing hGM-CSF. The protocol is based on a previously described method (54–56) with modifications:

1. PBMC obtained from patients or leukopacks are transferred into 50-ml centrifuge tube and sedimented at low speed.
2. The serum on the top of tube is collected into a clear tube as serum for cell culture. The blood cells at the bottom of the tube are resuspended with RPMI 1640 medium without serum (1:2 dilution).
3. The blood cells are gently laid on top of the tube containing a Ficoll density gradient.
4. Tubes are centrifuged at 300 × *g* for 20 min at room temperature.

5. After Ficoll density gradient centrifugation, the cells in the interface layer are collected into another tube with RPMI 1640 and 2% serum.
6. Cells are washed twice with serum-free medium, and the numbers of cells are counted.
7. Culture 1×10^6 cells/ml in RPMI 1640 containing 5% human serum, 2 mM L-glutamine, 100 U/ml penicillin, and 100 μg/ml streptomycin for 1 h in a humidified CO_2 incubator.
8. After 1 h culture, gently wash and remove the nonadherent cells. The adherent fraction is cultured in RPMI/AIM-V (1:1) medium with 1% of human serum, 1,000 U/ml hGM-CSF, and 500 U/ml hIL-4 for 5 days.
9. On day 3 of culture, cell clusters appear. Fresh medium with 1,000 U/ml of hGM-CSF is added if the color of medium becomes yellow.
10. On day 5 of culture, the loosely adherent cell or cell clusters are collected by gently dislodging the cells by pipetting and then the cells are counted (most cells are immature DCs).

3.2. Preparation of Tumor Cells

Tumor cells can be either freshly isolated from tumor samples or obtained from vials of frozen cell lines. The method described here is used to isolate and culture tumor cells from patient-derived breast or ovarian cancer sample under sterile condition.

1. The resected human tumor sample is weighted, minced to small pieces (1–3 mm), and digested in HBSS solution containing 1 mg/ml collagenase, 0.1 mg/ml DNase, 100 U/ml penicillin, and 100 μg/ml streptomycin.
2. The digested tumor tissue is then mashed through a sterile 50-μm nylon mesher under sterile conditions in a tissue culture hood.
3. Cells are washed twice with cold HBSS solution.
4. Single-tumor cell suspensions are obtained by passing through a cell strainer and the numbers of tumor cells counted.
5. Culture the tumor cells in high-glucose DMEM medium containing 10% human serum and antibiotics. Remove the nonadherent dead cells.
6. Incubate the cells at 37°C for 2–3 days. Cells are ready for fusion when they are in the logarithmic phase of growth.

3.3. Cell Fusion (see Note 1)

1. DCs generated from PBMC are cultured in 1,000 U/ml hGM-CSF medium for 5 days.
2. Tumor cells are maintained in DMEM supplemented with 10% heat-inactivated FCS, 2 mM L-glutamine, 100 U/ml penicillin, and 100 μg/ml streptomycin.

3. The DCs are mixed with tumor cells at a 10:1 ratio and the mixture is washed once with serum-free medium followed by low-speed spin (85 × *g*) to obtain cell pellets.
4. The mixed cell pellets are gently resuspended in prewarmed 50% PEG solution (1 ml per $1\text{–}5 \times 10^8$ cells) for 5 min at room temperature.
5. The PEG solution is diluted by slow addition and mixing of 1, 2, 4, 8, and 16-ml warm serum-free medium within 10 min.
6. The cell pellets obtained after centrifugation at 200 × *g* are resuspended in RPMI 1640 medium supplemented with 10% heat-inactivated FCS, 2 mM L-glutamine, 10 mM nonessential amino acids, 1 mM sodium pyruvate, 10% NCTC 109, 100 U/ml penicillin, 100 μg/ml streptomycin, and 500 U/ml hGM-CSF and further cultured for 5 days.
7. After 5 days, DC–tumor fusion cells are loosely adherent to the culture dish, whereas tumor–tumor fusions and unfused tumor cells are attached firmly to the dish. The loosely adherent fusion cells are obtained first by gentle pipetting.
8. The fusion efficiency is determined by dual expression of tumor antigens, such as MUC1 and DC markers (MHC class II molecules or costimulatory molecules).

3.4. Extraction of Hsp70 Peptide Complexes from DC–Tumor Fusion Cell Products (see Note 2)

1. DC–tumor fusion cells are collected and counted.
2. Resuspend the cell pellets with lysis buffer (50 mM Tris–HCl, pH 8.0, containing 50 mM NaCl, 1% NP40, 1 mM PMSF) (1-ml lysis buffer for 2×10^7 cells) on ice for 30 min.
3. Centrifuge the cell lysate at 16,000 × *g* for 15 min at 4°C.
4. After centrifuge, collect the supernatant into a clear tube.
5. Check protein concentration with standard procedure (Bio-Rad Bradford Protein Assay Kit):
   (a) Prepare dye reagent by diluting 1 part Dye Reagent Concentrate with 4 parts distilled, deionized water.
   (b) Prepare three to five dilutions of a protein standard, which is representative of the protein solution to be tested.
   (c) Pipette 100 μl of each standard and sample solutions into 5 ml of diluted dye reagent.
   (d) Incubate the test samples at room temperature for at least 5 min. Absorbance increases over time. Samples should be incubated at room temperature for no more than 15 min.
   (e) Measure absorbance at 595 nm. Calculate the protein concentration based on the standard curve.
6. The lysates are clarified by centrifugation, and the aqueous phase is collected and incubated with mAb against human

Hsp70 at concentration of 1:100, rotating through overnight at 4°C.

7. For protein analysis, the immunoprecipitates are dissolved in Laemmli SDS sample buffer (0.1 Tris–HCl, 4% SDS, 20% glycerol, 0.05% bromophenol blue, 5% 2-ME) and analyzed by immunoblotting.

   For binding of the immune complex:

   (a) Mix protein A sepharose and protein G sepharose at a 1:1 ratio followed by wash with lysis buffer once.

   (b) Spin down at 16,000×*g* for 1 min at 4°C, discard the supernatant, and resuspend the beads with 250-μl lysis buffer.

   (c) Pipette 100 μl A/G mixture beads into sample tubes and incubate for 2 h at 4°C.

   (d) After incubation, spin down at 16,000×*g* for 1 min at 4°C. Remove supernatant.

   (e) Wash beads with 0.5 ml of lysis buffer for 5 min (rotate at 4°C) followed by centrifugation at 16,000×*g* for 1 min at 4°C.

   (f) Wash beads with 0.5 ml of sterile PBS for 5 min (rotate at 4°C) followed by centrifugation at 16,000×*g* for 1 min at 4°C.

8. After extensive wash with lysis buffer, the immunoprecipitates are eluted with sterile high-salt elution buffer.

   (a) Elute the proteins with (500 mM NaCl, 100 μl) at RT for 2 h.

   (b) Centrifuge at 16,000×*g* for 1 min at 4°C.

   (c) Collect the supernatant and measure the protein concentration by Bradford protein assay.

9. The Hsp70.PC preparations are checked by LAL assay to ensure minimal contamination with endotoxins, aliquoted into 1.5-ml Eppendorf tubes and stored at −80°C.

4. Notes

1. The PEG solution is diluted by gradual addition and progressive mixing of 1, 2, 4, 8, and 16-ml warm serum-free medium. The cell pellets obtained after centrifugation at 200×*g* are resuspended in medium containing 10% heat-inactivated FCS and hGM-CSF. The variable factor for cell fusion is the length of time the cells are exposed to PEG. We have found that there

is some difference in the sensitivity of cells to PEG. It is desirable to perform a dose–response test to evaluate the conditions of PEG fusion for each type of tumor cell and to determine the optimal exposure time. Unlike electrofusion, DC–tumor fusion by PEG is an active and evolving process, and it is thus likely that the larger the initial contact surface between cells, the faster the integration of these cells. Fusion efficiency is lowest immediately after the fusion process is initiated, and a 1-week culture results in more than a tenfold increase in efficiency (Gong unpublished data). In addition, short-term culture gives the fusion cells sufficient time to integrate and display the antigen in the context of MHC molecules.

2. For protein concentration measurement, Bradford dye reagent absorbance increases over time. Samples should incubate at room temperature for at least 5 min, but no more than 15 min. After protein A/G sepharose binding with Hsp70 immunoglobulin, the beads should be gently mixed with lysis buffer to wash off nonspecific interactions. However, the use of vortex should be avoided since it may break the binding of sepharose beads and immunoglobulins. Background caused by actin contamination can be avoided by adding 10 mM ATP to lysis buffer. All steps should be performed at 4°C to reduce proteolysis and denaturation of antigens. This is especially important for the binding step which is typically incubated overnight (or at least for 2 h) at 4°C.

References

1. Lindquist S., Craig E. A. (1988) The heat shock proteins. *Ann. Rev. Genet.* **22**, 631–7.
2. Georgopolis C., Welch W. J. (1993) Role of the major heat shock proteins as molecular chaperones *Ann. Rev. Cell Biol.* **9**, 601–34.
3. Bukau B., Horwich A. L. (1998) The Hsp70 and Hsp60 chaperone machines *Cell* **92**, 351–66.
4. Tang D., Khaleque A. A., Jones E. R., Theriault J. R., Li C., Wong W. H., Stevenson M. A., et al. (2005) Expression of heat shock proteins and HSP messenger ribonucleic acid in human prostate carcinoma *in vitro* and in tumors *in vivo*. *Cell Stress Chaperones* **10**, 46–58.
5. Noessner E., Gastpar R., Milani V., Brandl A., Hutzler P. J., Kuppner M. C., Roos M., et al. (2002) Tumor-derived heat shock protein 70 peptide complexes are cross-presented by human dendritic cells *J Immunol* **169**, 5424–32.
6. Srivastava P. K., Amato R. J. (2001) Heat shock proteins: the "Swiss Army Knife" vaccines against cancers and infectious agents *Vaccine* **19**, 2590–7.
7. Nylandsted J., Brand K., and Jaattela M. (2000) Heat shock protein 70 is required for the survival of cancer cells *Ann N Y Acad Sci* **926**, 122–5.
8. Cornford P. A., Dodson A. R., Parsons K. F., Desmond A. D., Woolfenden A., Fordham M., Neoptolemos J. P., et al. (2000) Heat shock protein expression independently predicts clinical outcome in prostate cancer *Cancer Res* **60**, 7099–105.
9. Clark P. R., Menoret A. (2001) The inducible Hsp70 as a marker of tumor immunogenicity *Cell Stress Chaperones* **6**, 121–5.
10. Srivastava P. (2002) Interaction of heat shock proteins with peptides and antigen presenting cells: chaperoning of the innate and adaptive immune responses *Annu Rev Immunol* **20**, 395–425.
11. Srivastava P. (2003) Hypothesis: controlled necrosis as a tool for immunotherapy of human cancer. *Cancer Immun.* **3**, 4.
12. Srivastava P. K. (2000) Immunotherapy of human cancer: lessons from mice *Nat Immunol* **1**, 363–6.

13. Belli F., Testori A., Rivoltini L., Maio M., Andreola G., Sertoli M. R., Gallino G., et al. (2002) Vaccination of metastatic melanoma patients with autologous tumor-derived heat shock protein gp96-peptide complexes: clinical and immunologic findings *J Clin Oncol* **20**, 4169–80.
14. Mazzaferro V., Coppa J., Carrabba M. G., Rivoltini L., Schiavo M., Regalia E., Mariani L., et al. (2003) Vaccination with autologous tumor-derived heat-shock protein gp96 after liver resection for metastatic colorectal cancer *Clin Cancer Res* **9**, 3235–45.
15. Parmiani G., De Filippo A., Pilla L., Castelli C., and Rivoltini L. (2006) Heat shock proteins gp96 as immunogens in cancer patients *Int J Hyperthermia* **22**, 223–7.
16. Pilla L., Patuzzo R., Rivoltini L., Maio M., Pennacchioli E., Lamaj E., Maurichi A., et al. (2006) A phase II trial of vaccination with autologous, tumor-derived heat-shock protein peptide complexes Gp96, in combination with GM-CSF and interferon-alpha in metastatic melanoma patients *Cancer Immunol Immunother* **55**, 958–68.
17. Testori A., Richards J., Whitman E., Mann G. B., Lutzky J., Camacho L., Parmiani G., et al. (2008) Phase III comparison of vitespen, an autologous tumor-derived heat shock protein gp96 peptide complex vaccine, with physician's choice of treatment for stage IV melanoma: the C-100-21 Study Group *J Clin Oncol* **26**, 955–62.
18. Wood C., Srivastava P., Bukowski R., Lacombe L., Gorelov A. I., Gorelov S., Mulders P., et al. (2008) An adjuvant autologous therapeutic vaccine (HSPPC-96; vitespen) versus observation alone for patients at high risk of recurrence after nephrectomy for renal cell carcinoma: a multicentre, open-label, randomised phase III trial *Lancet* **372**, 145–54.
19. Enomoto Y., Bharti A., Khaleque A. A., Song B., Liu C., Apostolopoulos V., Xing P. X., et al. (2006) Enhanced immunogenicity of heat shock protein 70 peptide complexes from dendritic cell-tumor fusion cells *J Immunol* **177**, 5946–55.
20. Steinman R. M. (1991) The dendritic cell system and its role in immunogenicity *Annu Rev Immunol* **9**, 271–96.
21. Steinman R. M. (2001) Dendritic cells and the control of immunity: enhancing the efficiency of antigen presentation *Mt Sinai J Med* **68**, 106–66.
22. Gong J., Chen D., Kashiwaba M., and Kufe D. (1997) Induction of antitumor activity by immunization with fusions of dendritic and carcinoma cells *Nat Med* **3**, 558–61.
23. Gong J., Koido S., Chen D., Tanaka Y., Huang L., Avigan D., Anderson K., et al. (2002) Immunization against murine multiple myeloma with fusions of dendritic and plasmacytoma cells is potentiated by interleukin 12 *Blood* **99**, 2512–7.
24. Liu Y., Zhang W., Chan T., Saxena A., and Xiang J. (2002) Engineered fusion hybrid vaccine of IL-4 gene-modified myeloma and relative mature dendritic cells enhances antitumor immunity *Leuk Res* **26**, 757–63.
25. Lindner M., Schirrmacher V. (2002) Tumour cell-dendritic cell fusion for cancer immunotherapy: comparison of therapeutic efficiency of polyethylen-glycol versus electro-fusion protocols *Eur J Clin Invest* **32**, 207–17.
26. Homma S., Toda G., Gong J., Kufe D., and Ohno T. (2001) Preventive antitumor activity against hepatocellular carcinoma (HCC) induced by immunization with fusions of dendritic cells and HCC cells in mice *J Gastroenterol* **36**, 764–71.
27. Cao X., Zhang W., Wang J., Zhang M., Huang X., Hamada H., and Chen W. (1999) Therapy of established tumour with a hybrid cellular vaccine generated by using granulocyte-macrophage colony-stimulating factor genetically modified dendritic cells *Immunology* **97**, 616–25.
28. Wang J., Saffold S., Cao X., Krauss J., and Chen W. (1998) Eliciting T cell immunity against poorly immunogenic tumors by immunization with dendritic cell-tumor fusion vaccines *J Immunol* **161**, 5516–24.
29. Hayashi T., Tanaka H., Tanaka J., Wang R., Averbook B. J., Cohen P. A., and Shu S. (2002) Immunogenicity and therapeutic efficacy of dendritic-tumor hybrid cells generated by electrofusion *Clin Immunol* **104**, 14–20.
30. Xia J., Tanaka Y., Koido S., Liu C., Mukherjee P., Gendler S. J., and Gong J. (2003) Prevention of spontaneous breast carcinoma by prophylactic vaccination with dendritic/tumor fusion cells *J Immunol* **170**, 1980–6.
31. Kao J. Y., Gong Y., Chen C. M., Zheng Q. D., and Chen J. J. (2003) Tumor-derived TGF-beta reduces the efficacy of dendritic cell/tumor fusion vaccine *J Immunol* **170**, 3806–11.
32. Takeda A., Homma S., Okamoto T., Kufe D., and Ohno T. (2003) Immature dendritic cell/tumor cell fusions induce potent antitumour immunity *Eur J Clin Invest* **33**, 897–904.
33. Zhang J. K., Li J., Zhang J., Chen H. B., and Chen S. B. (2003) Antitumor immunopreventive and immunotherapeutic effect in mice induced by hybrid vaccine of dendritic cells and hepatocarcinoma in vivo *World J Gastroenterol* **9**, 479–84.

34. Li J., Holmes L. M., Franek K. J., Burgin K. E., Wagner T. E., and Wei Y. (2001) Purified hybrid cells from dendritic cell and tumor cell fusions are superior activators of antitumor immunity *Cancer Immunol Immunother* **50**, 456–62.
35. Xia D., Chan T., and Xiang J. (2005) Dendritic cell/myeloma hybrid vaccine *Methods Mol Med* **113**, 225–33.
36. Homma S., Kikuchi T., Ishiji N., Ochiai K., Takeyama H., Saotome H., Sagawa Y., et al. (2005) Cancer immunotherapy by fusions of dendritic and tumour cells and rh-IL-12 *Eur J Clin Invest* **35**, 279–86.
37. Kao J. Y., Zhang M., Chen C. M., and Chen J. J. (2005) Superior efficacy of dendritic cell-tumor fusion vaccine compared with tumor lysate-pulsed dendritic cell vaccine in colon cancer *Immunol Lett* **101**, 154–9.
38. Ogawa F., Iinuma H., and Okinaga K. (2004) Dendritic cell vaccine therapy by immunization with fusion cells of interleukin-2 gene-transduced, spleen-derived dendritic cells and tumour cells *Scand J Immunol* **59**, 432–9.
39. Akasaki Y., Kikuchi T., Homma S., Abe T., Kofe D., and Ohno T. (2001) Antitumor Effect of Immunizations With Fusions of Dendritic and Glioma Cells in a Mouse Brain Tumor Model *J Immunother* **24**, 106–13.
40. Scott-Taylor T. H., Pettengell R., Clarke I., Stuhler G., La Barthe M. C., Walden P., and Dalgleish A. G. (2000) Human tumour and dendritic cell hybrids generated by electrofusion: potential for cancer vaccines *Biochim Biophys Acta* **1500**, 265–79.
41. Tanaka H., Shimizu K., Hayashi T., and Shu S. (2002) Therapeutic immune response induced by electrofusion of dendritic and tumor cells *Cell Immunol* **220**, 1–12.
42. Siders W. M., Vergilis K. L., Johnson C., Shields J., and Kaplan J. M. (2003) Induction of specific antitumor immunity in the mouse with the electrofusion product of tumor cells and dendritic cells *Mol Ther* 7, 498–505.
43. Jantscheff P., Spagnoli G., Zajac P., and Rochlitz C. F. (2002) Cell fusion: an approach to generating constitutively proliferating human tumor antigen-presenting cells *Cancer Immunol Immunother* **51**, 367–75.
44. Goddard R. V., Prentice A. G., Copplestone J. A., and Kaminski E. R. (2003) In vitro dendritic cell-induced T cell responses to B cell chronic lymphocytic leukaemia enhanced by IL-15 and dendritic cell-B-CLL electrofusion hybrids *Clin Exp Immunol* **131**, 82–9.
45. Marten A., Renoth S., Heinicke T., Albers P., Pauli A., Mey U., Caspari R., et al. (2003) Allogeneic dendritic cells fused with tumor cells: preclinical results and outcome of a clinical phase I/II trial in patients with metastatic renal cell carcinoma *Hum Gene Ther* **14**, 483–94.
46. Trevor K. T., Cover C., Ruiz Y. W., Akporiaye E. T., Hersh E. M., Landais D., Taylor R. R., et al. (2004) Generation of dendritic cell-tumor cell hybrids by electrofusion for clinical vaccine application *Cancer Immunol Immunother* **53**, 705–14.
47. Suzuki T., Fukuhara T., Tanaka M., Nakamura A., Akiyama K., Sakakibara T., Koinuma D., et al. (2005) Vaccination of dendritic cells loaded with interleukin-12-secreting cancer cells augments in vivo antitumor immunity: characteristics of syngeneic and allogeneic antigen-presenting cell cancer hybrid cells *Clin Cancer Res* **11**, 58–66.
48. Trefzer U., Herberth G., Wohlan K., Milling A., Thiemann M., Sharav T., Sparbier K., et al. (2005) Tumour-dendritic hybrid cell vaccination for the treatment of patients with malignant melanoma: immunological effects and clinical results *Vaccine* **23**, 2367–73.
49. Shimizu K., Kuriyama H., Kjaergaard J., Lee W., Tanaka H., and Shu S. (2004) Comparative analysis of antigen loading strategies of dendritic cells for tumor immunotherapy *J Immunother* **27**, 265–72.
50. Phan V., Errington F., Cheong S. C., Kottke T., Gough M., Altmann S., Brandenburger A., et al. (2003) A new genetic method to generate and isolate small, short-lived but highly potent dendritic cell-tumor cell hybrid vaccines *Nat Med* **9**, 1215–9.
51. Hiraoka K., Yamamoto S., Otsuru S., Nakai S., Tamai K., Morishita R., Ogihara T., et al. (2004) Enhanced tumor-specific long-term immunity of hemagglutinating [correction of hemaggluttinating] virus of Japan-mediated dendritic cell-tumor fused cell vaccination by coadministration with CpG oligodeoxynucleotides *J Immunol* **173**, 4297–307.
52. Koido S., Ohana M., Liu C., Nikrui N., Durfee J., Lerner A., and Gong J. (2004) Dendritic cells fused with human cancer cells: morphology, antigen expression, and T cell stimulation *Clin Immunol* **113**, 261–9.
53. Galea-Lauri J., Darling D., Mufti G., Harrison P., and Farzaneh F. (2002) Eliciting cytotoxic T lymphocytes against acute myeloid leukemia-derived antigens: evaluation of dendritic cell-leukemia cell hybrids and other antigen-loading strategies for dendritic cell-based vaccination *Cancer Immunol Immunother* **51**, 299–310.
54. Gong J., Nikrui N., Chen D., Koido S., Wu Z., Tanaka Y., Cannistra S., et al. (2000) Fusions of human ovarian carcinoma cells with autologous or allogeneic dendritic cells induce antitumor immunity *J Immunol* **165**, 1705–11.

55. Gong J., Avigan D., Chen D., Wu Z., Koido S., Kashiwaba M., and Kufe D. (2000) Activation of antitumor cytotoxic T lymphocytes by fusions of human dendritic cells and breast carcinoma cells *Proc Natl Acad Sci USA* **97**, 2715–8.

56. Koido S., Nikrui N., Ohana M., Xia J., Tanaka Y., Liu C., Durfee J. K., et al. (2005) Assessment of fusion cells from patient-derived ovarian carcinoma cells and dendritic cells as a vaccine for clinical use *Gynecol Oncol* **99**, 462–71.

Chapter 20

Isolation of Heat Shock Protein Complexes

Fernando D. Cuello-Carrión, Mariel A. Fanelli, Niubys Cayado-Gutiérrez, Gisela Castro, and Daniel R. Ciocca

Abstract

Heat shock proteins (Hsp) are molecular chaperones with the capability to interact with a wide range of other proteins and are thus often found coupled with other heat shock and non-heat shock proteins. This can be an advantage to study specific interactions between a chaperone and other proteins and to generate an antitumoral immune response. In this chapter, we present two protocols to isolate Hsp. One involves column chromatography with hydroxyapatite and the other employs immunoprecipitation with antibodies coupled to magnetic beads. In both cases, we specifically want to isolate Hsp coupled with other proteins and use the Hsp complexes as intermediaries to present the coupled peptides/proteins to the immune system, or to explore the associations of a particular Hsp with other proteins.

Key words: Heat shock proteins, Hydroxyapatite, Chromatography, Immunoprecipitation

1. Introduction

Heat shock proteins (Hsp), like other proteins, can be isolated by column chromatography (1). However, Hsp are molecular chaperones having the capability to interact with several proteins; thus, the isolation of a "pure" Hsp might be difficult because they frequently appear coupled with other heat shock and non-heat shock proteins. This can be an advantage and one important example was the coprecipitation of p53 antioncoprotein with Hsp70 (2, 3), which shed light on the specific interaction between p53 with this chaperone (we now know that Hsp contribute to oncogen/antioncogen protein stabilization). Another of these protein complexes opened the door to new insights into the roles of Hsp complexes to generate an antitumoral immune response. Hsp70 and gp96 were identified by the group of Srivastava when they

Stuart K. Calderwood and Thomas L. Prince (eds.), *Molecular Chaperones: Methods and Protocols*,
Methods in Molecular Biology, vol. 787, DOI 10.1007/978-1-61779-295-3_20,

were testing affinity-purified proteins that stimulated the immune system to protect mice against tumor injection (4). These proteins elicited the immune response bringing unidentified tumor peptides to be presented to the immune system; in this manner, the tumor-derived Hsp-peptide complexes can be used for vaccination against malignancies (5, 6). The protocols to isolate both, the gp96-peptide complexes and grp170, from fresh tumor samples using affinity chromatography followed by anion-exchange chromatography, has been presented elsewhere (5, 7).

We are also interested in the approach to generate an anticancer vaccine (8); in this chapter, we present two protocols to isolate Hsp, one is by column chromatography with hydroxyapatite (a step that included the isolation of cell membranes) (8, 9) and the other is using immunoprecipitation with antibodies coupled to magnetic beads. In both cases, we specifically want to bring Hsp coupled with other proteins to use the Hsp together with other compounds as intermediaries to present the coupled peptides/proteins to the immune system, or to explore the associations of a particular Hsp with other proteins.

Hydroxyapatite (HA), also called hydroxylapatite, is a naturally occurring mineral form of calcium apatite with the formula $Ca_5(PO_4)_3(OH)$ (10), but is usually represented with the formule $Ca_{10}(PO_4)_6(OH)_2$ to denote that the crystal unit cell comprises two entities. HA is the hydroxyl end member of the complex apatite group. The OH^- ion can be replaced by fluoride, chloride, or carbonate, producing fluorapatite or chlorapatite. It crystallizes in the hexagonal crystal system. It has a specific gravity of 3.08 and is 5 on the Mohs hardness scale. Pure HA powder is white. Naturally occurring apatites can, however, also have brown, yellow, or green colorations, comparable to the discolorations of dental fluorosis.

HA is chemically similar to the mineral component of bones and hard tissues in mammals (10). Up to 50% of bone is made up of a modified form of the inorganic mineral HA (known as bone mineral). Carbonated calcium-deficient hydroxyapatite is the main mineral of which dental enamel and dentin are comprised. HA crystals are also found in the small calcifications (within the pineal gland and other structures) known as corpora arenacea or "brain sand."

HA is one of few materials classified as bioactive, meaning that it will support bone in-growth and osseous integration when used in orthopedic, dental, and maxillofacial applications. HA is commonly used as a filler to replace amputated bone or as a coating to promote bone ingrowths into prosthetic implants. Although many other phases exist with similar or even identical chemical makeup, the body responds much differently to them. Many modern implants, e.g., hip replacements and dental implants, are coated with HA.

2. Materials

2.1. Cells and Tissue Culture

1. HeLa cells (The HeLa cell line was provided by Dr. María Teresa Damiani (School of Medicine, National University of Cuyo, Argentina).
2. RPMI-1640 medium (Sigma Chemical Company, Argentina).
3. Fetal bovine serum (Internegocios S.A., Argentina).
4. Tissue culture incubator with a 5% CO_2-humidified atmosphere.

2.2. Hardware

1. Laminar flow tissue culture cabinet.
2. Pestel and steel mortar (made in Mendoza, Argentina).
3. Ultra-Turrax homogenizer (Janke & Kunkel, Germany).
4. Eppendorf microcentrifuge.
5. 1.5-ml microcentrifuge tubes.
6. Chromatography columns, Polyprep, Cat. 731–1550, Bio Rad.
7. Dynabeads M-280 Tosylactivated (DYNAL, Biotech Inc., Invitrogen, Argentina).
8. Immunoblot imager: ImageQuant LAS 4000 (Fuji, Japan).
9. Minigel and Western transfer apparatus (Bio Rad Corp., CA).

2.3. Buffers and Solutions

1. Homogenization buffer $NaHCO_3$ (30 mM, pH 7.0).
2. Sodium phosphate buffer (30 mM, pH 6.8).
3. Sucrose gradient (400 μL sucrose 40%, 400 μL sucrose 35%, and 400 μL sucrose 30%).
4. 50% ammonium sulfate.
5. 70% ammonium sulfate.
6. Hydroxyapatite, (Urodelia, St. Lys, France) equilibrated in sodium phosphate buffer.
7. 50 mM NaCl.
8. 100 mM NaCl.
9. 200 mM NaCl.
10. 300 mM NaCl.
11. 400 mM NaCl.
12. 600 mM NaCl.
13. Anti-gp96 antibody (NeoMarkers, Fremont, CA).
14. Anti-Hsp27 antibodies: rabbit polyclonal antibody against Hsp25/27 provided by Dr M. Gaestel (Germany) and mouse monoclonal antibody (StressGenn Biotechnologies Corp, Victoria, BC, Canada).

15. Silver stain Plus kit (Bio Rad, Argentina).
16. Ice-cold lysis buffer (2.5 mL) containing protease or phosphatase inhibitors (0.1 ng/mL aprotinin, 5 μg/mL leupeptin, 0.5 mM PMSF).
17. Buffer B (0.1 M borate buffer, pH 9.5).
18. Buffer C (phosphate-buffered saline, pH 7.4 with 0.1% BSA).
19. Buffer D (0.2 M Tris, pH 8.5 with 0.1% BSA).
20. Standard Laemmli sample buffer.
21. 12.5% denaturing SDS-polyacrylamide gels.
22. Nitrocellulose (13).
23. Chemiluminescence reagents (Dupont NEN, Boston, MA).

3. Methods

3.1. Chromatography: Protein Extraction from Human Tumor Tissues

3.1.1. Biopsy Collection

To obtain our antitumor therapeutic vaccine based in gp96 and tumor membranes, we begin with at least 1 cm^3 of tumor. The biopsy sample is immediately transported to the biochemistry laboratory in a sterile plastic container on ice. Once in the laboratory, to confirm the presence of tumor tissue, a small piece of the biopsy is processed for histopathological evaluation of the tumor (routine formalin fixation, paraffin embedding, and H&E staining). The remaining tissue is frozen and kept at −80°C for vaccine preparation.

3.1.2. Tissue Processing

The tumor tissue and all the materials used to prepare the vaccine were handled under sterile conditions under a laminar flow. The frozen tumor tissue was pulverized in a steel mortar (kept at −80°C) and the powder was transferred to a precipitation tube on ice, adding 750 μL of homogenization buffer. The tissue is then homogenized with an Ultraturrax apparatus and the solution transferred to Eppendorf microcentrifuge tubes (1.5 mL) and centrifuged at 8,000 ×*g* for 30 min at 4°C. The supernatant was saved for later use.

The pellet was resuspended in 400 μL of phosphate buffer and used for membrane preparation. Membranes were separated using two Eppendorf microcentrifuge tubes (1.5 mL), each containing a sucrose gradient (400 μL sucrose 40%, 400 μL sucrose 35%, and 400 μL sucrose 30%) (9). Then, 200 μL of the resuspended pellet were carefully layered onto the Eppendorf tubes, and the tubes were centrifuged in a microfuge (6,700 ×*g* for 30 min at 4°C). The material present at the 40 and 35% sucrose interface (which contains mainly plasma membranes, some endoplasmic reticulum membranes, and mitochondria) was recovered with a Pasteur pipette and placed in four sterile vaccine glass containers.

3.2. Precipitation

3.2.1. Ammonium Sulfate Precipitation

The supernatant was subjected to precipitation with cooled ammonium sulfate at 50% (pH 7) during 1 h in ice and then was centrifuged at 8,000×*g* for 30 min at 4°C. The pellet was resuspended with phosphate buffer (30 mM, pH 6.8) and then newly precipitated with ammonium sulfate at 70%. The last pellet was resuspended in 1 mL phosphate buffer.

3.3. Hydroxyapatite Chromatography

3.3.1. Hydroxyapatite: Characteristics and Properties

The HA was obtained by a precipitation method, then sintered before being sieved and/or spray-dried: the resulting powder consisted of more than 98% HA. The amount of CaO and TCP was inferior to 2%, and the powder had a negative surface charge; other powder characteristics have been reported elsewhere (11). The mechanism of HA chromatography is complicated and has been described as "mixed-mode" ion exchange. Protein–hydroxyapatite interactions are complex and not fully understood. Amino groups are attracted to crystal phosphates but repelled by crystal calcium ions. It is reversed for carboxyls. Amine binding to crystal phosphate is electrostatic.

The binding of carboxyl to crystal calcium ions involves the formation of coordination complexes between Ca and clusters of protein carboxyls. It may be difficult to predict the effectiveness of HA chromatography based on physical and chemical properties of the desired protein to be purified. For elution, a buffer with increasing phosphate concentration is typically used. The HA powder containing particles of about 80–160 μm was suspended in a sterile phosphate buffer solution (30 mM, pH 6.8) and autoclaved.

3.3.2. Column Preparation

The column was prepared with HA at 80–160 μm, 2 cm high, equilibrating with 10 volumes of phosphate buffer (30 mM pH 6.8) and then loaded with the resuspended pellet. After that, the column was eluted with 3 mL of the following solutions: 50, 100, 200, 300, 400, and 600 mM of NaCl. Fractions of 1 mL each were collected (3/each NaCl concentrations). To test the purity of fractions containing gp96, slot blots were performed followed by SDS-PAGE with silver staining (12) and Western blots (13) (Fig. 1).

3.4. Vaccine Preparation

The vaccine was prepared with two main approaches, one involving the purification of gp96 with HA chromatography columns. In the first purification step, the fractions collected from the HA column was tested with slot blot to identify gp96. The fractions collected with 100, 200, and 300 mM of NaCl were rich in gp96. To further characterize the fractions that contained more gp96, we performed SDS-PAGE. In the silver-stained SDS-PAGE gels, the 200-mM fractions showed abundant gp96 bands and additional lower MW unidentified bands with relatively little protein content (8). The isolation of gp96 by HA chromatography

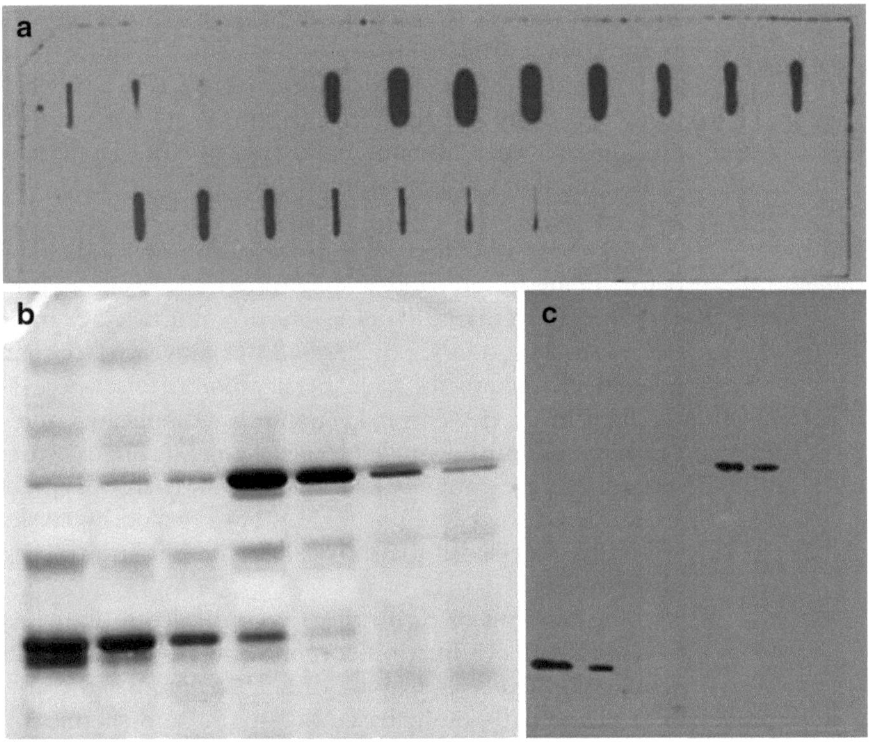

Fig. 1. The different steps of the purification of gp96 by column chromatography with hydroxyapatite are shown here: (**a**) slot blots to detect gp96 using different concentrations of the elution buffer, (**b**) SDS-PAGE with silver staining showing the fractions rich in gp96, and (**c**) Western blot to detect gp96.

columns were confirmed by Western blotting (8) (Fig. 2). We then tried to characterize the proteins/peptides present in the gp96 isolated fractions; β-catenin, P-cadherin, Her-2/neu, and survivin were identified by slot blots in these fractions (8) (data not shown). The other purification step was performed to isolate proteins from the cell membranes of the tumor cells by sucrose gradient centrifugation (9). Among the proteins recovered were Hsp70 as well as Hsp27, Her-2/neu, β-catenin, and P-cadherin (8). The fractions with higher gp96 content and purification were placed in the four glass vaccine containers mixing with the membranes and adding 0.5 mL of HA 45–80 μm in saline solution to each container (the vaccine was then ready to use; the protein concentration ranged from 1,100 to 1,400 μg/mL, and it was kept at −20°C).

Therefore our vaccine was composed of at least three heat shock proteins (gp96 was one of them, possibly with chaperoned proteins/peptides as shown in the slot blots), proteins from the cell membrane system, and HA particles.

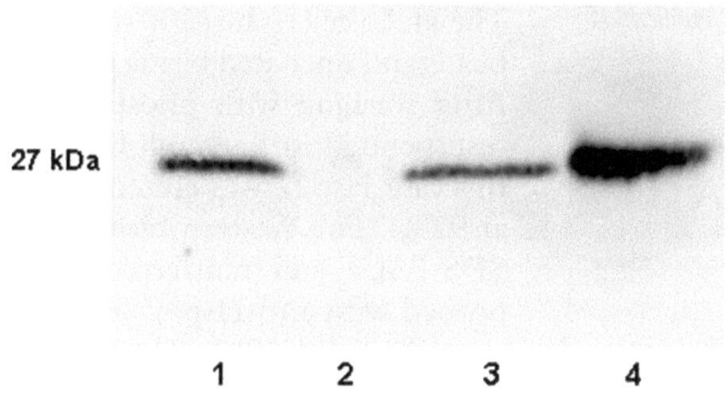

Fig. 2. Hsp27 immunoprecipitation (IP) followed by Western blot using a modified protocol. Lanes 1 and 3, HeLa cell lysate without IP (positive control). Lane 2, uncoated beads (negative control). Lane 4, Hsp27 IP using the beads coated with a mouse monoclonal anti-Hsp27 antibody.

3.5. Immunoprecipitation: Protein Extraction from HeLa Cell Line

3.5.1. Cell Culture

HeLa cell line was cultured in RPMI-1640 medium supplemented with 0.5% fetal bovine serum and antibiotics. Cells were grown in a 5% CO_2-humidified atmosphere at 37°C. The cell line was subcultured every 2–3 days reaching ~80% confluence.

3.5.2. Cell Lysate Preparation

Culture media was removed and HeLa cells were washed with ice-cold PBS and decanted. Then, we added ice-cold lysis buffer (2.5 mL) containing protease or phosphatase inhibitors directly to the flask. The cells were harvested by scraping and transferred to a conical tube. Finally, they were centrifuged for 10 min at 2,000 × g at 4°C. The supernatant was carefully collected and kept at −80°C. Protein determination was performed by Bradford method (14). The total protein concentration was 1 μg/μl.

3.6. Hsp27 Immunoprecipitation

3.6.1. Washing Procedure

Dynabeads M-280 Tosylactivated (DYNAL, Biotech INC., Invitrogen, Argentina) were resuspended using a pipette/vortex for approximately 1 min, avoiding foaming. We obtained 20 μl (0.6 mg) of beads and they were placed in the tube on a magnet until the beads had migrated to the side of the tube and the liquid turned clear. The supernatant was carefully pipetted off and the beads were resuspended in 1 mL of buffer B. This step was repeated one more time. At this point, the beads were ready for coating.

3.6.2. Coating Procedure

The dynabeads were homogenized and incubated with mouse anti-Hsp27 monoclonal antibody (StressGene, Victoria, Canada) during 48 h at 4°C with slow tilt rotation. The optimal amount of antibody was decided in each case; we used 12 μg per 4×10^7 dynabeads. After incubation, the tube was placed on the magnet and the supernatant was removed. The beads were washed on a roller with 1 mL of buffer D at 37°C/1 h, and Buffer C at RT/5–10 s twice. The beads were ready for use.

3.6.3. Hsp27 Purification Procedure

The HeLa cell lysate (500 μl) was incubated with anti-Hsp27-coated beads and uncoated beads for 1 h at 37°C with tilting and rotation. After washing with phosphate-buffer saline 1×, the beads were resuspended in Laemmli buffer and the target protein bound, in this case Hsp27 was eluted and concentrated by boiling for 5 min at 97°C. For Western blotting, proteins were separated in 12.5% SDS-PAGE and transferred to nitrocellulose (13). Blots were then probed with anti-Hsp27 (to check the success of the immunoprecipitation, 1:4,000). The specific protein band was detected by combining chemiluminescence and densitometry.

4. Notes

1. It is important to note that neither method can be used to isolate "pure" heat shock proteins. The two approaches presented here are useful, for example, in the development of vaccines that can be used against malignancies, but not in those cases where the goal is the complete purification of a pure HSP.

   Another methodological drawback is the purity of the hydroxyapatite, which is variable according to the manufacturer or supplier, and this may affect the results.
2. A major challenge in analytical scale protein purification is to selectively capture and isolate low abundance proteins from a complex sample with several thousands of other proteins. Immunoaffinity purification, or immunoprecipitation, is an ideal technique for capture and concentration of a broad range of proteins using specific antibodies. It can be used to confirm the identity of a protein, to quantify expression levels, to study the biochemical characteristics (such as protein/protein interactions), or to evaluate posttranslational modifications. Our method based on the use of immunoprecipitation with antibodies coupled to magnetic beads presents the disadvantage at the time of elution of the protein. This process must be optimized to obtain purified proteins/peptides coupled to Hsp27 that will be then presented to the immune system (next phase of our work).
3. Hsp27 elution from beads is presently not optimal and we are testing different elution buffers: our current standard is 0.2 M Glycine, pH 2.8 and 0.5 M NH_4OH with EDTA (0.5 mM), but the elution efficiency is currently not optimal.

Acknowledgments

The authors thank Urodelia (Saint Lys, France) for providing the HA.

Competing Interest Statement: The author DRC declare that a patent has been presented on the use of hydroxyapatite for isolation proteins to produce a vaccine. No competing interests are declared by the other authors.

Funding: This work was supported by the grants from the National Research Council (CONICET, PIP 2428), the National Agency for Scientific and Technological Promotion of Argentina (PICT 1047, 2007, Préstamo BID), and the Argentina Foundation for Cancer Research.

References

1. Dudich IV, Zav'yalov VP, Pfeil W, Gaestel M, Zav'yalova GA, Denesyuk AI, Korpela T (1995) Dimer structure as a minimum cooperative subunit of small heat-shock proteins. Biochemica et Biophysica Acta 1253:163–168
2. Pinhasi-Kimhi O, Michalovitz D, Ben-Zeev A, Oren M (1986) Specific interaction between the p53 cellular tumour antigen and major heat shock proteins. Nature 320:182–185
3. Stürzbecher H-W, Addison C, Jenkins JR (1988) Characterization of mutant p53-hsp72/73 protein-protein complexes by transient expression in monkey COS cells. Mol Cell Biol 8:3740–3747
4. Tamura Y, Peng P, Liu K, Daou M, Srivastava PK (1997) Immunotherapy of tumors with autologous tumor-derived heat shock protein preparations. Science 278:117–120
5. Janetzki S, Palla D, Rosenhauer V, Lochs H, Lewis JJ, Srivastava PK (2000) Immunization of cancer patients with autologous cancer-derived heat shock protein gp96 preparations: A pilot study. Int J Cancer 88:232–238
6. Eton O, Ross MI, East MJ, Mansfield PF, Papadopoulos N, Ellerhorst JA, Bedikian AY, Lee JE (2010) Autologous tumor-derived heat-shock protein peptide complex-96 (HSPPC-96) in patients with metastatic melanoma. J Transl Med 8:9
7. Wang XY, Kazim L, Repasky EA, Subjeck J (2003) Immunization with tumor-derived ER chaperone grp170 elicits tumor-specific CD8+ T-cell responses and reduces pulmonary metastatic disease. Int J Cancer 105: 226–231
8. Ciocca DR, Frayssinet P, Cuello-Carrion FD (2007) A pilot study with a therapeutic vaccine based on hydroxyapatite ceramic particles and self-antigens in cancer patients. Cell Stress & Chaperones 12:33–43
9. Iyengar R, Rojas F, Rich KA, Birnbaumer L (1991) Membrane receptors: criteria and selected methods of study. In: Hormone Action and Molecular Endocrinology, ed Hughes MR, Schrader WT, O'Malley BW Baylor College of Medicine, Houston, TX, 3–18.
10. Williams RJP (1989) The functional forms of biominerals. Mann S, Webb J, Williams RJP (Eds), VCH Publisher, Weinheim, 1.
11. Laquerriere P, Grandjean-Laquerriere A, Jallot E, Balossier G, Frayssinet P, Guenounou M (2003) Importance of hydroxyapatite particles characteristics on cytokines production by human monocytes *in vitro*. Biomaterials 24: 2739–2747.
12. Bollag DM, Rozycki MD, Edelstein SJ (1996) Gel electrophoresis under denaturing conditions in Protein Methods (2nd ed) a John Wiley & Sons, Inc, Publication.
13. Fanelli MA, Cuello-Carrión FD, Dekker J, Schoemaker J, Ciocca DR (1998) Serological detection of heat shock protein hsp27 in normal and breast cancer patients. Cancer Epidemiol Biomarkers Prev 7:791–795.
14. Bradford, M. M. (1976) A Rapid and Sensitive Method for the Quantitation of Microgram Quantities of Protein Utilizing the Principle of Protein-Dye Binding. Anal. Biochem. 72:248–254.

Chapter 21

Enhancing Antigen Cross-Presentation and T-Cell Priming by Complexing Protein Antigen to Recombinant Large Heat-Shock Protein

Xiang-Yang Wang, Huanfa Yi, Xiaofei Yu, Damin Zuo, and John R. Subjeck

Abstract

Large heat-shock proteins (HSPs), including hsp110 and grp170, are unique immunochaperones capable of carrying and introducing antigens into professional antigen-presenting cells for efficient cross-presentation. Therefore, reconstituted chaperone complexes of large HSPs and protein antigen may be exploited for augmentation of an antigen-specific immune response. The methods for the preparation of the recombinant protein antigen chaperone complex and characterization of its T-cell priming capability in both in vitro and in vivo settings are described.

Key words: Large heat-shock protein, hsp110, grp170, Chaperone vaccine, Antigen presentation, T-cell priming

1. Introduction

Heat-shock proteins (HSPs) are among the most abundant and ubiquitous intracellular proteins. As molecular chaperones, they are actively involved in almost every aspect of protein homeostasis, e.g., folding/refolding, assembly, translocation, and degradation (1). The studies during the last decade have shown that certain tumor-derived HSPs can serve as effective cancer vaccines (2–4), which has been attributed to an HSP-carried peptide antigenic "fingerprint" of the tumor (5). Indeed, it has been documented that exogenous HSPs are highly efficient in directing associated antigens into antigen-presenting cells (APCs) through the interactions with surface receptors (6–9), resulting in the cross-presentation of antigens on MHC class I molecules. In addition to promoting

Stuart K. Calderwood and Thomas L. Prince (eds.), *Molecular Chaperones: Methods and Protocols*, Methods in Molecular Biology, vol. 787, DOI 10.1007/978-1-61779-295-3_21, © Springer Science+Business Media, LLC 2011

antigen processing and presentation, HSP interactions with certain signaling receptors, such as toll-like receptors, facilitate phenotypic and functional maturation of professional APCs, e.g., dendritic cells (DCs) or monocytes (10, 11). Thus, the properties of HSPs as antigen carriers and as activators of innate immune cells enable these chaperone molecules to be utilized as physiological adjuvants for development of various immunotherapeutic approaches against cancer or infectious diseases (12).

Large HSPs, called hsp110 and grp170, exhibit similar albeit distinct structural and functional features compared to other chaperone molecules (13). In light of their exceptional client protein-holding capacity and superior immunostimulatory activity (4, 14–17), we have created novel recombinant heat-shock vaccines by complexing clinically relevant tumor protein antigens to these large HSPs in vitro. Since it has long been understood that HSPs chaperone full-length protein substrates, these generated complexes are believed to resemble natural intracellular HSP–substrate chaperone complexes. We have demonstrated that these chaperone vaccines exhibit potent antitumor activities in various tumor models (16, 18–21), and large HSPs are significantly more potent than complete Freund's adjuvant (CFA) as an adjuvant (19). The "chaperoning" approach that we have developed clearly provides several advantages over autologous vaccines, including no requirement for a surgical tumor specimen, unlimited quantities of off-the-shelf vaccines with uniformity, broad applicability, and easy immunomonitoring using well-defined antigens.

Here, we describe the methods for preparing recombinant large HSPs and protein antigens using the baculovirus protein expression system, generating large HSP–protein antigen chaperone complexes by heat shock, and assessing the complex-stimulated T-cell activation using both in vitro and in vivo systems.

2. Materials

2.1. Preparation of Recombinant Large HSPs and Protein Antigen

1. BacPAK baculovirus expression system (Clontech).
2. Baculovirus rapid titer kit (Clontech).
3. Ni^{2+}-nitrilotriacetic acid (Ni-NTA)-agarose (Qiagen).
4. Lysis buffer: 20 mM Tris–HCl (pH7.9), 0.5 M NaCl, 5 mM Imidazole (Sigma), 0.1% Nonidet P-40, and protease inhibitor cocktail tablets (Roche Molecular Biochemicals).
5. Binding buffer: 20 mM Tris–HCl (pH7.9), 0.5 M NaCl, 5 mM Imidazole.

6. Wash buffer: 20 mM Tris–HCl (pH 7.9), 0.5 M NaCl, 20–50 mM Imidazole.
7. Elution buffer: 20 mM Tris–HCl (pH 7.9), 0.5 M NaCl, 300 mM Imidazole.

2.2. Chaperone Complex Formation In Vitro

1. Luciferase aggregation assay buffer: 25 mM HEPES (pH 7.9), 5 mM magnesium acetate, 50 mM KCl, and 5 mM β-mercaptoethanol (β-ME).
2. Complexing buffer: Phosphate-buffered saline (PBS) containing 20 mM HEPES, pH 7.2, 20 mM NaCl.
3. GelCode blue stain reagent (Pierce).
4. Bicinchoninic acid (BCA) protein assay kit (Pierce).
5. Enhanced chemiluminescence detection system (Amersham Pharmacia).

2.3. Measuring Antigen Cross-Presentation In Vitro Using Dendritic Cells

1. Complete bone marrow-derived DC (BMDC) medium: RPMI-1640 medium containing 10% heat-inactivated fetal bovine serum (FBS), 10 mM HEPES (pH 7.3, Invitrogen), 20 ng/ml GM-CSF (R&D system), 2 mM L-glutamine, 100 U/ml penicillin, and 100 μg/ml streptomycin and 50 μM β-ME (Sigma).
2. Red blood cell lysis buffer: 0.15 M NH_4Cl, 10 mM $KHCO_3$, 0.1 M Na_2EDTA, pH 7.2.
3. TruStain FcX™ (anti-mouse CD16/32) antibody (Clone 93, BioLegend).
4. FITC BrdU Flow kits (BD Pharmingen).
5. PE Rat anti-mouse CD8 antibody (clone 53–6.7, BioLegend).
6. PerCP/Cy5.5 anti mouse CD90.1 (clone OX-7, BioLegend).
7. FACS staining buffer: PBS containing 0.1% (m/v) bovine serum albumin, 0.1% NaN_3.

2.4. Measuring the Immunogenicity of Chaperone Complex Vaccines In Vivo

1. Red blood cell lysis buffer: 0.15 M NH_4Cl, 10 mM $KHCO_3$, 0.1 M Na_2EDTA, pH 7.2.
2. RPMI-1640 complete medium containing 10% FBS, 2 mM L-glutamine, 100 U/ml penicillin, and 100 μg/ml streptomycin and 50 μM β-ME.
3. 96-well nitrocellulose-backed microtiter plates (Millipore).
4. Rat anti-mouse IFN-γ (clone R4-6A2, BD Pharmingen).
5. Wash solution: PBS containing 0.05% (v/v) Tween 20.
6. Biotinylated IFN-γ antibody (clone XMG1.2, BD Pharmingen).
7. Avidin-alkaline phosphatase D (Vector Laboratories).
8. 5-bromo-4-chloro-3-indolyl phosphatase/Nitro Blue Tetrazolium (BCIP/NBT) (Boehringer Mannheim).

3. Methods

3.1. Preparation of Recombinant Large Heat-Shock Protein and Protein Antigen

The baculovirus-insect cell expression system, which not only facilitates proper protein folding and posttranslational processing, but also greatly reduces endotoxin contamination, is used to generate recombinant large HSPs (i.e., hsp110 and grp170) and protein antigens (e.g., gp100). The full-length cDNAs for large HSPs or targeted antigens are first subcloned into a baculovirus transfer vector (pBacPAK) containing the promoter for transcription as well as the sequences for homologous recombination and selection. The 6 × histidine (His) affinity tag is introduced into the N-terminus or C-terminus of the proteins to facilitate protein binding to Ni^{2+}-NTA agarose for purification. The constructed plasmids are cotransfected along with replication-deficient baculovirus DNA into Sf21 insect cells for virus packaging.

The Bacfectin–DNA mixture is prepared by diluting transfer vector (0.5 μg) with linear viral DNA (0.1 μg) and the Bacfectin (4 μl) in 100 μl sterile nuclease-free water and then added dropwise to a 35-mm culture dish containing Sf21 cells while gently swirling the dish to mix. 1.5 ml of BacPak complete medium is then added after culture at 27°C for 5 h. When the signs of infection, such as irregular shapes and increased volumes, appear (approximately 5 days posttransfection), culture supernatants are harvested for plaque assay. Freshly prepared sf21 cells in 35-mm dish are incubated with the serially diluted supernatant (10^{-2}~10^{-4}) at room temperature for 1 h. After removing the virus inoculums, prewarmed 1% SeaPlaque agarose solution (FMC Bioproducts) is gently overlaid on the infected cell monolayer, and BacPAK complete medium is subsequently added on the solidified agarose. After culture in a humidified incubator for 5–7 days, the plates are stained with 0.03% neutral red solution (Sigma) in PBS and the well-isolated viral plaques are picked using sterile Pasteur pipettes. Vortex of agarose plug in the medium allows the viruses to diffuse out of the agarose plug. 0.1 ml of inoculum from each virus plaque is amplified for 3–4 days using fresh sf21 cells until the cells appear grainy with irregularly shaped membranes. The supernatants are collected after centrifugation at 1,000 × *g* for 5 min at room temperature (passage I virus stock). Expression of large HSPs and targeted antigen should be determined at this step using the infected remaining cells by SDS-PAGE and immunoblotting analysis. The positive recombinant virus plaques are further amplified (0.1 ml passage I stock into 1.5×10^7 cells/30 ml medium in a 150-mm plate) for 4–6 days to produce passage II virus stock. Several aliquots should be kept at −70°C for long-term storage and the remainder kept at 4°C as the working stock. The titration of virus should be performed using passage II virus so that subsequent infections can be optimized to produce the maximal yield of recombinant proteins. Virus titers can be determined using a baculovirus rapid titer Kit (Clontech) and calculated as follows: the titer of the virus stock (pfu/ml) = (average plaques per dish) × 10 × (dilution factor)$^{-1}$. Passage II virus stocks for large HSPs and protein antigens (e.g., gp100) usually have a high titer of 5×10^7 pfu/ml.

Recombinant proteins are produced in large quantities by infecting insect cells growing in suspension using higher multiplicity of infection (MOI, ~5 to 10). Adding glucose (4.5 mg/ml) to culture media was seen in our hands to increase the glycosylation of certain glycoproteins, such as grp170 and gp100. Three to four days after virus infection, a small aliquot of cells is subjected to SDS-PAGE to ensure protein expression before cells are collected for the scale-up of protein production. Cells are sonicated in lysis buffer (10^9 cells per 100 ml lysis buffer) and incubated for 30 min on ice. The supernatant of postcentrifugation at 10,000 × g for 1 h is incubated with Ni^{2+}-NTA agarose beads under native conditions at 4°C (volume ratio ~20:1) and the resins are packed onto columns the next day. The columns are washed first with tenfold bed volume of binding buffer and subsequently with wash buffer to remove nonspecific protein binding. His-tagged recombinant proteins are eluted from column with 5–10 ml of elution buffer. The eluted proteins are dialyzed against PBS using Slide-A-Lyzer (Pierce) and concentrated with Centriplus (Milipore) or Vivaspin (Vivascience) ultrafiltration columns. The recovered proteins are quantified using BCA protein assay with bovine serum albumin as a standard. The estimated total protein yield is 1–2 mg per 10^9 cells.

3.2. Complex Formation of Large Heat-Shock Protein and Protein Antigen

It is recommended that aggregation protection assays using luciferase as a reporter protein are performed to assess the chaperoning capability of recombinant large HSPs prior to the complex reconstitution (22). 150 nM luciferase and hsp110 or grp170 at a molar ratio of 1:1 are incubated in the aggregation assay buffer at 43°C for 30 min, and protein aggregation is monitored by measuring the optical density at 320 nm using a spectrophotometer that is connected to a water bath for temperature control. Hsp110 or grp170 should be able to protect 70–90% of luciferase from heat shock-induced denaturation (Fig. 1). For the generation of chaperone vaccines, recombinant large HSPs and targeted protein antigens (e.g., melanoma antigen gp100) are incubated under heat-shock conditions for 30 min and then incubated at 37°C for another 1 h. Since the "melting" temperatures differ for individual protein antigens, a pilot study should be carried out to determine the approximate temperature at which the protein starts to denature as indicated by aggregation and precipitation of the protein. The addition of hsp110 or grp170 to the protein antigen should then inhibit aggregation as a result of complex formation. While this

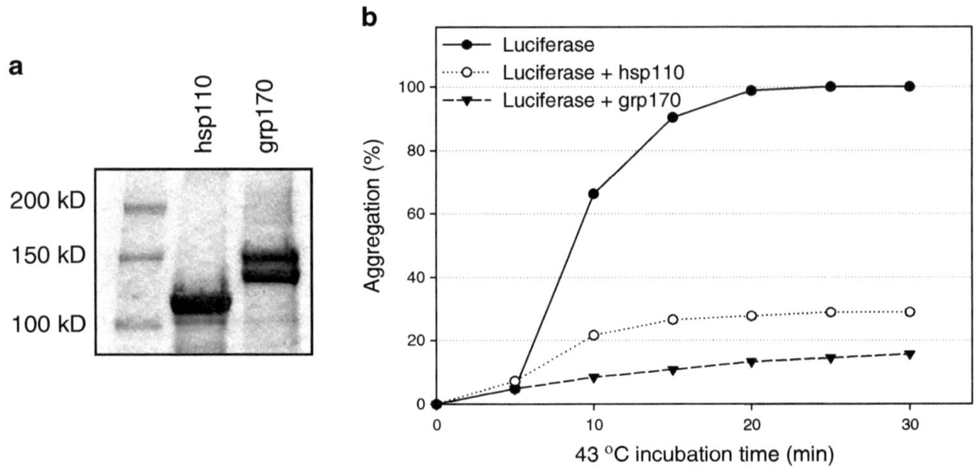

Fig. 1. In vitro chaperoning activity assay using luciferase as a reporter protein. (**a**) Purified hsp110 and grp170 were analyzed by SDS-PAGE and subsequent gel staining. (**b**) Luciferase was incubated in the presence or absence of hsp110 or grp170 at 43°C. The optical density of reaction samples at 320 nm was monitored using a spectrophotometer. Aggregation of heated luciferase alone was set as 100%.

model is generally applicable, there is an important caveat. There is an upper temperature limit of approximately 65°C that can be used to induce aggregation since hsp110 itself begins to aggregate at about 70°C. While most proteins examined do aggregate in the available temperature range and therefore adhere to this simple model, not all do. Some proteins are simply thermostable and not suitable for complexing, e.g., ovalbumin. However, we have had experience with protein antigens, which were purified from bacterial inclusion bodies, that did not aggregate but readily complexed with hsp110 at higher temperature as indicated by co-immunoprecipitation (e.g., the intracellular domain of Her-2/neu[18]), yielding potent vaccines. Therefore, while a simple precipitation assay is an initial way to set parameters of complexing, failure of this assay should not be the reason to abandon the antigen as not suitable. Most protein antigens that we have examined do aggregate in the available temperature range and complex with hsp110 or grp170. In addition, the molar ratios of large HSPs and the antigen can be adjusted to achieve the maximal complexing efficiency, although a one-to-one molar ratio has been found to be effective in most instances.

The complex formation is confirmed using immunoprecipitation assays as previously described (18). Anti-hsp110 (1:200) or -grp170 (1:100) antibodies are incubated with the complexes to pull down chaperone proteins. Normal rabbit sera are used as negative controls. The immune complexes are then precipitated by Protein-A Sepharose CL-4B (Amersham Pharmacia) and subjected to SDS-PAGE followed by either Gel-blue staining (Pierce) or immunoblotting analysis with antibodies against the targeted antigen (e.g., gp100). Visualization of coprecipitated hsp110 (or grp170) and the protein antigen by Gel-blue staining also allows for an estimation of the molar ratio of complexing. This can be quantitated from a gel scan by adjusting for molecular weight differences and by assuming similar amino acid compositions. For example, gp100 has a molecular weight of about 75 kDa, so a one-to-one molar ratio with hsp110 would be seen as a ratio of about 0.70 based on Gel-blue band intensities. In practice, this ratio is closer to 0.60 indicating that most hsp110 molecules are occupied with gp100 protein antigens. However, there has been no analysis of this ratio versus vaccine efficiency and it is expected that a lower ratio would also be fine. In our experience, the simple visualization of both bands by Gel-blue staining using standard loading conditions appears to guarantee sufficient complexing to generate a significant immune response.

3.3. Measuring the Chaperone Complex-Facilitated Antigen Cross-Presentation In Vitro

In order to test the ability of large HSPs to enhance the cross-presentation of chaperoned protein antigen, mouse BMDCs are prepared and used as APCs. The mouse tibiae and femurs are removed and cut with scissors to expose the marrow cavity, and the bone marrow cells are flushed out from the bones using a 3 cc syringe (VWR) attached to a 27-G needle (Becton Dickinson) with ice-cold serum-free RPMI1640. Clusters within the marrow suspension should be disintegrated by vigorous pipetting, followed by the removal of red blood cells from the cell suspension using lysis buffer. Cells are washed and passed through a cell strainer (Becton Dickinson) to remove small pieces of bone and debris. Yields are routinely $5{\sim}7\times10^7$ mononuclear cells per mouse (two tibiae and two femurs) with a viability of >99%. Cells in complete BMDC culture medium are seeded at 2×10^6 cells/well into 12-well plates (Corning) on day 0. The loosely adherent granulocytes should be carefully depleted on day 3 and replaced with fresh complete medium. On days 5 and 7, half of the culture supernatant is harvested and centrifuged. The cell pellet is resuspended in fresh medium and given back into the original wells. After 7 day cultures, the clustering adherent DCs dislodge from the stromal cells and float in the culture medium. On day 9, the non- and semiadherent cells are collected by gentle pipetting using a Pasteur pipet. The cells contain at least 80% of DCs as identified by surface marker CD11c, MHC class II, and B7.1/B7.2.

In vitro antigen-specific T-cell proliferation assay is used to measure large HSP-enhanced antigen cross-presentation. Bromodeoxyuridine (BrdU), a synthetic nucleoside that is an analogue of thymidine, can be incorporated into the newly synthesized DNA of dividing cells. Therefore, BrdU incorporation assay involving immunofluorescent staining of incorporated BrdU and flow cytometric analysis provides an alternative nonradioisotope-based

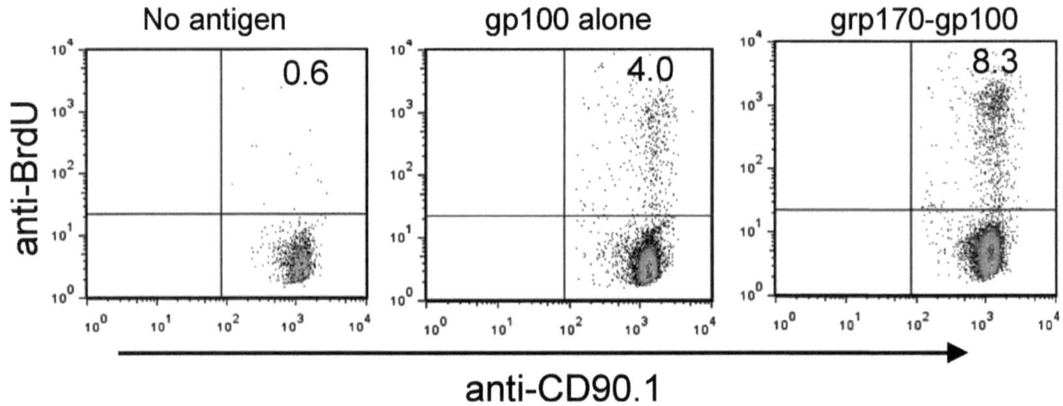

Fig. 2. Recombinant grp170 promotes gp100 cross-presentation and activation of gp100-specific T-cells in vitro. Purified CD8+ Pmel T-cells were cocultured with BMDCs pulsed with gp100 or grp170–gp100 chaperone complexes at a molar ratio of 10:1 (T-cells:DCs) for 72 h. BrdU (10 μM) was added at the beginning of the coculture. Cells were stained with anti-CD8, CD90.1, and BrdU antibodies conjugated with various fluorophores. BrdU+ cells were analyzed using FACS gating in CD8+CD90.1+ cells.

method for assessing the complex-promoted T-cell activation. The procedure described here is to determine the frequency of BrdU incorporating T-cells following coculture with BMDCs that have been loaded with recombinant large HSP-protein antigen complexes (e.g., grp170–gp100).

BMDCs are incubated with recombinant gp100 protein, grp170–gp100 protein complexes or left untreated in 1 ml RPMI complete medium overnight at 37°C. BMDCs are washed and seeded into 96-well U-bottom cell culture cluster as stimulators ($1–2\times10^4$/well). Gp100-specific CD8 T-cells are purified from Pmel17 TCR transgenic mice (Jackson Laboratory) as responders using mouse CD8 cell recovery column kit (Cedarlane laboratories Limited) and cocultured with BMDCs at different ratios in the 96-well plates. BrdU solution is added at a final concentration of 10 μM. After a 72-h culture in a humidified 37°C, 5% CO_2 incubator, cells are washed and blocked with CD16/CD32 antibodies in FACS staining buffer, followed by staining using anti-CD8-PE and CD90.1-PerCP/Cy5.5 antibodies at 4°C for 30 min. Cells are fixed, treated with DNase I (300 μg/ml), and stained with FITC-conjugated anti-BrdU antibodies (BD Pharmingen). Cells are analyzed on a flow cytometer (e.g., BD FACS Calibur) for the frequency of BrdU+CD90.1+ T-cells (Fig. 2).

3.4. Measuring the Chaperone Complex-Induced T-Cell Activation In Vivo

The most reliable readout for the immunogenicity of protein antigen-targeted chaperone complex vaccines is to determine their ability to generate functional T-cell responses in vivo. Several T-cell assays have been established, which include enzyme-linked immunosorbent spot (ELISPOT) assay, in vitro cytolytic assay, in vivo CTL killing assay, intracellular cytokine staining, and MHC class I peptide tetramer assay. Here, we briefly discuss the ELISPOT assay, which measures the frequency of cytokine producing T-cells in response to antigen stimulation on a single-cell level. The ELISPOT takes advantage of the relatively high concentration of the cytokines in the environment surrounding the cytokine-secreting cell, which can be captured and detected using high-affinity cytokine antibodies. A total of 25–30 μg of large HSP–protein antigen (e.g., gp100) complexes is injected in a volume of 100 μl intradermally to naïve mice. Mice immunized with large HSP alone, protein antigen with or without heat-shock treatment, serve as controls. One week later, a second vaccination is given to boost the immune responses. Spleen or draining lymph nodes are collected after an additional week and a single-cell suspension is prepared. Red blood cells are removed by lysis buffer. A 96-well, nitrocellulose-backed microtiter plates (Millipore) are precoated with 10 μg/ml rat anti-mouse IFN-γ in PBS overnight at 4°C or for 2 h at room temperature. Splenocytes, lymph node cells, or purified $CD8^+$ T-cells purified using magnetic beads (Myltenyi Biotec) are plated at a concentration of $2–10 \times 10^5$ cells/well in RPMI1640 complete medium containing antigens (e.g., 20 μg/ml gp100 protein or 1 μg/ml $gp100_{25–32}$ peptide) on a level surface. Following culture in a humidified 37°C, 5% CO_2 incubator for 24 h and extensive rinse with wash buffer, the plates are incubated sequentially with 5 μg/ml biotinylated IFN-γ antibody and 0.2 U/ml avidin–alkaline phosphatase D. Spots are developed by adding BCIP/NBT to each well and incubating at room temperature until color develops. The number of the spots is counted with a Zeiss ELISPOT reader and presented as the number of IFN-γ spots per 1×10^6 cells.

4. Notes

1. Given the inhibitory effect of FBS on the transfection, it is necessary to wash the insect cell once with BacPAK Grace basic medium (GIBCO) and replace the normal medium with serum-free medium before adding the Bacfectin–DNA mixture to the cells.
2. To achieve maximal protein expression, log phase Sf21 cells that are at least 98% viable should be used for infections. Both the quality of virus plaques and the level of protein production are highly dependent on the viability of the cells. In addition, sometimes a protein antigen can be engineered to improve solubility and expression, e.g., removal of a transmembrane domain as in the case of melanoma antigen gp100.
3. Including a low concentration of imidazole in the lysis and wash buffers can help minimize nonspecific binding. Drying of

the columns should be always avoided during washes, since this results in the early elution of His-tagged proteins.

4. Male mice are preferred for the generation of DCs because of their larger bone size compared to female mice. We flush out the bone marrow cells using ice-cold serum-free RPMI 1640 instead of the complete medium because FBS in the complete medium can generate a lot of foam during the repeated flushes that may be harmful to the cells. To increase the purity of DCs, the loosely adherent granulocytes should be removed thoroughly without disturbing the large clusters of DCs.
5. A diffuse darkening of the nitrocellulose membrane is commonly observed when too many cells per well are seeded. Inadequate removal of cells from the ELISPOT plates after antibody incubation may cause a high background. Washing the plate at least once with distilled water helps reduce the "noise".

Acknowledgments

This work was supported by NIH research grants CA129111, CA154708, CA099326, and ACS RSG-08-187-01-LIB.

References

1. Welch WJ. Heat shock proteins functioning as molecular chaperones: their roles in normal and stressed cells. Philos Trans R Soc Lond B Biol Sci 1993;339:327–33.
2. Tamura Y, Peng P, Liu K, Daou M, Srivastava PK. Immunotherapy of tumors with autologous tumor-derived heat shock protein preparations. Science 1997;278:117–20.
3. Graner M, Raymond A, Romney D, He L, Whitesell L, Katsanis E. Immunoprotective activities of multiple chaperone proteins isolated from murine B-cell leukemia/lymphoma. Clin Cancer Res 2000;6:909–15.
4. Wang XY, Kazim L, Repasky EA, Subjeck JR. Characterization of heat shock protein 110 and glucose-regulated protein 170 as cancer vaccines and the effect of fever-range hyperthermia on vaccine activity. J Immunol 2001;166:490–7.
5. Srivastava P. Interaction of heat shock proteins with peptides and antigen presenting cells: chaperoning of the innate and adaptive immune responses. Annu Rev Immunol 2002;20: 395–425.
6. Binder RJ, Han DK, Srivastava PK. CD91: a receptor for heat shock protein gp96. Nat Immunol 2000;1:151–5.
7. Delneste Y, Magistrelli G, Gauchat J, et al. Involvement of LOX-1 in dendritic cell-mediated antigen cross-presentation. Immunity 2002;17: 353–62.
8. Berwin B, Hart JP, Rice S, et al. Scavenger receptor-A mediates gp96/GRP94 and calreticulin internalization by antigen-presenting cells. Embo J 2003;22:6127–36.
9. Facciponte JG, Wang XY, Subjeck JR. Hsp110 and Grp170, members of the Hsp70 superfamily, bind to scavenger receptor-A and scavenger receptor expressed by endothelial cells-I. Eur J Immunol 2007;37:2268–79.
10. Asea A, Kraeft SK, Kurt-Jones EA, et al. HSP70 stimulates cytokine production through a CD14-dependant pathway, demonstrating its dual role as a chaperone and cytokine. Nat Med 2000;6:435–42.
11. Vabulas RM, Ahmad-Nejad P, Ghose S, Kirschning CJ, Issels RD, Wagner H. HSP70 as endogenous stimulus of the Toll/interleukin-1

receptor signal pathway. J Biol Chem 2002;277: 15107–12.

12. Wang XY, Facciponte JG, Subjeck JR. Molecular chaperones and cancer immunotherapy. Handb Exp Pharmacol 2006;172:305–29.
13. Easton DP, Kaneko Y, Subjeck JR. The hsp110 and Grp1 70 stress proteins: newly recognized relatives of the Hsp70s. Cell Stress Chaperones 2000;5:276–90.
14. Oh HJ, Chen X, Subjeck JR. Hsp110 protects heat-denatured proteins and confers cellular thermoresistance. J Biol Chem 1997;272: 31636–40.
15. Park J, Easton DP, Chen X, MacDonald IJ, Wang XY, Subjeck JR. The chaperoning properties of mouse grp170, a member of the third family of hsp70 related proteins. Biochemistry 2003;42:14893–902.
16. Park J, Facciponte JG, Chen X, et al. Chaperoning Function of Stress Protein grp170, a Member of the hsp70 Superfamily, Is Responsible for its Immunoadjuvant Activity. Cancer Res 2006;66:1161–8.
17. Wang XY, Arnouk H, Chen X, Kazim L, Repasky EA, Subjeck JR. Extracellular targeting of endoplasmic reticulum chaperone glucose-regulated protein 170 enhances tumor immunity to a poorly immunogenic melanoma. J Immunol 2006;177:1543–51.
18. Manjili MH, Henderson R, Wang XY, et al. Development of a recombinant HSP110-HER-2/neu vaccine using the chaperoning properties of HSP110. Cancer Res 2002;62:1737–42.
19. Wang XY, Chen X, Manjili MH, Repasky E, Henderson R, Subjeck JR. Targeted immunotherapy using reconstituted chaperone complexes of heat shock protein 110 and melanoma-associated antigen gp100. Cancer Res 2003;63:2553–60.
20. Manjili MH, Wang XY, Chen X, et al. HSP110-HER2/neu chaperone complex vaccine induces protective immunity against spontaneous mammary tumors in HER-2/neu transgenic mice. J Immunol 2003;171:4054–61.
21. Kim H, Sun X, Subjeck J, Wang X-Y. Evaluation of renal cell carcinoma vaccines targeting carbonic anhydrase IX using heat shock protein 110. Cancer Immunology, Immunotherapy 2007;56:1097–105.
22. Oh HJ, Easton D, Murawski M, Kaneko Y, Subjeck JR. The chaperoning activity of hsp110. Identification of functional domains by use of targeted deletions. J Biol Chem 1999;274:15712–8.

Chapter 22

Investigating Receptors for Extracellular Heat Shock Proteins

Ayesha Murshid, Jimmy Theriault, Jianlin Gong, and Stuart K. Calderwood

Abstract

Extracellular heat shock proteins (HSP) play important roles in cell signaling and immunity. Many of these effects are mediated by cell surface receptors expressed on a wide range of cell types. We have investigated the nature of such proteins by cloning candidate receptors into cells (CHO-K1) with the rare property of being null for HSP binding. Using this approach, we have discovered that Hsp70 binds to a least two classes of receptor: c-type lectin receptors (CLR) and scavenger receptors (SR). However, the nature of the receptor–ligand interactions is not yet clear. Hsp70 can bind to LOX-1 (a member of both the CLR and SR), with the c-type lectin binding domain (CTLD) as well as the SR family members SREC-I and FEEL-1/CLEVER-1/STABILIN-1, which by contrast have arrays of EGF-like repeats in their extracellular domains. In this chapter, we discuss (1) methods for determining HSP receptors, (2) approaches to study of individual receptors in cells that contain multiple such receptors, and (3) methods for investigating HSP receptor function in vivo.

Key words: Extracellular, Heat, Shock, Protein, Scavenger, Receptor, Immunity

1. Introduction

It is apparent that heat shock proteins (HSP) play significant signaling roles in the extracellular microenvironment (1). HSP have been found in human serum particularly after disease or stress (2, 3). The 70-kDa HSP (Hsp70) has been shown to be released from cells after acute stress as well as being secreted after exposure to a number of stimuli (4, 5). Extracellular HSPs may thus be able to play the role of danger signal (danger activated molecular pattern or DAMP) (6). In this context, they may interact with pattern recognition receptors (PRR) such as Toll-like receptors (TLR) and

Stuart K. Calderwood and Thomas L. Prince (eds.), *Molecular Chaperones: Methods and Protocols*,
Methods in Molecular Biology, vol. 787, DOI 10.1007/978-1-61779-295-3_22, © Springer Science+Business Media, LLC 2011

activate proinflammatory signaling and transcription (7, 8). Proteins including Hsp60, Hsp70, GP96 have been implicated as DAMPs (9, 10). However, interpretation of such experiments requires caution and careful control as some HSPs have the ability to bind to prokaryotic molecules that activate TLR signaling, such as lipopolysaccharides (11). In addition, many members of the HSP family can participate in adaptive immunity by binding to antigenic peptides and transporting them into antigen-presenting cells (APC) (12, 13). HSPs mediate the process of ***antigen cross presentation*** (14) by facilitating internalization of antigens and permitting their delivery to major histocompatibility class I (MHC class I) molecules. MHC I-peptide complexes can then stimulate cognate T cell receptors on T lymphocytes and initiate the activation of clones of such powerful immune effectors. HSP may thus play a versatile role in antitumor immunity by activating the innate and adaptive arms. HSP can additionally activate natural killer cells and lead to tumor cell killing and CD25+ immune regulatory T cells (15–17). HSP can thus upregulate or downregulate immunity depending on context.

Many studies have suggested that HSP activate immunity by binding to receptors on the cell surface (18–26). HSP binding is saturable and competed for by unlabeled ligand, properties of receptor-mediated signaling. However, this is where consensus seems to end, and some controversy exists as to the most significant HSP receptors. We have attempted to address this issue by screening the various contenders for binding to Hsp70 and Hsp90.

2. Materials

2.1. Plasmids

1. pET23 *hsp90a* plasmid.
2. pDEST™10.
3. AC-to-BAC Baculovirus transfection kit.
4. pCDNA3.1 eukaryotic expression vector.

2.2. Cell Lines

1. Chinese Hamster Ovary-K1 cells (CHO-K1).
2. A375 human melanoma.
3. MISA human breast carcinoma cells.
4. Sf9 insect cells.
5. MC38 cells stably expressing the MUC1 tumor antigen.
6. B16 melanoma cells.
7. B16 melanoma cells stably expressing the MUC1 tumor antigen.

2.3. Mouse Systems and Primary Murine Cells

1. Wild type mice C57BL/6 and *tlr2*$^{-/-}$/*tlr4*$^{-/-}$ double knockouts.
2. Primary mouse bone marrow dendritic cells were prepared from C57BL/6 as in text.
3. Splenocytes and/or lymph node cells (LNC) were isolated from mice immunized with Hsp70.PC fusion vaccine as described (27).

2.4. Chromatography

1. Ni-NTA purification system (Qiagen).
2. 10 ml Sephadex G-25 in PD10 column (Sigma-Aldrich).
3. 5.0 ml ADP-agarose column (Sigma-Aldrich).
4. 20 ml DEAE-cellulose anion exchange column (Pierce Chemicals).

2.5. Buffers and Reagents

1. *Hypotonic Buffer*. 10 mM $NaHCO_3$, 0.5 mM PMSF, pH 7.1.
2. *Buffer D*. 20 mM Tris–acetate, 20 mM NaCl, 15 mM b-mercaptoethanol, 3 mM $MgCl_2$, 0.5 mM PMSF, pH 7.5.
3. *ADP-agarose Elution buffer*. 3.0 mM ADP in buffer D.
4. *FPLC buffer*. 20 mM sodium mono and diphosphate, 20 mM NaCl, pH 7.0.
5. *DEAE-cellulose elution buffer*. 150 mM NaCl in FPLC buffer.
6. *Hsp70 binding buffer (PFNC)*. 0.5% FBS, 0.05% NaN_3 and 1 mM $CaCl_2$.
7. Hanks' Buffered Saline Solution.

2.6. Antibodies

Anti-Hsp70 antibody (SPA-810, Assay Designs Inc).

2.7. Chromophores

Alexa488 (Molecular Probes).

2.8. shRNA to SREC-I

MISSION™ shRNA plasmids (shRNA) were purchased (Sigma-Aldrich, St. Louis, MI) and the Lentivirus generation and transduction were performed according the manual of ViralPower™ Lentiviral Expression Systems (Invitrogen).

3. Methods

3.1. Screening for HSP Receptors

We have screened receptors for HSP binding in the context of cell surface expression, by expressing candidate receptors in cells null for Hsp70 binding (see Note 1). A number of primary and tissue culture cells were, therefore, screened for lack of capacity to bind to Hsp70. We screened both primary cells and established cell lines (Table 1). Maintenance of established cell lines is described previously (28). Human Umbilical Vein Endothelial Cells (HUVEC) were maintained in Endothelial Basal Medium-2 (EBM-2) supple-

Table 1
Binding to Hsp70

Cell type	Species	Hsp70 binding
THP1, monocyte	Human	+
RAW264.7, macrophage	Mouse	++
Primary macrophage	Mouse	++
Primary dendritic cell	Mouse	++
HEK293 embryonic kidney	Human	+
Vacular endothelial	Human	++
PC-3, prostate carcinoma	Human	+
HeLa, cervical carcinoma	Human	++
Hela S3 cervical carcinoma	Human	+
MCF7, mammary Cancer	Human	+
IMR90, fibroblast	Human	–
K562, pluripotent leukemia	Human	–
A375, melanoma	Human	–
CHO K1, ovarian cells	Chinese hamster	–

mented with Clonetics™ SingleQuot® (Cambrex/Biowittaker). Isolation of peritoneal macrophages was carried out as previously described (29). Briefly, peritoneal macrophages were isolated from 6- to 10-week-old C57BL/6 background mice. The mice were injected intraperitoneally with 3 ml of thioglycollate, and after 4 days peritoneal exudate cells were harvested by lavage with 10 ml of RPMI 1640 and cultured in RPMI 1640 medium supplemented with 10% heat-inactivated FBS and penicillin–streptomycin. Bone marrow-derived dendritic cells (BMDCs) were generated from the femur and tibiae of C57BL/6 mice. The bone marrow was flushed out and cultured in RPMI 1640 supplemented with 10% heat-inactivated fetal bovine serum (FBS) and 40 ng/ml GMCSF for 6 days. On day 3, a third of the media was replaced with fresh growth media.

3.1.1. Alexa 488-Labeled Purified HSP70 Preparation

Human melanoma cells A375-MEL or mouse MISA cells were used as starting material for Hsp70 preparation because high endogenous HSC70 and/or HSP70 levels were detected in these cell types (J. Theriault and SK Calderwood, unpublished). In addition, for some experiments we used minced mouse liver as an abundant source for Hsp70. The Hsp70 purification protocol was based on previous studies (30). Briefly, a 10-ml cell pellet of tumor cells or minced liver was homogenized in 40 ml hypotonic buffer by Dounce homogenization. The homogenate was then spun at 10,000 × *g* for 30 min, and the supernatant was further treated for 60 min at 100,000 × *g*. The sample buffer was changed to buffer D using a PD-10 desalting column (Amersham-Biosciences). The material was then applied directly to a 5-ml ADP-agarose column preequilibrated with buffer D. Hsp70 was eluted from the ADP-agarose column with 3 mM ADP in buffer D. The sample buffer was then changed to FPLC buffer with PD-10 column. The supernatant was applied to a DEAE anion exchange column equilibrated with FPLC buffer (Amersham-Biosciences). Hsp70 was eluted with the FPLC buffer containing 150 mM NaCl. Protein concentrations were determined by *Bradford* assay. Purified Hsp70 was then labeled with fluorophore Alexa 488 according to the manufacturer's instructions (Molecular Probes, USA). Intactness and purity of the labeled Hsp70 was checked by SDS-PAGE, Coomassie staining, and the presence of Hsp70 in the preparation was confirmed by Western blotting using a mouse monoclonal antibody specific for HSP70.

3.1.2. Alexa 488-Labeled Purified Hsp90 Preparation

Hsp90 alpha DNA was prepared by PCR amplification from the pET23 plasmid and cloned into pDEST™10. Overexpression of Hsp90 alpha in Sf9 cells was achieved according to the BAC-to-BAC Transfection kit protocol of *Invitrogen*. Transfer vector was transformed into DH10BAC competent cells containing bacmid DNA. Later, colonies containing recombinant bacmid were identified and prepared. The bacmid DNA was then transfected into Sf9 cells using CellFECTIN (Invitrogen) to make recombinant baculovirus according to manufacturer's protocol.

Sf9 cells were grown in Sf900II serum-free medium (Invitrogen) supplemented with 100 U/ml penicillin–streptomycin and 2 mM of L-glutamine in suspension cultures with continuous shaking at 150 rpm at 27°C in a nonhumidified environment. The insect cultures were infected in the log phase of growth with recombinant baculovirus. Cells were harvested 48-h post infection and washed with Hank's buffered saline solution, and the protein was purified using the Ni-NTA purification system according to the manufacturer's protocol (Invitrogen). AcTEV protease was used to cleave the 6× His tag from the fusion protein generated using pDEST™10 after purifying the recombinant protein on a nickel chelating resin. Purified Hsp90 was then labeled with Alexa 488 as above. Intactness and purity of the labeled Hsp90 was checked by SDS-PAGE, Coomassie Blue staining, and the presence of Hsp90 in the preparation was confirmed by immunoblot.

3.1.3. HSP Binding Assay

Cells were first screened by binding to Alexa 488-labeled HSP in vitro and analysis by flow cytometry. 2×10^5 nontrypsinized cells were washed twice in PFNC buffer and incubated with 150 nM Alexa 488-labeled BSA (negative control), Hsp70, or Hsp90 for 30 min on ice with gentle shaking. The cells were washed twice in PFNC buffer and Alexa 488-labeled protein binding was monitored by flow cytometry (Becton Dickinson).

Experiments utilizing flow cytometry were next confirmed by confocal fluorescence microscopy. Alexa Fluor conjugated BSA, Hsp70, or Hsp90 were prepared as above. Cells were labeled with ligand for 20–30 min on ice. Cells were later washed with ice-cold stripping buffer to remove unbound Hsp90.PC. Cells were then fixed with 4% para-formaldehyde and permeabilized with 0.1% Triton X-100. Fluorescence was then visualized using a Zeiss 510 confocal microscope (Carl Zeiss GmbH, Jena, Germany). Fluorophores were visualized using 488-nm excitation and a band pass 505–530 emission filter for Alexa 488. Images were taken using a 63× numerical aperture (NA) 1.4 oil immersion objective lens.

To assay individual receptors for HSP binding, we selected CHO-Kl cells as null for HSP binding in the wild-type state. Cells were then transfected with expression plasmids for individual receptors following the protocol used for study of LOX-1.

2.5×10^5 CHO-Kl cells were transiently transfected with 5 μg of empty vector (pCDNA3) or pCDNA3 plasmids encoding Myc-tagged LOX-1, for 48 h using the Superfect transfection reagent according to the manufacturer's instructions (QIAGEN). Expression of recombinant proteins was analyzed after transfection by SDS-PAGE and immunoblot with the mouse monoclonal anti-Myc antibody (clone 9E10, Stratagene, USA). Cell lines were maintained by selection with neomycin and checked routinely for expression of Myc-tagged product. In addition, we examined the cell surface location of the candidate receptors using antibodies to the extracellular domains of such proteins.

Using this approach, we have examined a number of candidate receptors. As previous studies had suggested a role for LOX-1 in immune responses to Hsp70, we began with the study of this protein and have confirmed that Hsp70 binds avidly to CHO-LOX-1 cells (Table 2) (21, 31). LOX-1 has been assigned to at least two distinct protein families, the c-type lectins and the scavenger receptors (SR) (31, 32). C-type lectin receptors (CLR) are a large family of receptors characterized by the possession of a common binding domain – the Ca^{++} dependent carbohydrate binding motif (CTLD) (33, 34). Binding to protein ligands can be inhibited by use of hapten sugars that differ between different CLR family members. In the case of LOX-1, fucoidin is a hapten sugar that interacts with its CTLD domain and inhibits Hsp70 binding to LOX-I (21). Scavenger receptors (SR) have been studied mostly in endothelial

Table 2
Candidate HSP receptors

Receptor	Type	Expressed in	Hsp70	Hsp90
TLR2	Signaling	APC etc.	–	ND
TLR4	Signaling	APC etc.	–	ND
CD14	Signaling	APC etc.	–	ND
CD40	Signaling	APC	–	ND
CD91	Internalizing	many	–	ND
LOX-1	Scavenger/CTL	Endo, APC	+	++
DC-SIGNscavenger/CTL	APC	–	ND	
Dectin 1	Scavenger	APC	–	ND
CLEC-1	Scavenger	APC	–	ND
CLEC2	Scavenger	APC	–	ND
SREC-1	Scavenger	APC	++	++
FEEL-1	Scavenger	APC	++	++
NKG2A	CTL	NK	++	ND
NKG2C	CTL	NK	++	ND
NKG2D	CTL	NK, T cell	++	ND

Binding to receptors was assayed in CHO transfectants with two exceptions, which are DC-SIGN, which was in K-562 (also HSP binding null), and the TLR, which were in HEK293. Ability to compete with 25-fold XS cold HSP70 is indicated in the last column

cells but are expressed in dendritic cells and macrophages also (J. Gong, A. Murshid and SK Calderwood, in preparation). SR are a group of proteins that are clustered according to their function in cells – their ability to interact with chemically modified proteins in the extracellular fluid, as exemplified by binding oxidized low density lipoprotein and acetylated or maleylated bovine serum albumin (BSA) (35–38). Binding of HSP to SR can initially be screened by competition assay using known SR binding proteins such as maleylated BSA, oxidized LDL, acetylated LDL, apolipoprotein B, or polyanions such as polyinosine (21). In this approach, one ligand is labeled (fluorescently for flow cytometry) and the other one remains unlabeled. The competition assay is done with a constant concentration of labeled HSP and varying concentration of unlabeled ligands such as mBSA, AcLDL. The basis for this interaction as well as HSP binding is not well understood as the extracellular domains of individual SR are highly divergent (38).

We screened members of these receptor families and confirmed that Hsp70 binds to LOX-1 and as well as two other SR family members, SREC-I and FEEL-1/Stabilin1/CLEVER-1 when expressed in CHO-K1 (Table 2) (39–41). Others have subsequently shown that another SR family member – scavenger receptor A can interact with HSP family members, which, however, fail to bind CD36, MARCO, and CLA-1 (21, 42). Hsp70 also binds to other members of the CLR family such as NKG2D that are expressed in natural killer cells (Table 2) (39) (see Note 2). However, in our studies, Hsp70 failed to bind to a number of major CLR family members including Dectin-1, DC-SIGN, CLEC-1, and CLEC-2 (Table 2) (39). Some studies also indicate a role for the LDL receptor-related protein (LRP) or CD91 in HSP binding. LRP/CD91 contains four clusters of binding repeats that mediate association with at least 30 different ligands including apolipoprotein E, α2-macroglobulin, pro-urokinase, and others (43, 44). Most ligands bind specifically to two of these clusters of binding repeats within domains II and IV (43). However, we could not detect Hsp70 binding to either domain II or IV when expressed in CHO-K1 cells, and in addition, LRP null cells appeared to bind Hsp70 as well as wild-type cells (Table 2) (20, 39, 45). Endocytosis of the molecular chaperone calreticulin was also not decreased in $CD91^{-/-}$ cells casting some doubt on CD91 as a universal endocytic receptor for HSPs (46). By contrast, scavenger receptor SRA has been shown to be required for a large proportion of gp96 and calreticulin uptake (19). In the case of Hsp90, neither fucoidin (LOX-1 agonist) nor α2-macroglobulin (LRP/CD91 agonist) was able to block representation of a peptide bound to Hsp90 (47). Nonetheless, others have shown that inactivation of CD91 can lead to loss of antigen representation ability in cells exposed to gp96/peptide complexes (48). The large HSPs Hsp110 and Grp170 that have potent immune properties can bind to SRA and SREC-I, and there is also some evidence for binding to LRP/CD91 (22). There is additional evidence indicating a possible role for LRP/CD91 in Hsp70 binding to macrophages (21). Inhibition by α2-macroglobulin competition has often been used as a criterion for HSP binding to LRP/CD91 (22). In addition, our unpublished experiments indicate that levels of SREC-I and LOX-1 are very low in unstimulated murine macrophages in which CD91 may play a significant role. The usage of HSP receptors could thus vary with the nature of the chaperone ligand, immune cell type, and the activation state of the cell. Our studies in vivo indicate that TLR signaling is essential for SREC-I expressing DC to traffic to afferent lymph nodes after vaccination with Hsp70 vaccine (27). Hsp70 can itself induce SREC-I expression in TLR pathway proficient murine DC, suggesting a feed-forward mechanism in which Hsp70 induces its own receptors (SR) and amplifies immune effects of Hsp70-peptide complexes (27). There are also indications that

Hsp70 activates signaling receptors such as TLR2, TLR4, and CD40 and may be involved in inducing inflammation and innate immunity (9, 10, 49). However, using the CHO transfection system described above, we were unable to confirm direct binding of Hsp70 to these molecules (Table 2). At least in the case of TLR2, indirect activation of this receptor downstream of both LOX-1 and SREC-I is observed after receptor binding to bacterial protein OMP1 (50).

Figure 1 shows a cartoon indicating that at least four HSP receptors exist and could potentially be coexpressed in a single cell. In addition, a range of other ligands can interact with such receptors. The existence of multiple receptors, therefore, complicates interpretation of experiments probing the function of extracellular HSP. This difficulty is exacerbated by the findings that while SREC-I and LOX-1 have proimmune functions, SRA-1 appears to be inhibitory to the immune response by inhibiting the activity of TLR4 (21, 27, 42, 51). The receptors may thus have both additive and confounding effects (see Note 3).

With this plethora of receptors, the nature of the HSP–receptor interaction is still in some doubt. However, the crystal structure of ligand-bound LOX-1 has recently been determined. These studies indicate that the ligand (oxidized LDL) binding surface is hydrophobic except for a basic spine composed of arginine residues essential for ligand binding (52, 53). These positively charged arginine residues together with the hydrophobic residues appear to

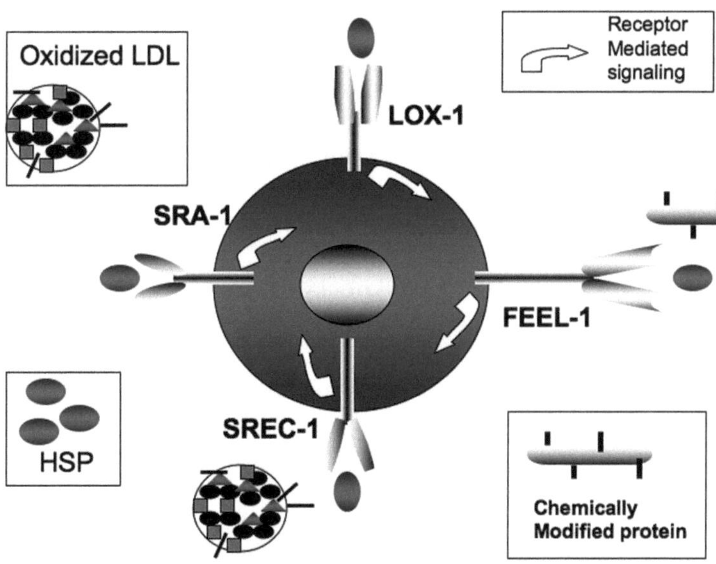

Fig. 1. A cell expressing four HSP receptors that could bind simultaneously to extracellular HSP. We also depict three of the potential ligands in the extracellular fluid, including HSP, oxidized LDL particles, and chemically modified proteins. We depict relative lengths of the intracellular and extracellular domains of the proteins (FX).

confer the specificity of LOX-1 for negatively charged lipids and lipoproteins (52). LOX-1 binds to its ligands as a homodimer with an intramolecular disulfide bond (53). It is not clear to what degree the ligand binding properties of SREC-I and FEEL-1/CLEVER1 resemble those of LOX-1. The extracellular domains of these two SR do not contain CTLD, but consist mostly of multiple EGF-like repeats (54). These repeat domains have a length of approximately 40 amino acids and are characterized by conserved arrangement of six cysteine residues found in EGF itself (55, 56). It has been shown that at least four tandem repeats of EGF-like regions are required for the FEEL-1 homolog FEEL-2 (STABILIN-2) to bind the acidic lipid phosphatidylserine (57). It is apparent, therefore, that much is left to be learnt regarding the specificity and sequence requirements for HSP binding to these candidate receptors.

3.2. Studying HSP–SREC-I Interaction In Vivo

As we currently have not developed SREC-I knockout mice, we took the approach of knocking down SREC-I in dendritic cells by shRNA. Our studies had indicated that in mice with TLR2/TLR4 knockdown, responses to Hsp70 vaccine (Hsp70-FC) were strongly inhibited (27, 58). We utilized this finding to study the significance of SREC-I in responses to Hsp70 in vivo. To address this issue, we first determined whether the decreased immunity in TLR knockout cells could be reversed by adoptive transfer of wild-type DC (WT-DC) into *tlr2*$^{-/-}$/*tlr4*$^{-/-}$ mice (27). Indeed, transfer of Hsp70-FC pulsed DC was able to induce immunity to tumors in naïve WT mice and support tumor specific cytotoxicity (27). We next showed that immunization of *tlr2*$^{-/-}$/*tlr4*$^{-/-}$ mice with WT DC that had been pulsed with Hsp70-FC also increased the cytotoxic lymphocyte (CTL) activity and CTL frequency of such mice (27). This implies that WT donor DC were able to compensate for the endogenous, inactive DC in *tlr2*$^{-/-}$/*tlr4*$^{-/-}$ mice. Groups of such knockout mice were next immunized with DC generated from WT mice that had been infected with SREC-I shRNA or control shRNA constructs and then pulsed with Hsp70-FC. SREC-I knockdown inhibited the ability of WT DC to compensate for TLR knockout, indicating that this is a viable approach for studying SREC-I function in vivo. These experiments, therefore, suggested that SREC-I is essential for antigen presentation by DC exposed to Hsp70.PC. These findings were confirmed by assessing the frequency of antigen-specific T cells induced using an MHC-class-I/peptide tetramer (iTAg). The 8-mer peptide (SAPDTRPA) is a dominant epitope from MUC1 that binds to C57BL/6 MHC class I, H-2K^{b} (59). MUC1 is one of the most prominent tumor antigens present in the Hsp70-FC vaccine (27). The MUC1-8 iTAg was used to identify and assess the tetramer-positive T cells. The numbers of MUC1-8 iTAg-positive T cells from *tlr2*$^{-/-}$/*tlr4*$^{-/-}$ immunized with SRECI knockdown DC were significantly decreased compared with those from mice transferred with DC infected with control virus. Immunization of *tlr2*$^{-/-}$/*tlr4*$^{-/-}$ mice either with Hsp70-FC pulsed DC after SREC-I knockdown or after infection with control virus resulted 1.45 and 3.51% CD8 T cells positive for MUC1-8, indicating that SRECI plays an important role in the induction of antigen-specific T cells by Hsp70-based vaccines.

3.2.1. shRNA Directed Against SRECI

MISSION™ shRNA plasmids (shRNA) were purchased (Sigma-Aldrich, St. Louis, MI) and the Lentivirus generation and transduction were performed according the manual of ViralPower™ Lentiviral Expression Systems (Invitrogen). The effectiveness of shRNA for knockdown of murine SRECI (five clones: TRCN0000067873; TRCN0000067874; TRCN0000067875; TRCN0000067876, and TRCN0000067877) was examined Real-Time PCR and immunoblot. The most effective construct for mSREC-I mRNA knockdown (TRCN0000067875) was used. The Lentivirus (insert sequence was 5′-CCGCAGGT-ATGCACGCGT-3′, which does not target any mouse genes, but activates RISC and the RNAi pathway in the cells) was used as negative control.

Three-day-old DC were collected, purified, placed in 96-well round bottom plates with 1×10^6 cell per well for O/N culture in medium containing GM-CSF. On the second day, half of the medium (100 μl) was removed and 100 μl Lentivirus supernatant (1 virus :1 DC) was added. After 20 h, 150 μl of medium from each well was replaced with fresh GM-CSF medium for additional culture. DC infected with SREC-I shRNA or control constructs or uninfected DC were collected for SREC1 expression and T cell stimulation assay. Significant SRECI knockdown at the protein level was achieved in 90–100 h.

4. Notes

1. *Purified HSP.* Most HSPs are encoded by multigene families. This is particularly true for Hsp70 in which there are, depending on species at least 12 closely related gene family members each of which may bind to cells with differing affinities (39). This complication could be remedied by the use of recombinant proteins. However, if recombinant proteins are produced in *Escherichia coli*, great care must be taken to remove contaminating PAMPs such as lipopolysaccharides or peptidoglycans (11). We have prepared recombinant HSPs using the baculovirus/sf9 system to avoid these problems. However, LPS in particular can be introduced into protein preparations from contaminated glassware and we have been very careful to assay all batches of HSP by the *limulus* assay.

2. HSP bind with only moderate affinity to the receptors uncovered so far and these are promiscuous in terms of binding partners. In most cases, the structural basis for HSP–receptor binding is not known. Thus, much needs to be learned regarding HSP–receptor interaction. Future in vitro studies should address questions of three-dimensional structure of HSP–receptor complexes and the exact affinities of the interactions.
3. Many cells express multiple HSP receptors, and great care must be taken to isolate the properties of individual receptors. It seems likely that other as yet unknown receptors exist.

Acknowledgments

This work was supported by NIH research grants RO-1CA047407, R-O1CA119045, and RO-1CA094397.

References

1. Calderwood SK, Mambula SS, Gray PJ, Jr., Theriault JR. Extracellular heat shock proteins in cell signaling. FEBS Lett 2007;581(19): 3689–94.
2. Pockley AG. Heat shock proteins, inflammation, and cardiovascular disease. Circulation 2002; 105(8):1012–7.
3. Pockley AG, Shepherd J, Corton JM. Detection of heat shock protein 70 (Hsp70) and anti-Hsp70 antibodies in the serum of normal individuals. Immunol Invest 1998;27(6):367–77.
4. Mambula SS, Calderwood SK. Heat induced release of Hsp70 from prostate carcinoma cells involves both active secretion and passive release from necrotic cells. Int J Hyperthermia 2006; 22(7):575–85.
5. Mambula SS, Calderwood SK. Heat shock protein 70 is secreted from tumor cells by a nonclassical pathway involving lysosomal endosomes. J Immunol 2006;177(11):7849–57.
6. Matzinger P. The danger model: a renewed sense of self. Science 2002;296(5566):301–5.
7. Vabulas RM, Ahmad-Nejad P, da Costa C, et al. Endocytosed HSP60s use toll-like receptor 2 (TLR2) and TLR4 to activate the toll/interleukin-1 receptor signaling pathway in innate immune cells. J Biol Chem 2001;276(33): 31332–9.
8. Vabulas RM, Ahmad-Nejad P, Ghose S, Kirschning CJ, Issels RD, Wagner H. HSP70 as endogenous stimulus of the Toll/interleukin-1 receptor signal pathway. J Biol Chem 2002;277(17):15107–12.
9. Asea A, Rehli M, Kabingu E, et al. Novel signal transduction pathway utilized by extracellular HSP70: role of toll-like receptor (TLR) 2 and TLR4. J Biol Chem 2002;277(17):15028–34.
10. Asea A, Kraeft SK, Kurt-Jones EA, et al. HSP70 stimulates cytokine production through a CD14-dependant pathway, demonstrating its dual role as a chaperone and cytokine. Nat Med 2000;6(4):435–42.
11. Henderson B, Calderwood SK, Coates AR, et al. Caught with their PAMPs down? The extracellular signalling actions of molecular chaperones are not due to microbial contaminants. Cell Stress Chaperones 2009.
12. Singh-Jasuja H, Toes RE, Spee P, et al. Cross-presentation of glycoprotein 96-associated antigens on major histocompatibility complex class I molecules requires receptor-mediated endocytosis. J Exp Med 2000;191(11):1965–74.
13. Srivastava P. Interaction of heat shock proteins with peptides and antigen presenting cells: chaperoning of the innate and adaptive immune responses. Annu Rev Immunol 2002;20: 395–425.
14. Rock KL. The ins and outs of cross-presentation. Nat Immunol 2003;4(10):941–3.
15. Multhoff G. Activation of natural killer cells by heat shock protein 70. Int J Hyperthermia 2002;18(6):576–85.
16. Multhoff G, Hightower LE. Cell surface expression of heat shock proteins and the immune response. Cell Stress Chaperones 1996;1(3):167–76.

17. van Eden W, van der Zee R, Prakken B. Heat-shock proteins induce T-cell regulation of chronic inflammation. Nat Rev Immunol 2005;5(4): 318–30.
18. Berwin B, Delneste Y, Lovingood RV, Post SR, Pizzo SV. SREC-I, a type F scavenger receptor, is an endocytic receptor for calreticulin. J Biol Chem 2004;279(49):51250–7.
19. Berwin B, Hart JP, Rice S, et al. Scavenger receptor-A mediates gp96/GRP94 and calreticulin internalization by antigen-presenting cells. Embo J 2003;22(22):6127–36.
20. Binder RJ, Han DK, Srivastava PK. CD91: a receptor for heat shock protein gp96. Nat Immunol 2000;1(2):151–5.
21. Delneste Y, Magistrelli G, Gauchat J, et al. Involvement of LOX-1 in dendritic cell-mediated antigen cross-presentation. Immunity 2002;17(3):353–62.
22. Facciponte JG, Wang XY, Subjeck JR. Hsp110 and Grp170, members of the Hsp70 superfamily, bind to scavenger receptor-A and scavenger receptor expressed by endothelial cells-I. Eur J Immunol 2007;37(8):2268–79.
23. Gross C, Hansch D, Gastpar R, Multhoff G. Interaction of heat shock protein 70 peptide with NK cells involves the NK receptor CD94. Biol Chem 2003;384(2):267–79.
24. Kettner S, Kalthoff F, Graf P, et al. EWI-2/CD316 Is an Inducible Receptor of HSPA8 on Human Dendritic Cells. Mol Cell Biol 2007;27(21):7718–26.
25. Sondermann H, Becker T, Mayhew M, Wieland F, Hartl FU. Characterization of a receptor for heat shock protein 70 on macrophages and monocytes. Biol Chem 2000;381(12): 1165–74.
26. Whittall T, Wang Y, Younson J, et al. Interaction between the CCR5 chemokine receptors and microbial HSP70. Eur J Immunol 2006; 36(9):2304–14.
27. Gong J, Zhu B, Murshid A, et al. T Cell Activation by Heat Shock Protein 70 Vaccine Requires TLR Signaling and Scavenger Receptor Expressed by Endothelial Cells-1. J Immunol 2009.
28. Theriault JR, Mambula SS, Sawamura T, Stevenson MA, Calderwood SK. Extracellular HSP70 binding to surface receptors present on antigen presenting cells and endothelial/epithelial cells. FEBS Lett 2005;579(9): 1951–60.
29. Mambula SS, Sau K, Henneke P, Golenbock DT, Levitz SM. Toll-like receptor (TLR) signaling in response to Aspergillus fumigatus. J Biol Chem 2002;277(42):39320–6.
30. Peng P, Menoret A, Srivastava PK. Purification of immunogenic heat shock protein 70-peptide complexes by ADP-affinity chromatography. J Immunol Methods 1997;204(1):13–21.
31. Mehta JL, Chen J, Hermonat PL, Romeo F, Novelli G. Lectin-like, oxidized low-density lipoprotein receptor-1 (LOX-1): a critical player in the development of atherosclerosis and related disorders. Cardiovasc Res 2006;69(1): 36–45.
32. Chen M, Masaki T, Sawamura T. LOX-1, the receptor for oxidized low-density lipoprotein identified from endothelial cells: implications in endothelial dysfunction and atherosclerosis. Pharmacol Ther 2002;95(1):89–100.
33. Zelensky AN, Gready JE. The C-type lectin-like domain superfamily. Febs J 2005;272(24): 6179–217.
34. Drickamer K. C-type lectin-like domains. Curr Opin Struct Biol 1999;9(5):585–90.
35. Adachi H, Tsujimoto M. Endothelial scavenger receptors. Prog Lipid Res 2006;45(5): 379–404.
36. Rigotti A. Scavenger receptors and atherosclerosis. Biol Res 2000;33(2):97–103.
37. van Berkel TJ, Out R, Hoekstra M, Kuiper J, Biessen E, van Eck M. Scavenger receptors: friend or foe in atherosclerosis? Curr Opin Lipidol 2005;16(5):525–35.
38. Krieger M. The other side of scavenger receptors: pattern recognition for host defense. Curr Opin Lipidol 1997;8(5):275–80.
39. Theriault JR, Adachi H, Calderwood SK. Role of scavenger receptors in the binding and internalization of heat shock protein 70. J Immunol 2006;177(12):8604–11.
40. Adachi H, Tsujimoto M. Characterization of the human gene encoding the scavenger receptor expressed by endothelial cell and its regulation by a novel transcription factor, endothelial zinc finger protein-2. J Biol Chem 2002; 277(27):24014–21.
41. Politz O, Gratchev A, McCourt PA, et al. Stabilin-1 and -2 constitute a novel family of fasciclin-like hyaluronan receptor homologues. Biochem J 2002;362(Pt 1):155–64.
42. Wang XY, Facciponte J, Chen X, Subjeck JR, Repasky EA. Scavenger receptor-A negatively regulates antitumor immunity. Cancer Res 2007;67(10):4996–5002.
43. Herz J, Strickland DK. LRP: a multifunctional scavenger and signaling receptor. J Clin Invest 2001;108(6):779–84.
44. Newton CS, Loukinova E, Mikhailenko I, et al. Platelet-derived growth factor receptor-beta (PDGFR-beta) activation promotes its association with the low density lipoprotein receptor-related protein (LRP). Evidence for co-receptor function. J Biol Chem 2005; 280(30):27872–8.

45. Obermoeller-McCormick LM, Li Y, Osaka H, FitzGerald DJ, Schwartz AL, Bu G. Dissection of receptor folding and ligand-binding property with functional minireceptors of LDL receptor-related protein. J Cell Sci 2001;114 (Pt 5):899–908.
46. Walters JJ, Berwin B. Differential CD91 dependence for calreticulin and Pseudomonas exotoxin-A endocytosis. Traffic 2005;6(12): 1173–82.
47. Kurotaki T, Tamura Y, Ueda G, et al. Efficient cross-presentation by heat shock protein 90-peptide complex-loaded dendritic cells via an endosomal pathway. J Immunol 2007; 179(3):1803–13.
48. Binder RJ, Srivastava PK. Peptides chaperoned by heat-shock proteins are a necessary and sufficient source of antigen in the cross-priming of CD8+ T cells. Nat Immunol 2005;6(6): 593–9.
49. Becker T, Hartl FU, Wieland F. CD40, an extracellular receptor for binding and uptake of Hsp70-peptide complexes. J Cell Biol 2002;158(7):1277–85.
50. Jeannin P, Bottazzi B, Sironi M, et al. Complexity and complementarity of outer membrane protein A recognition by cellular and humoral innate immunity receptors. Immunity 2005;22(5):551–60.
51. Kottke T, Pulido J, Thompson J, et al. Antitumor Immunity Can Be Uncoupled from Autoimmunity following Heat Shock Protein 70-Mediated Inflammatory Killing of Normal Pancreas. Cancer Res 2009.
52. Sawamura T. LOX-1 unlocked. Structure 2005;13(6):834–5.
53. Ohki I, Ishigaki T, Oyama T, et al. Crystal structure of human lectin-like, oxidized low-density lipoprotein receptor 1 ligand binding domain and its ligand recognition mode to OxLDL. Structure 2005;13(6):905–17.
54. Pluddemann A, Neyen C, Gordon S. Macrophage scavenger receptors and host-derived ligands. Methods 2007;43(3):207–17.
55. Appella E, Weber IT, Blasi F. Structure and function of epidermal growth factor-like regions in proteins. FEBS Lett 1988; 231(1):1–4.
56. Shibata M, Ishii J, Koizumi H, et al. Type F scavenger receptor SREC-I interacts with advillin, a member of the gelsolin/villin family, and induces neurite-like outgrowth. J Biol Chem 2004;279(38):40084–90.
57. Park SY, Kim SY, Jung MY, Bae DJ, Kim IS. Epidermal growth factor-like domain repeat of stabilin-2 recognizes phosphatidylserine during cell corpse clearance. Mol Cell Biol 2008;28(17):5288–98.
58. Enomoto Y, Bharti A, Khaleque AA, et al. Enhanced immunogenicity of heat shock protein 70 peptide complexes from dendritic cell-tumor fusion cells. J Immunol 2006;177(9): 5946–55.
59. Apostolopoulos V, Yu M, Corper AL, et al. Crystal structure of a non-canonical low-affinity peptide complexed with MHC class I: a new approach for vaccine design. J Mol Biol 2002;318(5):1293–305.

Chapter 23

Analysis of Cellular Migration Using a Two-Chamber Methodology

Michelle M. Kouspou and John T. Price

Abstract

Directed cell migration is fundamental to both physiological and pathophysiological processes such as embryogenesis, wound healing, and cancer metastasis. A complex series of events are required for directional cell migration, which is initiated by a migration-promoting or chemotactic stimulus, resulting in cellular polarization and entry into a cyclical pattern of leading edge protrusion, adhesion, and retraction of the trailing edge allowing cell movement. Heat shock proteins such as Hsp27, Hsp90, alphaB-crystallin, as well as heat shock transcription factors, are important players in both physiological and pathophysiological cell migration. A variety of techniques are currently available to assess cell migration, and among the most commonly utilized are those that employ a two-chamber methodology, such as that developed by Stephen Boyden in the early 1960s. Herein, we describe the use of a multiwell microchemotaxis migration assay that has the advantages of being reusable, inexpensive, highly reproducible, and requiring only a small volume of reagents.

Key words: Migration, Heat shock, Stress, Chemotaxis, Adhesion, Metastasis

1. Introduction

Cell migration is an essential component of many physiological and pathophysiological processes. The migration of a cell is fundamental to normal embryonic development, efficient wound healing, and a functional immune system (1–4). It also contributes to many pathologies such as chronic inflammatory diseases, vascular disease, tumor formation, and metastasis (5–8).

Cell migration consists of a complex cascade of events that involve the integration of localized, transient signaling events with changes in cellular architecture. For migration to occur efficiently, cells must have an asymmetric or a polarized morphology, with a

Stuart K. Calderwood and Thomas L. Prince (eds.), *Molecular Chaperones: Methods and Protocols*,
Methods in Molecular Biology, vol. 787, DOI 10.1007/978-1-61779-295-3_23, © Springer Science+Business Media, LLC 2011

defined leading and trailing edge (9, 10). Cellular signaling at the leading edge of the cell results in cell protrusion and adhesion, and upon contraction of the cell with detachment of its trailing edge, the cell migrates forward (11). This sequence of steps of polarization, protrusion, adhesion, contraction, and detachment, known as the cell motility cycle, occurs in a wide variety of cell types in response to a wide variety of factors.

Cell migration can be stimulated by many different cues that are often asymmetric in nature, thus not only providing a stimulus of the basic migratory machinery but also affording directionality (9, 11, 12). The type of stimulus that mediates the cell migration is often used to define the form of migration. For example, cells undergo chemotaxis in response to soluble gradient cues (12), and haptotaxis in response to anchored guidance cues or altered adhesion gradients in the extracellular matrix (13, 14). Although the overall basic mechanisms of cell migration have been largely defined, the complexities of the process and the vast array of modulators are still being revealed within physiological and pathophysiological settings.

In recent years, stress proteins, such as members of the heat shock protein (Hsp) family, have been shown to play both negative and positive roles in the regulation of cell migration. For example, in cancer, Hsps have long been associated with tumor cell growth and survival (15–17); however, it is becoming apparent that Hsps also play a prominent role in cell migration pathways. Hsp27, Hsp70, Hsp90, and alphaB crystallin have all been shown to modulate cancer cell migration through a variety of mechanisms, including modulation of actin polymerization and stabilization of cell signaling molecules fundamental to cell migration pathways (18–26). In addition to Hsps, heat shock transcription factor 1 (HSF1), the major transcriptional regulator of the heat shock response and Hsps, has also been shown to play a role in cell migration (27).

A variety of in vitro cell migration assays are currently available to examine the various modes of migration such as chemokinesis, chemotaxis, and haptotaxis. Of these, the most commonly used are two-chamber based assays, which include the Boyden chamber (28) and variations of this apparatus, such as transwell cell culture inserts and multi-well chamber assays (29). These consist of two chambers separated by a porous filter, with chemoattractants and cells being placed in the lower and upper chambers, respectively. During incubation, cells migrate through the pores of the filter toward the chemoattractant or haptotactic gradient and adhere to the underside of the filter, providing a quantitative measure of their migratory capacity. Within this chapter, we describe a protocol for the use of a 48-well microchemotaxis chamber available from Neuro Probe, Inc. We have used this successfully to examine chemotaxis, haptotaxis, and chemokinesis in a variety of cancer cell lines and identify novel mediators of cancer cell migration and cell signaling pathways associated with the process (26, 29–31). The advantages of this assay is that it is relatively cheap, reproducible, and a high number of samples can be run simultaneously with minimal handling unlike that of the single well Boyden chamber and cell culture inserts. Another advantage of the system is that due to the small volumes required for the assay, the amounts of expensive reagents such as inhibitors, antibodies, and growth factors is reduced.

This protocol is specifically designed for the human breast cancer cell line, MDA-MB-231, using EGF and fibroblast conditioned media as chemoattractants. However, this assay can be adopted for a wide variety of cell types in response to not only chemotactic stimuli but also chemokinetic and haptotactic stimuli.

2. Materials

1. Standard 48-well microchemotaxis chamber and associated apparatus including the following: lower chamber, silicone gasket, upper chamber, thumbnuts, curved nose, fine forceps for handling filters, and large filter clamps to keep filters extended widthwise (Fig. 1) (Neuro Probe, Inc., Gaithersburg, MD, USA).
2. Standard Polyvinylpyrrolidone (PVP)-free, 25 × 80 mm 8-μm polycarbonate track-etch (PCTE) filters (cat. no. PFB8, Neuro Probe, Inc., Gaithersburg, MD, USA).
3. MDA-MB-231 cells (cat. no. HTB-26; ATCC).
4. Dulbecco's modified Eagles's medium (DMEM), high glucose, pH 7.4.

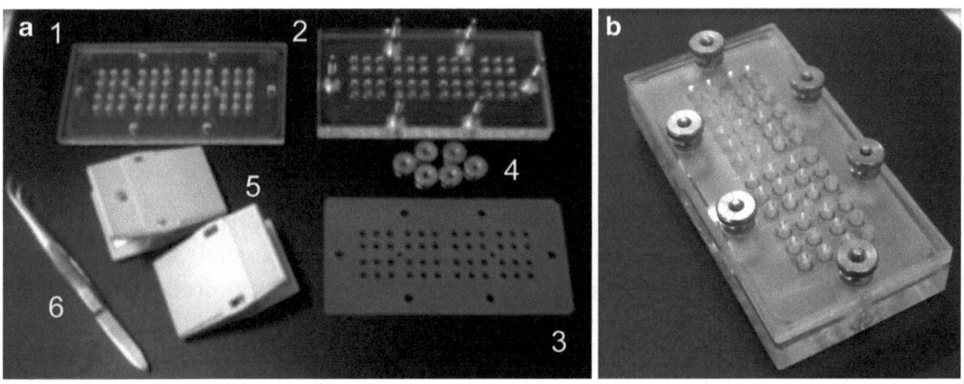

Fig. 1. The apparatus associated with the multiwell chemotaxis assay (**a**) showing the upper plate (1), lower plate (2), silicone gasket (3), the thumbnuts that hold the apparatus together (4), large clamps (5), and fine curved forceps (6) for manipulating the polycarbonate filters. The assembled multiwell microchemotaxis chamber (**b**).

5. Fetal bovine serum (FBS).
6. Bovine serum albumin (BSA): cell culture grade, endotoxin-free. Aliquots of 20% stock solution in DMEM allow uniform supplementation of each solution in each assay and uniformity over many assays.
7. Base assay medium (DMEM containing 0.1% BSA).
8. Phosphate-buffered saline, pH 7.4 (cat. no. 10010-031; Invitrogen, Carlsbad, CA, USA).
9. Versene (cat. no. 15040066; Invitrogen, Carlsbad, CA, USA).
10. Epidermal Growth Factor (EGF; cat. no. 354052; BD Biosciences, San Jose, CA., USA).
11. Human Collagen Type I (cat no C5533; Sigma-Aldrich Corp. St. Louis, MO, USA) for coating PCTE filters.
12. Methanol.
13. Diff-Quik staining solutions I and II, or equivalent.
14. Large glass microscope slides for mounting filters (50 × 70 mm).
15. Cell culture dishes (150 mm; cat. no. 351013; BD Biosciences, San Jose, CA., USA) for placing chambers in during incubation.
16. Tergazyme™ (cat. no. 1325; Alconox, White Plains, NY, USA) for cleaning the chambers.

3. Methods

3.1. Assay Preparation Overview

To efficiently perform the assay, there are a number of steps that should be completed before preparing the cells for analysis.

1. To remove any potential particle buildup (e.g., dust), which increases the likelihood of air bubble generation within the assay, chambers and gaskets should be washed in running distilled water. Shake off any excess water from the chambers and gaskets, allowing them to thoroughly dry in a laminar flow hood.
2. Chemoattractants and background controls should be made up prior to the assay being commenced, decanted into an Eppendorf, and stored on ice until required. Just prior to use, the chemoattractants are allowed to equilibrate to room temperature to prevent any potential temperature differential within the chambers, which causes increased air bubble formation.
3. Filters should be precoated with the extracellular matrix component of choice, completely dry and free of dust, with the left-hand corner removed for orientation purposes.

3.2. Chemoattractants

A wide variety of both organic and inorganic substances have been shown to act as powerful chemoattractants and include growth factors, cytokines, and peptides. Many factors can also act as chemokinetic and haptotactic agents and although we only examine chemotactic agents within this protocol, this assay can be modified to also examine chemokinesis and haptotaxis (see Note 1).

Depending upon the cell type being studied, relevant chemoattractants can be chosen. For example, EGF acts as a strong chemo attractant in many epithelial cancer cell lines, and we have used this to good effect in our studies (29). A single source of chemoattractant has the advantage that downstream signaling is less complex and can be easily examined; however, the use of a mixed chemoattractant source, such as fibroblast conditioned medium (Fb-CM), has the advantage that it can stimulate robust cell migration, is less concentration dependent, and is applicable to a wide variety of cell types. In addition to single sources of chemoattractants, we also run Fb-CM as a positive control within our migration assays.

3.2.1. Positive Controls

We have used Fb-CM prepared from near-confluent monolayers of NIH-3T3 cells or human primary fibroblasts as a positive control.

1. To generate conditioned media, near-confluent monolayers of fibroblasts are rinsed in PBS and then DMEM + 10% FBS is added.
2. After 48 h of incubation, the Fb-CM is harvested under sterile conditions and centrifuged in a benchtop centrifuge at full speed for 20 min to remove any cellular debris.
3. The clarified supernatants are pooled, decanted to fresh tubes, and frozen at −80°C. Repeated freezing–thawing should be avoided for uniformity.
4. Prior to its use in the microchemotaxis assay, the Fb-CM should be supplemented with BSA (20% stock) to a final concentration of 0.1%.

An alternative to Fb-CM is to use FBS-supplemented medium. Although not as robust as Fb-CM, we have effectively used 10%-FBS supplemented medium as a chemoattractant. If FBS is to be used, it should be a uniform batch that the cells are growing in, and again should be aliquoted and not thawed and refrozen, to keep uniformity between assays. As with Fb-CM, the FBS/medium should be supplemented with BSA to a final concentration of 0.1% prior to the assay.

3.2.2. Background Migration Controls

Appropriate background controls that measure a cells intrinsic unstimulated random migratory potential should be run for each cell type being examined. The base assay medium, which acts as the background control, consists of medium, in this case, DMEM with

0.1% BSA. This control provides a consistent measure of cell migration for a particular cell type and provides an indication of interassay variation. Running this control can aid in determining whether the chamber is becoming contaminated, if background migration increases between assays, and can also indicate if the source of chemoattractant has a loss of activity (see Note 2).

3.3. Coating Filters

The clear polycarbonate filters used within the assay have a shiny side and a dull side, which becomes useful for orientation purposes. We always load the filters into the chamber with the shiny side down, allowing us to determine the side of the filter where the migrated cells are located. The filters are available with varying pore sizes, and a suitable filter should be chosen for the cell type of interest; we generally use 8-μm filters for most human epithelial cancer cell lines.

To provide a consistent and definable surface to which the cells can attach and migrate, we consistently coat the filters with ECM components, most commonly collagen type I or IV.

1. To coat filters, submerge them in a solution of collagen type I (20 μg/ml in 10 mM acetic acid) or collagen type IV (40 μg/ml in 0.2 mM acetic acid) and gently agitate overnight at room temperature.
2. After incubation, remove coated filters and dry by attaching large filter clamps (Fig. 1) to each end of the filter. The filters are then hung vertically from one end until dry.
3. After drying, and with the shiny side of the filter facing down, the top left-hand corner is removed. This allows for the correct orientation of the filter during subsequent assays.
4. The coated filters are then stored at room temperature, in a sealed container to protect them from dust, until required.

3.4. Preparation of Cells

3.4.1. Cell Culture

The cells should be passaged uniformly prior to the assay and should be in logarithmic growth. Cellular confluency at the time of harvesting impacts upon the migratory potential of many cell types and should be kept consistent between assays, usually between 70 and 80%. To overcome any potential variability that may come from culture conditions and confluency, we routinely harvest cells ~48 h prior to assay, reseeding cell culture flasks at a defined cellular seeding density that provides 70–80% confluence at the time of harvest.

3.4.2. Harvesting of Cells

To prevent loss of adhesion molecules through proteolysis, cells are harvested from cell culture flasks nonenzymatically.

1. Cell culture medium is removed and the adherent cells are washed twice with versene solution.
2. A sufficient volume of versene solution is then used to cover the cells and the flask is incubated at 37°C, 5% CO_2 for 5–10 min.

At regular intervals, the flasks are removed and "gently" tapped to dislodge cells until all the cells have lost adhesion.

3. DMEM + 10% FBS is added after which time the cells are washed twice with base assay medium (DMEM/0.1% BSA).

If the cells that are to be used are highly adherent, then trypsin can be used (see Note 3).

3.4.3. Counting and Verifying Cell Viability

1. During the second wash (see Subheading 3.3, step 2), the cells are stained with trypan blue and counted.
2. The cells are resuspended with base assay medium at a concentration of 1.0×10^6 viable cells/mL (see Note 4). Only cell populations with >95% viability should be used in the assay.
3. To facilitate easier loading of the cells into the chambers, as well as to minimize the chance of introducing air bubbles within the assay, the cells are transferred to Eppendorf tubes. These cells are ready to be assayed and should be stored on ice if a protracted period (>30 min) will elapse before use.

3.4.4. Treatment of Cells

Depending upon the investigations being performed, some assays will require prior and/or continued treatment of the cells with agents of choice, and this should be factored into the protocol. For example, the effects of the heat shock response (HSR) upon cell migration can be examined by incubating cells with stimulators of the HSR (30 min to 2 h) a day prior to the assay, and allowing their recovery for 18–22 h prior to determining their cell migration. We have also used inhibitors of the HSR by incubating these reagents for 18–22 h prior to analysis of cell migration. In the case of inhibiting of cell signaling pathways, after washing, counting, and viability-checking the cells, we pretreat them with the appropriate inhibitory agents for 30–45 min at 37°C, 5% CO_2. The cells are then loaded into the assays in the presence of the agents.

3.5. Assembling the Chamber

3.5.1. Preparation of the Components

At this stage of the assay, the cells have been harvested, checked for viability, counted, and resuspended at 1×10^6 cells/mL in base assay medium. The chemoattractants have been prepared in base assay medium at appropriate concentrations (e.g., EGF; 10 ng/mL) (see Note 5), and the positive, and background migration controls have been prepared in Eppendorf tubes. The filters have been appropriately coated and the left-hand corner removed. All of the components are at room temperature to prevent any potential temperature differential within the chamber that will increase the likelihood of air bubble formation.

3.5.2. Chambers

Setting up and loading the chambers is performed on the lab bench as most assays run between 4 and 8 h. Set the chambers up in an area that is not prone to drafts, an important consideration when placing the filter onto the chambers. It is important that the chambers are thoroughly dry before adding the chemoattractant into the lower chamber, as surface tension is required to keep the chemoattractant welled up in the lower chamber such that air bubbles are avoided when placing the filter.

3.5.3. Addition of Chemoattractant to the Lower Chamber

1. Orientate the lower chamber so that the NP trademark is in the upper left corner. Adjust a 100- or 200-μL micropipette so that the volume expelled up to the first stop of the pipette forms a slight positive meniscus when the well is filled, usually ~27–29 μL (Fig. 2) (see Note 6).
2. Expel the chemoattractant smoothly but quickly into the well to avoid air bubbles. Depressing the plunger to the first stop of the pipette will retain fluid within the pipette tip; however, this action prevents the introduction of air bubbles into the bottom chamber which would occur if the plunger was fully depressed.
3. Complete the filling of the lower chambers in no more than 5 min to prevent evaporation.

3.5.4. Assembling Filter, Silicone Gasket, and Top Chamber

1. After the lower chamber is filled, place a filter membrane on the chamber by lifting the filter by the ends with two forceps, with the shiny side of the filter facing down. Orientate the filter so that the cut top left-hand corner corresponds to the NP trademark. Hold the filter over the filled wells and lower the filter, allowing the middle portion to make contact first (Fig. 3). The filter position can be adjusted at this point; however, excessive movement will cause contamination between the wells.
2. Apply the silicone gasket with the cut corner in the top left-hand corner (Fig. 3).

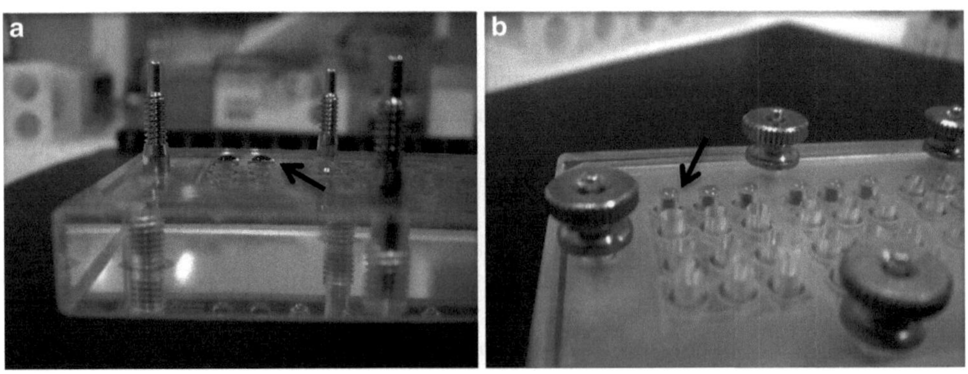

Fig. 2. The positive menisci (*arrows*) that form in the lower wells (**a**) and the upper wells (**b**) when appropriate volumes of chemoattractant and cell suspension are added, respectively.

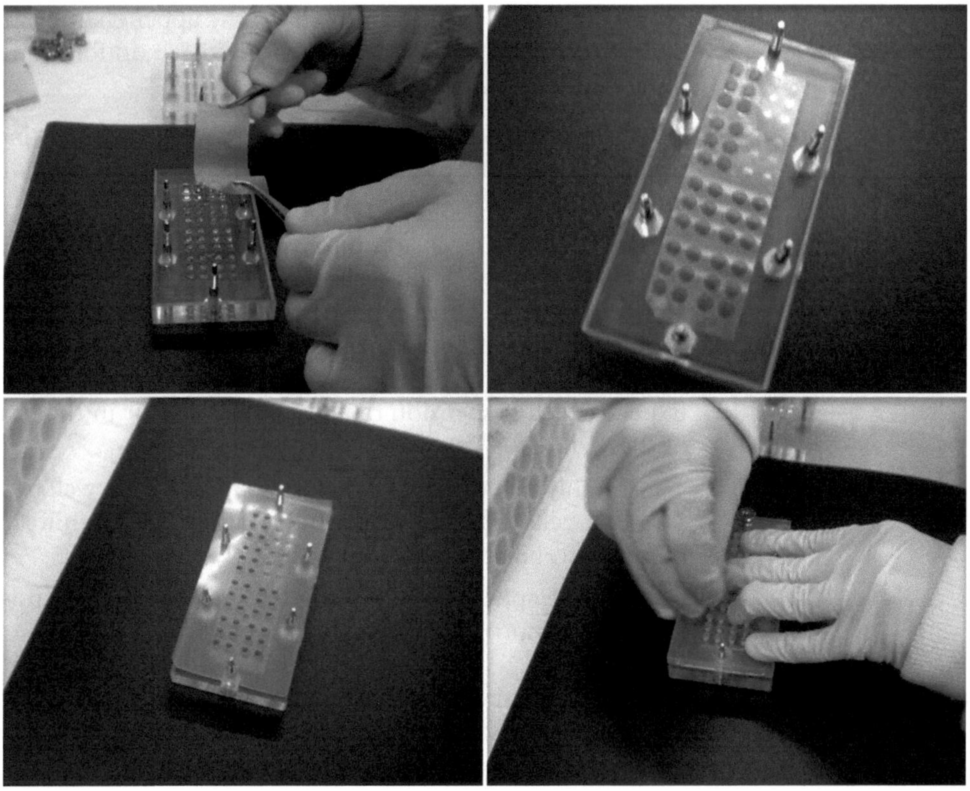

Fig. 3. The procedure for placing the filter upon the lower plate after the chemoattractants have been loaded (*upper left panel*), and the final resting position of the filter on the lower plate (*upper left panel*) prior to placing the gasket on the lower plate (*lower left panel*). Image of the top plate being held evenly and firmly while the thumbnuts are being screwed into position (*lower right panel*).

3. Place the top plate with the NP trademark orientated at the top left-hand corner. Push the top plate down against the bottom plate and hold it firmly with one hand, while applying the thumbnuts with the other (Fig. 3). It is important to maintain even pressure on the assembly while tightening the thumbnuts; this prevents air bubbles from being introduced into the bottom wells.
4. Check for air bubbles that may have been introduced into the bottom chamber. To do this, invert the apparatus and tap it lightly on the bench; this action will dislodge any air bubbles that have been trapped at the filter–chemoattractant interface.
5. Slowly turn the apparatus right-side up. If the air bubbles are small, they often remain lodged at the bottom of the well and do not rise to interfere with the filter and chemoattractant interface. However, large air bubbles will rise to the interface and be clearly visible. These wells should be discounted in any future analysis.

3.5.5. Addition of Cells to the Upper Chamber

The major concerns when adding cells to the upper chamber are to avoid introducing air bubbles and to ensure that a uniform number of cells are added to each chamber.

1. Gently mix the cell suspension by drawing into the pipette several times, but be careful not to introduce any air bubbles into the suspension.
2. Pipette the cell suspension into each upper well, adjusting the volume so that the filled wells have a slight positive meniscus, ~54–56 μL (Fig. 2)(see Note 6). To achieve this without introducing air bubbles, hold the pipette at a steep angle so that the end of the pipette tip rests against the top rim of the well. Expel the cell suspension rapidly to dislodge air in the bottom of the well.
3. Thoroughly mix the cell suspension between each well to make sure that cells are evenly distributed allowing for uniform cell numbers to be added to each well.
4. Check for trapped air bubbles in the upper wells, as this will seriously compromise the assay (see Note 7).

3.6. Incubation

1. Place the chamber in a 150-mm cell culture dish containing 30–50 mL of distilled water and cover.
2. Place the dish containing the chamber at 37°C, 5% CO_2 for 4–8 h. The incubation time will vary depending upon the cell line that is being used, and the end user should determine optimal times (see Note 8).

3.7. Preparation of Staining Solutions

Before harvesting the assay, prepare the fixative and staining solutions in preparation for staining the migrated cells upon the filter.

1. Decant methanol, Diff-Quik solution I, and Diff-Quick solution II into 250-mL beakers sufficient to cover the filter when the filter is vertically placed in the solution (see Note 9).
2. Fill two 500-mL beakers with distilled water for washing the filters after the staining procedure.

3.8. Harvesting and Staining the Assay

3.8.1. Disassembly of the Chamber

1. Remove the chamber from the incubator and the cell culture dish.
2. Unscrew the thumbnuts while holding down the top plate and then invert the entire chamber onto a paper towel.
3. Hold the four corners of the top plate, which is now situated at the bottom, and push down evenly so that it remains level as it is lowered. If the rubber gasket gets caught on the post hardware, carefully push it down evenly onto the plate. Do not touch the filter, which will be stuck to the gasket. The migrated cells are now facing up on the filter.

4. Lift up the edge of the filter that previously had the corner removed with a pair of forceps and attach a large wide blade clamp to 1 mm of the filter. Lift the clamp and slowly peel the filter away from the gasket. Hold the filter vertically with the migrated cells facing away from you.

3.8.2. Fixing and Staining of the Cells

1. Holding the clamp, and with the cells that have migrated through the filter facing away from you, i.e., the shiny side, fix the cells by gently lowering the filter into the methanol for 2 min. Make sure that the side of the filter with the migrated cells does not touch the side of the beaker.
2. Remove the filter from the methanol and immediately submerge the filter into Diff-Quik solution I (1 min) followed by Diff-Quick II (2 min).
3. Remove excess stain by dipping the filter repeatedly in distilled water until all the excess stain is removed.
4. Mount the filter onto a large microscope slide by submerging the filter vertically in distilled water and placing a large microscope slide against the side of the filter that contains the migrated cells, i.e., the shiny side. Remove the slide and the filter simultaneously and vertically from the water; this results in the filter being mounted onto the slide. Carefully remove the clamp from the filter and reposition the filter while wet.
5. While the filter is still wet, wipe the filter with a wet Kimwipe or tissue paper to remove cells that have not migrated. The filter is then allowed to completely dry.

3.9. Quantitation of Cell Migration

A number of options exist for counting the cells that have migrated. However, whether counting manually or mechanically, at least five random fields on each well should be counted and each experimental condition should be done in at least triplicate within the assay. Although manual counting is laborious, it is often the most accurate method. Owing to the presence of cell debris, partially migrated cells, and cell debris filling the filter pores, and variations in nuclear sizes among migrated cells, automated counting by image analysis is a difficult proposition. We routinely capture color images from a standard light microscope (200× or 400× magnification) (Fig. 4) using a digital camera and import these images into ImageJ (NIH, USA). The cells are counted manually using the point tool to label counted cells in ImageJ.

3.10. Washing the Chambers

After the assay, rinse all parts in running distilled water and soak in distilled water overnight. The components are then allowed to dry at room temperature under a protective paper towel. Proteins and other contaminants accumulate in the chambers, and this can lead to erratic results and the background migration being unusually high. Periodic soaking of all of the components in Tergazyme™

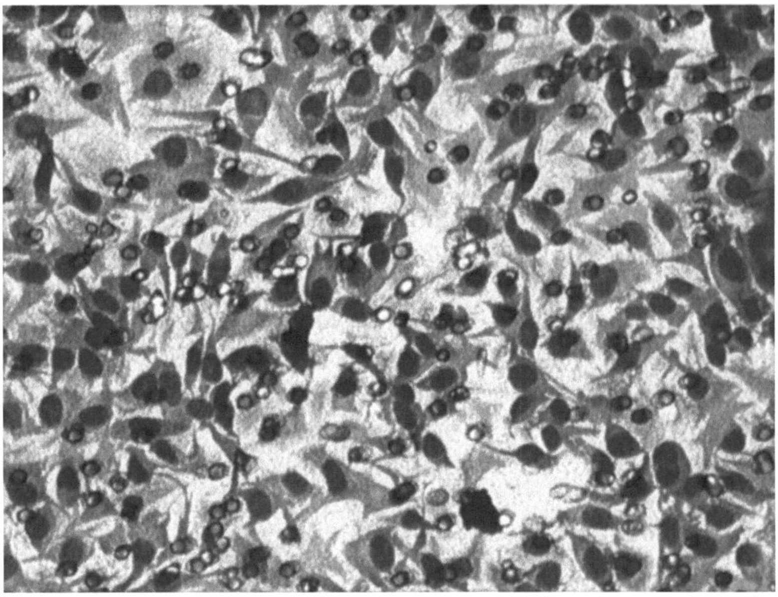

Fig. 4. Image of fixed and stained MDA-MB-231 human breast cancer cells that have migrated through the 8-μM pores of a collagen I coated polycarbonate filter membrane toward EGF (10 ng/mL).

(an enzyme active detergent available form most laboratory suppliers) for 1–3 h at 55–60°C removes the contaminants. This incubation is followed with thorough rinsing under running distilled water and soaking in distilled water overnight. The chambers can be cleaned in this way as often as necessary.

4. Notes

1. Agents that are known to be chemokinetic can be used to prestimulate cells in culture for a defined period prior to analysis within the assay. In addition, the chemokinetic nature of agents can be determined by placing equal concentrations of the agent in both the lower and top wells. Haptotactic migration can be observed by coating the underside of the filter membrane with an ECM component such as vitronectin. Alternatively, a solution of the ECM component can be added to the lower wells in base assay medium, at a defined concentration. During incubation, the ECM coats the underside of the filter establishing a haptotactic gradient.
2. We have found that although growth factors, such as EGF, are stored appropriately, their activity decreases significantly in cell migration assays over time. Therefore, care needs to be taken that assays are performed with reagents that have not had substantial loss of activity.

3. If trypsin is used to remove adherent cells from the cell culture flask, 10 mL of DMEM + 10% FBS should be added to inactivate the trypsin and then the cells should be incubated for 45–60 min at 37°C, 5% CO_2 to allow the recovery of cell surface adhesion molecules. The cells are then washed twice with base assay medium.
4. Depending upon the migratory nature of the cells under study, we vary the concentration of cells between 5×10^5 cells/mL for highly migratory cells, and 2×10^6 cells/mL for those that have low migration rates. This ensures that sufficient cell numbers migrate to allow accurate quantitation of cell migration by random field counting.
5. The concentration of the chemoattractant is an important consideration in the assay due to the relationship of migrated cell numbers and chemoattractant concentration being bell-shaped in nature, rather than sigmoidal. Therefore, an optimal concentration of chemoattractant needs to be determined for the chemoattractant and the cell line of interest.
6. The volume required to fill the wells in the lower chamber is approximately between 27 and 29 μL. However, as the chambers can slightly vary the exact volume for any given chamber should be determined. The volume required to fill the upper chamber is approximately 55–56 μL and will form a slight positive meniscus.
7. An easy way of checking for air bubbles is to look for the reflection of overhead lights in the menisci. Often, a well with an abnormally large positive meniscus usually has a trapped air bubble. If an air bubble is present, remove the contents of the well completely and refill it.
8. Due to factors such as the breakdown of the chemotactic gradient within the chamber, the loss of adhesion of migrated cells to the underside of the filter, and confounding cell proliferation, we seldom use incubation times of greater than 12 h.
9. Other cell stains that can be used include Giesma stain, hematoxylin and eosin, and crystal violet. Depending upon the number of assays we run, we routinely reuse the Diff-Quik stains for several months; however, although we reuse the methanol, we tend to change this more frequently.

Acknowledgments

We would like to thank Ryan Chai, Benjamin Lang, and Reece Lim for the critical reading of this manuscript. This work was supported by an NHMRC R Douglas Wright fellowship, NHMRC project grant 606549, and Cancer Council Victoria grant 545969CA to (to J.T. Price).

References

1. Aman A, Piotrowski T Cell migration during morphogenesis. Dev Biol 341 (1):20–33.
2. Ulrich F, Heisenberg CP (2009) Trafficking and cell migration. Traffic 10 (7):811–818.
3. Mrass P, Weninger W (2006) Immune cell migration as a means to control immune privilege: Lessons from the cns and tumors. Immunol Rev 213:195–212.
4. Yoneyama H, Matsuno K, Matsushimaa K (2005) Migration of dendritic cells. Int J Hematol 81 (3):204–207.
5. Sprague AH, Khalil RA (2009) Inflammatory cytokines in vascular dysfunction and vascular disease. Biochem Pharmacol 78 (6):539–552.
6. Jin T, Xu X, Hereld D (2008) Chemotaxis, chemokine receptors and human disease. Cytokine 44 (1):1–8.
7. Price JT, Thompson EW (2002) Mechanisms of tumour invasion and metastasis: Emerging targets for therapy. Expert Opin Ther Targets 6 (2):217–233.
8. Yang J, Weinberg RA (2008) Epithelial-mesenchymal transition: At the crossroads of development and tumor metastasis. Dev Cell 14 (6):818–829.
9. Ridley AJ, Schwartz MA, Burridge K, Firtel RA, Ginsberg MH, Borisy G, Parsons JT, Horwitz AR (2003) Cell migration: Integrating signals from front to back. Science 302 (5651):1704–1709.
10. Vicente-Manzanares M, Webb DJ, Horwitz AR (2005) Cell migration at a glance. J Cell Sci 118 (Pt 21):4917–4919.
11. Petrie RJ, Doyle AD, Yamada KM (2009) Random versus directionally persistent cell migration. Nat Rev Mol Cell Biol 10 (8): 538–549.
12. Van Haastert PJ, Devreotes PN (2004) Chemotaxis: Signalling the way forward. Nat Rev Mol Cell Biol 5 (8):626–634.
13. Chia J, Kusuma N, Anderson R, Parker B, Bidwell B, Zamurs L, Nice E, Pouliot N (2007) Evidence for a role of tumor-derived laminin-511 in the metastatic progression of breast cancer. Am J Pathol 170 (6):2135–2148.
14. Sotobori T, Ueda T, Myoui A, Yoshioka K, Nakasaki M, Yoshikawa H, Itoh K (2006) Bone morphogenetic protein-2 promotes the haptotactic migration of murine osteoblastic and osteosarcoma cells by enhancing incorporation of Integrin Beta1 into lipid rafts. Exp Cell Res 312 (19):3927–3938.
15. Whitesell L, Lindquist S (2009) Inhibiting the transcription factor Hsf1 as an anticancer strategy. Expert Opin Ther Targets 13 (4):469–478.
16. Garrido C, Brunet M, Didelot C, Zermati Y, Schmitt E, Kroemer G (2006) Heat shock proteins 27 and 70: Anti-apoptotic proteins with tumorigenic properties. Cell Cycle 5 (22):2592–2601.
17. Mahalingam D, Swords R, Carew JS, Nawrocki ST, Bhalla K, Giles FJ (2009) Targeting Hsp90 for cancer therapy. Br J Cancer 100 (10):1523–1529.
18. Annamalai B, Liu X, Gopal U, Isaacs JS (2009) Hsp90 is an essential regulator of EphA2 receptor stability and signaling: Implications for cancer cell migration and metastasis. Mol Cancer Res 7 (7):1021–1032.
19. Di K, Wong YC, Wang X (2007) Id-1 promotes TGF-beta1-induced cell motility through Hsp27 activation and disassembly of adherens junction in prostate epithelial cells. Exp Cell Res 313 (19):3983–3999.
20. Garg M, Kanojia D, Seth A, Kumar R, Gupta A, Surolia A, Suri A Heat-shock protein 70–2 (Hsp70-2) expression in bladder urothelial carcinoma is associated with tumour progression and promotes migration and invasion. Eur J Cancer 46 (1):207–215.
21. Golembieski WA, Thomas SL, Schultz CR, Yunker CK, McClung HM, Lemke N, Cazacu S, Barker T, Sage EH, Brodie C, Rempel SA (2008) Hsp27 mediates sparc-induced changes in glioma morphology, migration, and invasion. Glia 56 (10):1061–1075.
22. Hedges JC, Dechert MA, Yamboliev IA, Martin JL, Hickey E, Weber LA, Gerthoffer WT (1999) A role for p38(mapk)/hsp27 pathway in smooth muscle cell migration. J Biol Chem 274 (34):24211–2421.
23. Lesko E, Gozdzik J, Kijowski J, Jenner B, Wiecha O, Majka M (2007) Hsp90 antagonist, geldanamycin, inhibits proliferation, induces apoptosis and blocks migration of rhabdomyosarcoma cells in vitro and seeding into bone marrow in vivo. Anticancer Drugs 18 (10):1173–1181.
24. Luo XG, Zou JN, Wang SZ, Zhang TC, Xi T (2010) Novobiocin decreases smyd3 expression and inhibits the migration of MDA-MB-231 human breast cancer cells. IUBMB Life 62 (3):194–199.
25. Moyano JV, Evans JR, Chen F, Lu M, Werner ME, Yehiely F, Diaz LK, Turbin D, Karaca G, Wiley E, Nielsen TO, Perou CM, Cryns VL (2006) AlphaB-crystallin is a novel oncoprotein that predicts poor clinical outcome in breast cancer. J Clin Invest 116 (1):261–270.
26. Price JT, Quinn JM, Sims NA, Vieusseux J, Waldeck K, Docherty SE, Myers D, Nakamura

A, Waltham MC, Gillespie MT, Thompson EW (2005) The heat shock protein 90 inhibitor, 17-allylamino-17-demethoxygeldanamycin, enhances osteoclast formation and potentiates bone metastasis of a human breast cancer cell line. Cancer Res 65 (11):4929–4938.

27. O'Callaghan-Sunol C, Sherman MY (2006) Heat shock transcription factor (Hsf1) plays a critical role in cell migration via maintaining map kinase signaling. Cell Cycle 5 (13):1431–1437.
28. Boyden S (1962) The chemotactic effect of mixtures of antibody and antigen on polymorphonuclear leucocytes. J Exp Med 115:453–466
29. Price JT, Tiganis T, Agarwal A, Djakiew D, Thompson EW (1999) Epidermal growth factor promotes mda-mb-231 breast cancer cell migration through a phosphatidylinositol 3′-kinase and phospholipase c-dependent mechanism. Cancer Res 59 (21):5475–5478
30. Pereira JJ, Meyer T, Docherty SE, Reid HH, Marshall J, Thompson EW, Rossjohn J, Price JT (2004) Bimolecular interaction of insulin-like growth factor (IGF) binding protein-2 with alphavbeta3 negatively modulates IGF-I-mediated migration and tumor growth. Cancer Res 64 (3):977–984
31. Ackland ML, Newgreen DF, Fridman M, Waltham MC, Arvanitis A, Minichiello J, Price JT, Thompson EW (2003) Epidermal growth factor-induced epithelio-mesenchymal transition in human breast carcinoma cells. Lab Invest 83 (3):435–448

INDEX

Stuart K. Calderwood and Thomas L. Prince (eds.), *Molecular Chaperones: Methods and Protocols*, Methods in Molecular Biology, vol. 787, DOI 10.1007/978-1-61779-295-3, © Springer Science+Business Media, LLC 2011